Engineering Modeling

Engineering Modeling

Edited by **Tommy Haynes**

Willford Press

New York

Published by Willford Press,
118-35 Queens Blvd., Suite 400,
Forest Hills, NY 11375, USA
www.willfordpress.com

Engineering Modeling
Edited by Tommy Haynes

International Standard Book Number: 978-1-68285-034-3 (Hardback)

Printed in the United States of America.

Contents

Preface

This book has been a concerted effort by a group of academicians, researchers and scientists, who have contributed their research works for the realization of the book. This book has materialized in the wake of emerging advancements and innovations in this field. Therefore, the need of the hour was to compile all the required researches and disseminate the knowledge to a broad spectrum of people comprising of students, researchers and specialists of the field.

The emerging discipline of modeling and simulation in engineering is influenced by developments in system theories, artificial intelligence and software intelligence. Engineering modeling is an interdisciplinary field which includes concepts and techniques from mathematics, computer science, and management. It describes the design and conceptualization of large mechanical systems using complex non-linear models. It is useful to apprehend efficient numerical modeling along with model verification and validation. This book focuses upon the modeling theory and methodology of complex systems, simulation technology and approximation theory. It encompasses advanced and innovative modeling approaches and numerical strategies that would provide a comprehensive overview of the field to students, researchers and engineers.

At the end of the preface, I would like to thank the authors for their brilliant chapters and the publisher for guiding us all-through the making of the book till its final stage. Also, I would like to thank my family for providing the support and encouragement throughout my academic career and research projects.

Editor

1

Modeling elastic-plastic buckling of sandwich axisymmetric shells: on the limits of "shell" models and analytical solutions

Alain G Combescure

Correspondence:
alain.combescure@insa-lyon.fr
LaMCoS, INSA-Lyon, Université de Lyon, CNRS UMR 5259, 18-20 rue des Sciences, F69621 Villeurbanne, France

Abstract

Background: The objective of this paper is to answer the question: "Can a 'shell' model always be used to predict the elastic buckling of a shell?"

Method: This paper shows that such a model leads to significantly overestimated critical loads in the case of sandwich shells and gives an explanation for this overestimation.

Results: A dependable model is proposed and applied to a few structures of revolution, for which it is shown that shell analyses are sometimes overly on the unsafe side.

Conclusion: Of course, in such cases, 3D analysis is possible, but the associated computation cost is several orders of magnitude higher than that of the Fourier series analysis proposed in this paper.

Background

Computational modeling techniques to analyze the buckling of anisotropic and multilayered shells are very well developed, so that further advances may not seem necessary. However the quest for lighter structures, which has induced many studies for decades, has led to the development of sandwich structures in order to manufacture robust, yet lightweight, mechanical parts. Some of these parts are subjected to compression and their buckling strength must be assessed. The prediction of the critical buckling loads of these structures seems to be simple: it suffices to model these objects as multilayered shells, thus leading to an immediate prediction of their critical loads. However, a comparison of the critical loads predicted from calculations with experimental results reveals significant differences. Some remarkable works on buckling [1-3] were published at the turn of the century, and quasi-analytical solutions have been proposed for cylinders in axial compression [4]. These solutions take into account two types of buckling: global shell buckling and local skin buckling (wrinkling). Many works dealing with the study of crashworthiness optimization (e.g. [5-7]) are also available. Some authors have also worked on the interaction between the two types of buckling [8-11] and shown that in some cases this interaction leads to a critical load which is smaller than that predicted by calculating the critical loads of each mode independently. The objective of the present work is to examine the finite element modeling of this problem in the case of axisymmetric structures and propose a modeling strategy which leads to good results in all situations. The particular case of plastic buckling is also analyzed in depth.

The paper is structured as follows:

- first, the modeling of elastic-plastic buckling and its application to structures of revolution is reviewed;
- then, three sample applications are presented(a cone under internal pressure, a cone under external pressure, and a sphere under external pressure).

Method : A short review of the modeling of the elastic-plastic buckling of structures of revolution

Elastic buckling of structures of revolution

The structures being considered in this work are structures of revolution subjected to axisymmetric loading. Their elastic response (which can involve geometric and material nonlinearities) is also axisymmetric. However, when studying the stability of these objects, the loss of this symmetry of revolution must be envisaged. Because of the axisymmetric nature of both the structure and the constraints, the instability mode can be decomposed into a Fourier series. One can easily show (Koiter [12], Bushnell [13,14]) that in the case of a perfectly axisymmetric structure subjected to axisymmetric loading the instability involves only a single Fourier mode. This type of problem can be modeled using axisymmetric finite elements and a Fourier series decomposition. This is the chosen approach because it is quite efficient. 3D shells, or even solid elements, could also be used, but these would lead to much longer computation times. Two means of modeling these sandwich shells will be compared: the multilayer shell model called COQMULT, whose kinematics requires that the distance between layers remain constant during the deformation, and an axisymmetric volume model denoted QUA8 (in our case, an 8-node element enabling an exact representation of a linearly varying strain across the thickness), which removes all kinematic constraints between the two skins. Let us consider the local coordinate system s, θ, n which is tangent to the shell at any point of the mid-surface (Figure 1).

The membrane strains ϵ_{ij}^m and bending strains ϵ_{ij}^b for shells are well-known [15,16]. They are recalled below in the case of the Kirchhoff-Love theory for the nth Fourier mode:

$$\epsilon_{ss}^m = \frac{\partial u}{\partial s} - \frac{w}{R} \tag{1}$$

$$\epsilon_{\theta\theta}^m = \frac{1}{r}\left(\frac{\partial v}{\partial \theta} + w\, cos\phi + u\, sin\phi\right) \tag{2}$$

$$\epsilon_{s\theta}^m = \frac{1}{2}\left(\frac{\partial v}{\partial s} + \frac{1}{r}\left(\frac{\partial u}{\partial \theta} - v\, sin\phi\right)\right) \tag{3}$$

$$\epsilon_{ss}^b = -\frac{\partial^2 w}{\partial s^2} - \frac{1}{R}\frac{\partial u}{\partial s} \tag{4}$$

$$\epsilon_{\theta\theta}^b = \frac{1}{r^2}\left(-\frac{\partial^2 w}{\partial \theta^2} - r\, sin\phi\frac{\partial w}{\partial s} + cos\phi\frac{\partial v}{\partial \theta} - \frac{r}{R}u\, sin\phi\right) \tag{5}$$

$$\epsilon_{s\theta}^b = \frac{1}{2r^2}\left(-2r\frac{\partial^2 w}{\partial s\partial \theta} + 2r\, cos\phi\frac{\partial v}{\partial s} - \frac{r}{R}\frac{\partial u}{\partial \theta} - v\, sin\phi\, cos\phi + \frac{\partial w}{\partial \theta}sin\phi\right) \tag{6}$$

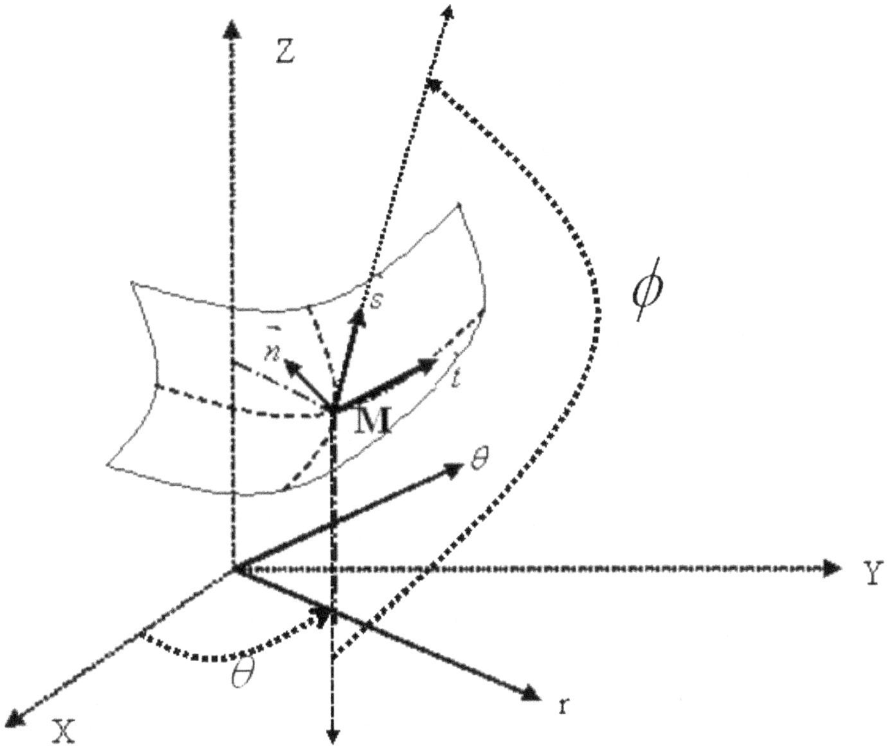

Figure 1 Definition of the local coordinate system for an axisymmetric shell.

where u, v, w denote respectively the axial, circumferential and normal displacements of the shell. r is the distance to the axis of revolution, R the axial curvature of the shell, and ϕ the angle with direction $-z$ ($+z$ being the axis of revolution).

The quadratic membrane strains , which measure the variation of the shell strain state in an infinitely small increment, and thus enable one to calculate buckling are given by:

$$\epsilon_{ss}^{q} = \frac{1}{2}\left[\left(\frac{\partial w}{\partial s} + \frac{u}{R}\right)^2 + \left(\frac{\partial u}{\partial s} - \frac{w}{R}\right)^2 + \left(\frac{\partial v}{\partial s}\right)^2\right] \quad (7)$$

$$\epsilon_{\theta\theta}^{q} = \frac{1}{2r^2}\left[\lambda_\phi^2 + \mu_\phi^2 + \eta_\phi^2\right] \quad (8)$$

$$\epsilon_{s\theta}^{q} = \frac{1}{2r}\left[\lambda_\phi\left(\frac{\partial w}{\partial s} + \frac{u}{R}\right) + \left(\frac{\partial u}{\partial s} - \frac{w}{R}\right)\mu_\phi + \frac{\partial v}{\partial s}\eta_\phi\right] \quad (9)$$

where $\lambda_\phi = \frac{\partial w}{\partial \theta} - vcos\phi$, $\mu_\phi = \frac{\partial u}{\partial \theta} - vsin\phi$ and $\eta_\phi = \frac{\partial v}{\partial \theta} + wcos\phi + usin\phi$. For the volume elements ($QUA8$), the linear strains are:

$$\epsilon_{rr} = \frac{\partial u_r}{\partial r} \quad (10)$$

$$\epsilon_{zz} = \frac{\partial u_z}{\partial z} \quad (11)$$

$$\epsilon_{\theta\theta} = \frac{1}{r}\left(u_r + \frac{\partial v}{\partial \theta}\right) \tag{12}$$

$$\epsilon_{rz} = \frac{1}{2}\left(\frac{\partial u_r}{\partial z} + \frac{\partial u_z}{\partial r}\right) \tag{13}$$

$$\epsilon_{r\theta} = \frac{1}{2}\left(\frac{\partial v}{\partial r} + \frac{1}{r}\left(\frac{\partial u_r}{\partial \theta} - v\right)\right) \tag{14}$$

$$\epsilon_{z\theta} = \frac{1}{2}\left(\frac{\partial v}{\partial z} + \frac{1}{r}\frac{\partial u_z}{\partial \theta}\right) \tag{15}$$

In these expressions, u_r, u_z, v denote respectively the radial displacement, the vertical displacement and the circumferential displacement. The quadratic strains are:

$$\epsilon^q_{rr} = \left[\left(\frac{\partial u_r}{\partial r}\right)^2 + \left(\frac{\partial u_z}{\partial r}\right)^2 \left(\frac{\partial v}{\partial r}\right)^2\right] \tag{16}$$

$$\epsilon^q_{zz} = \left[\left(\frac{\partial u_r}{\partial z}\right)^2 + \left(\frac{\partial u_z}{\partial z}\right)^2 \left(\frac{\partial v}{\partial z}\right)^2\right] \tag{17}$$

$$\epsilon^q_{\theta\theta} = \frac{1}{r^2}\left[\left(\frac{\partial u_r}{\partial \theta} - v\right)^2 + \left(\frac{\partial u_z}{\partial \theta}\right)^2 + \left(u_r + \frac{\partial v}{\partial \theta}\right)^2\right] \tag{18}$$

$$\epsilon^q_{rz} = \frac{1}{2}\left(\frac{\partial u_r}{\partial r}\frac{\partial u_r}{\partial z} + \frac{\partial u_z}{\partial r}\frac{\partial u_z}{\partial z} + \frac{\partial v}{\partial r}\frac{\partial v}{\partial z}\right) \tag{19}$$

$$\epsilon^q_{r\theta} = \frac{1}{2r}\left[\left(\frac{\partial u_r}{\partial \theta} - v\right)\frac{\partial u_r}{\partial r} + \frac{\partial u_z}{\partial \theta}\frac{\partial u_z}{\partial r} + \left(\frac{\partial v}{\partial \theta} + u_r\right)\frac{\partial v}{\partial r}\right] \tag{20}$$

$$\epsilon^q_{z\theta} = \frac{1}{2r}\left[\left(\frac{\partial u_r}{\partial \theta} - v\right)\frac{\partial u_r}{\partial z} + \frac{\partial u_z}{\partial \theta}\frac{\partial u_z}{\partial z} + \left(\frac{\partial v}{\partial \theta} + u_r\right)\frac{\partial v}{\partial z}\right] \tag{21}$$

Finally, buckling analysis consists of two steps: first the calculation of the axisymmetric prestress and then a series of buckling calculations for each Fourier mode *j*. The critical load and the associated critical mode correspond to the Fourier mode *j* which leads to the smallest load.

The shell finite elements chosen for the calculations are plane shell elements ($\frac{1}{R} = 0$) which possess only four degrees of freedom per node (three translations plus the rotation about the circumferential direction) [15].

The volume elements chosen are 8-node isoparametric axisymmetric elements with three Fourier degrees of freedom.

Nonlinear buckling of structures of revolution

Nonlinear instability may develop in any non linear analysis: for a given load level, they are characterized by the existence of more than one equilibrium state. For an axisymmetric structure the unstable state can be either axisymmetric or non-axisymmetric. If the structure remains axisymmetric, well-known incremental calculation techniques with control (such as the arc-length control method [17,18]) can generally be used to predict the critical load, except when the loading is exactly orthogonal to the instability mode. However, if the instability mode is not axisymmetric, this instability cannot be predicted using incremental analysis alone. In that case, there are two possible modeling strategies for predicting instability.

- The first strategy consists in meshing the structure of revolution in 3D and using this mesh to seek the instability. However, one should note that if the mesh satisfies the symmetry of revolution and the loading is axisymmetric the 3D analysis leads to an axisymmetric deformed shape and, thus, "misses" the non-axisymmetric instability. In this case, one must either introduce a defect (by perturbing the mesh of the structure or the loading) or use a mesh which does not satisfy the symmetry of revolution, hoping that this "imperfection" will solicit the instability sufficiently for it to develop "naturally."
  One can also use the specific technique of axisymmetric elements with non-axisymmetric defects ([16,19]).
- The second strategy consists in calculating the nonlinear response of the structure of revolution (which, thus, remains axisymmetric), then checking the stability of each resulting nonlinear state k. In such a stability study, for each load step k, one examines all the possible Fourier modes: the instability point is the first step k whose load multiplier is equal to 1. This Fourier analysis with uncoupled modes is still possible even though the response of the structure is nonlinear (in terms of geometry or material) because the preloads remain axisymmetric.

The geometrically nonlinear case

In the case of a pure geometrically nonlinear response, one calculates the axisymmetric equilibrium states using standard nonlinear incremental techniques. Thus, one obtains a sequence of m equilibrium states, denoted C_k, which are characterized by two variable fields: the displacement field between the initial structure and the current structure u_k, and the equilibrated Cauchy stress field σ_k. Once these m states have been found, one studies their stability. In order to do that, one calculates the stability of the deformed shape which corresponds to each state C_k by examining all the possible Fourier modes to find the mode j which leads to the smallest critical load λ_j^k. The first state k which satisfies $\lambda_j^k = 1$ is the critical state: then, buckling occurs following Fourier mode j. One should stress the fact that this buckling is usually associated with a loss of symmetry of revolution (if $j \neq 0$).

The elastic-plastic case

In this case, modeling the instability is a little trickier because of the unilateral nature of the elastic-plastic constitutive relation. The calculation of the equilibrium states follows the same scheme as in the geometrically nonlinear case: one calculates a sequence of statically and plastically admissible states, again denoted C_k. These states are characterized by three quantities: the displacements and stresses (as in elasticity), plus the internal variables A_k. One shall now consider the special case of Von Mises elastic-plastic case, but the present approach is the same for any irreversible elastic-plastic material. In case of damage model, this procedure is still valid but must be applied in a slightly modified manner. Let us consider the case of a tension curve with Young's modulus E, tangent modulus E_T and secant modulus E_s, as shown in Figure 2 below.

In order to calculate the instability using the same method as in nonlinear elasticity, one must linearize the constitutive relation during the buckling around state C_k. The natural way to linearize the material's behavior around the current state is simply to use the tangent constitutive law for all Gauss points on the loading surface, *i.e.* to consider the tangent constitutive equation:

$$\underline{\underline{H_t}} = \underline{\underline{H}}\left(\underline{\underline{I}} - \frac{\underline{\underline{A}}\left[\underline{s} \otimes \underline{s}^T\right]\underline{\underline{AH}}}{s^{*2}\, E_h + \underline{s}^T \underline{\underline{AHA}}\underline{s}}\right) \tag{22}$$

In this equation, H is the elastic constitutive matrix, s the stress deviator and E_h the elastic-plastic slope, which is related to the tangent modulus through the equation:

$$E_h = \frac{E\,E_T}{E - E_T} \tag{23}$$

s^* is Von Mises' equivalent stress. A, the operator which leads to Von Mises' stress knowing the components of the tensor, is given by the relation:

$$s^* = \sqrt{\underline{\sigma}^T \underline{\underline{A}}\underline{\sigma}} \tag{24}$$

Thus, the equilibrium states can be calculated incrementally. When the loading increases monotonically, the iterations converge rapidly. In the case of unloading, or when the direction of loading changes abruptly, the increment in the local stress tensor is very different from the current state and this tangent behavior operator, which is calculated using one's knowledge of the current stress state, is far from predicting the correct increment. For example, if one considers a stress state over the loading surface and the stress increment is tangent to the plasticity convex set, this approach leads to an elas-

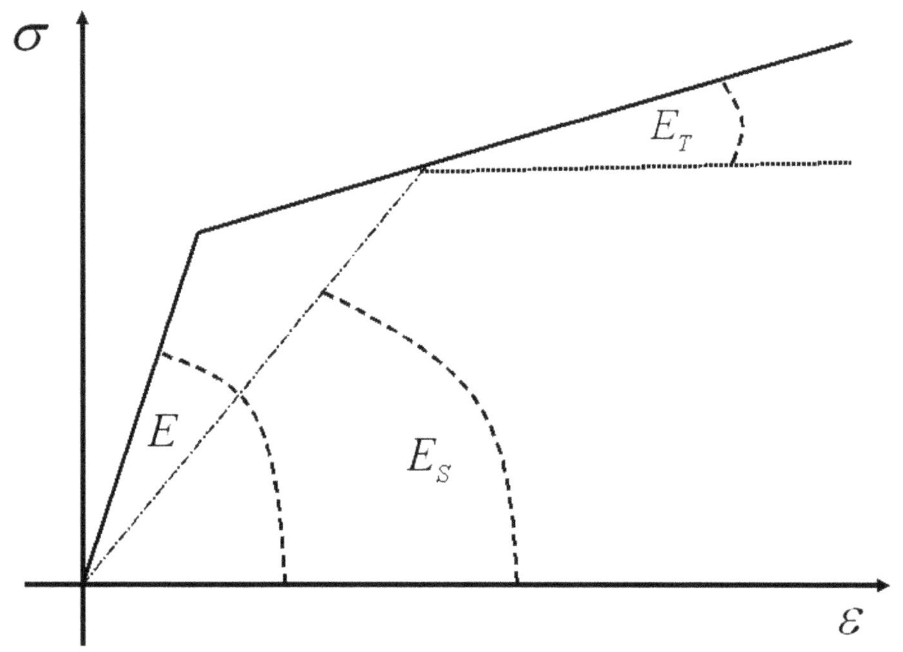

Figure 2 The stress-strain curve and definition of the different moduli.

tic increment. In the case of incremental analyses, this approximation is corrected by the constraint that the final state must be plastically admissible. This remark shows that if one evaluates the stability of a state using for state C_k the standard tangent constitutive matrix based on incremental plasticity theory (which is what is done in most programs) one tends to overestimate the stiffness, which leads to an unsafe prediction of the instability. This is an unconservative method of predicting plastic instability. The most pessimistic approximation consists, for any stress state on the plasticity convex set, in choosing the approximation of the tangent modulus which uses the following equation for the tangent constitutive matrix:

$$\underline{\underline{H_t}} = \frac{E_T}{E}\underline{\underline{H}} \tag{25}$$

The middle-of-the-road method consists in using finite plasticity theory, in which, at any point of the loading surface, the tangent constitutive law is approximated by Equation 26. These three ways of calculating the tangent stiffness in buckling can be applied in the general case, whether one uses a COQMULT model or a QUA8 continuum mechanics model.

$$\underline{\underline{H_t}} = \left[\left(\frac{1}{E_T} - \frac{1}{E_S}\right)\frac{\underline{\underline{A}}\underline{s} \otimes \underline{s}^T\underline{\underline{A}}}{s^{*2}} + \underline{\underline{H}}^{-1} + \left(\frac{1}{E_S} - \frac{1}{E}\right)\underline{\underline{A}}\right]^{-1} \tag{26}$$

Results and discussions

Now let us apply the buckling models to the prediction of elastic and elastic-plastic buckling for three types of sandwich shells:

- a half-sphere fixed at the base and subjected to uniform external pressure;
- three cone frustra fixed at their bases and subjected to uniform external pressure;
- the same three cone frustra, this time subjected to uniform internal pressure.

Geometries and materials

All the shells considered consisted of three layers. The two skins were the same thickness (t = 1 mm) and the core layer was filled with a material which was generally weaker than the skin. Three core layers with thicknesses e equal to 1, 9 and 49 mm, respectively (making their overall thicknesses equal to 3, 11 and 51 mm) are examined. The chosen generic geometries and the properties of the materials, which were assumed to be isotropic, are defined below:

The half-sphere

The half-sphere had a radius R equal to 1 m and was fixed at the base.

The cones

Three types of cones, with angles 30°, 45° and 60° from the horizontal, were studied. All three had the same radius at the base R_{min} equal to 1 m and the same maximum radius R_{max} equal to 2 m. The bases were fixed. Thus, 9 cases under internal pressure and 9 cases under external pressure were studied.

The materials

The materials were assumed to be elastic or perfectly elastic-plastic. The yield strain was 0.001. The same Poisson's coefficient is chosen for all materials (equal to 0.3). The Young's modulus of the skins was $E_{skin} = 200{,}000$ MPa. For the core layer of the sandwich shells, decreasing values of the Young's modulus were chosen among the following list: $E_{core} =$ 200,000 MPa, 20,000 MPa, 2,000 MPa, 200 MPa and 20 MPa. For the nonlinear analyses, the moduli $E_{core} = 200{,}000$ MPa, 2,000 MPa and 20 MPa alone were studied. Hereafter, the ratio $\frac{E_{core}}{E_{skin}}$ will be denoted β.

The finite element models

All these cases were modeled as axisymmetric problems in two ways. The first model will be referred to as the COQMULT model. The mid-line was meshed using p + 1 nodes, *i.e.* p elements. Each layer was assumed to have its own offset, its own thickness and its own material. The second model will be referred to as the QUA8 model. The thickness was represented by 3 quadrangular isoparametric elements with 8 nodes and 9 integration points. The generatrix was meshed using q elements. Thus, the final mesh contained 3q elements and $10q + 7$ nodes. The stability analysis was carried out in the Fourier basis by seeking the Fourier mode leading to the lowest critical load.

In the case of the QUA8 mesh of the cone, rectangular elements were chosen. Thus, the mesh represented the shell's geometry exactly. A preliminary convergence study was performed for each model in order to predict the linear buckling pressure of a single layer under external or internal pressure. This study enabled us to choose the mesh size for each case. The coarsest mesh which enabled the critical pressure to converge to within 1% of the exact solution has been chosen. A decomposition into 100 regular elements along the mid-line ensured the convergence of the elastic buckling load.

Regarding the meshing of the cone, 50 shell elements were sufficient to achieve convergence. Table 1 summarizes the meshes which were chosen for the calculations.

Linear elastic buckling

Now let us compare the linear elastic buckling loads which were obtained for each case.

A formula for calculating the change in critical pressure for each case when replacing a homogenous shell by a sandwich shell

A simple method for predicting the buckling pressure of a sandwich shell modeled using shell theory if one knows the result for a shell of the same geometry, but made of a single material whose properties are those of the skin will be presented. Let us compare the membrane and bending stiffnesses of the simple shell with those of the sandwich shell. The membrane stiffness of a shell of thickness h and Young's modulus E is (Eh). The associated bending stiffness is $D = E\frac{h^3}{12(1-\nu^2)}$. Let $(Eh)^*$ and D^* denote respectively the homogenized membrane and bending stiffnesses of the multilayer shell. All the formulas giving the theoretical critical loads of the shells considered here are proportional to the product of these two stiffnesses raised to some power m:

$$k = \left(D^*(Eh)^*\right)^m \quad (27)$$

Table 1 The meshes chosen for the sphere and cone cases

Geometry	p COQMULT	q QUA8
Sphere	100	100
Cone	50	100

The ratio of the two critical loads can be expressed in general form as:

$$r_{gen} = \left(\frac{D^*(Eh)^*}{DEh}\right)^m \tag{28}$$

Now let us define the variable $x = \frac{2t}{e}$ where t is the skin thickness and e the core thickness. One has the relation $h = e(1 + x)$. With these notations, one has:

$$(Eh)^* = eE_{skin}(\beta + x) \tag{29}$$

$$D^* = e^3\frac{E_{skin}}{12(1 - \nu^2)}\left[\beta + x\left(3 + 3x + x^2\right)\right] \tag{30}$$

Therefore, the buckling load of the sandwich shell is obtained by multiplying the critical load of the homogenous shell with the same overall thickness by the coefficient r given by the equation:

$$r = \frac{f_{case}(\beta, x)}{(1 + x)^m} \tag{31}$$

The function f and the power m depend on the case being considered. Here, one has three cases:

- The sphere under external pressure. The critical buckling pressure is given by the formula:

$$P_E^{sphere} = \frac{2E}{\sqrt{3(1 - \nu^2)}}\left(\frac{h}{R}\right)^2 \tag{32}$$

  This critical load is found by applying the Volmir method [20], based on Donnell's shell model. The equation giving the stability of a sphere of radius R and thickness h made of a single material (Young's modulus E, Poisson's coefficient ν) under an external pressure p is:

$$D\nabla^6 w + p\frac{R}{2}\nabla^4 w + \frac{Eh}{R^2}\nabla^2 w = 0 \tag{33}$$

  In Equation (33), w is the displacement normal to the shell. The solution is found by assuming that the buckling mode w satisfies $\nabla^2 w = -z^2 w$, where z is a parameter to be determined. Then, one has the following equation in w:

$$\left(Dz^4 + p\frac{R}{2}z^2 + \frac{Eh}{R^2}\right)z^2 w = 0 \tag{34}$$

  Equation 34 is satisfied if the pressure p satisfies:

$$p = \frac{2Eh}{R^3 z^2} + \frac{2D}{R}z^2 \tag{35}$$

  The buckling pressure is obtained for the value of z^2 which minimizes that function. One gets:

$$z_{min}^2 = \sqrt{\frac{Eh}{DR^2}} = \frac{1}{Rh}\sqrt{12(1 - \nu^2)} \tag{36}$$

Substituting z given by Equation 36 into Equation 35, one retrieves Equation 32. Now let us extend this analytical solution to the case of multilayer spherical shells, which will enable us to predict the analytical critical pressure of the sphere under the assumption that that shell still satisfies shell theory. In order to do that, one replaces the membrane and bending stiffnesses Eh and D in Equation (33) by their homogenized values $(Eh)^*$ and D^*. Hence:

$$p^*_{crit} = \frac{4}{R^2}\sqrt{D^*\,(Eh)^*} \tag{37}$$

This is a very general formula which is valid for any kind of multilayer shell. One can immediately note that the exponent m of interest is equal to $\frac{1}{2}$.
In our case, one gets:

$$r_{sphere}\,(\beta, x) = \frac{\sqrt{\beta^2 + \beta x\,(4 + 3x + x^2) + x^2\,(3 + 3x + x^2)}}{(1 + x)^2} \tag{38}$$

Now let us describe some properties of this ratio. When the Young's modulus of the core layer tends toward zero (the modular ratio of core to skin β is then 0.), the limit of the ratio r_{sphere} is:

$$r_{sphere}\,(0, x) = \frac{x\sqrt{3 + 3x + x^2}}{(1 + x)^2} \tag{39}$$

When $\beta = 1$ one gets back to r = 1.

- The cone under external pressure. The buckling mode of the cone under external pressure fixed along its smaller diameter and free along its larger diameter is similar to that of a cylinder of finite length under external pressure. The proof which leads to the critical loads shows that the buckling pressure is proportional to $\left[(D^*)^3\,(Eh)^*\right]^{2.5}$. After some calculations, the ratio of the critical loads is found to be:

$$r_{cone-pext}\,(\beta, x) = \frac{\left[\left(\beta + x\left[3 + 3x + x^2\right]\right)^3(\beta + x)\right]^{0.25}}{(1 + x)^{2.5}} \tag{40}$$

Here, the exponent m is equal to $\frac{5}{2}$. Indeed, one can note that if $\beta = 1, r = 1$. If β equals zero, one gets the lower critical pressure from the equation:

$$r_{cone-pext}\,(0, x) = \frac{\sqrt{x}\,(3 + 3x + x^2)^{\frac{3}{4}}}{(1 + x)^{2.5}} \tag{41}$$

- The cone under internal pressure. In the case of the cone under internal pressure, the critical pressures are estimated by replacing the cone by an equivalent cylinder subjected to uniform axial compression at its base. The proof of the theoretical formulas giving the critical load of the cylinder [12] is more complex than, but similar to that given above for the sphere under external pressure. One can show that the critical load is still governed by $\sqrt{D^*(Eh)^*}$. Therefore, this case is similar to that of the sphere under external pressure and the reduction factor is the same:

$$r_{cone-pint}\,(\beta, x) = r_{sphere}\,(\beta, x) \tag{42}$$

The sphere under external pressure

The results of the two finite element models are compared to the values predicted by Equation (32) in Table 2 below.

Table 2 The calculated buckling loads for the 3 thicknesses

Overall thickness mm	P_E (Koiter) (MPa)	COQMULT (MPa)	QUA8 (MPa)
3	2.179	2.179	2.179
11	29.29	29.12	29.01
51	630	612	587

The agreement was very good for the two thinnest shells. The results of the calculation for the thickest shell were still close to the theoretical formula, but less accurate. Let us observe that the thicker the shell, the further the result given by the solid finite element calculation from that given by the COQMULT calculation. The latter also departed from the theoretical solution obtained with Sanders Donnell's simplified shell theory. Let us note that in the case of the thick shell the slenderness ratio ($\frac{R}{h}$) was only 20: the shell was not really thin and thin shell assumptions were no longer relevant.

Table 3 below compares, for each case, the critical pressures obtained with the two models. In all the calculations, the critical loads of Fourier modes 0 to 5 differed by less than 3%.

The calculated critical loads are given in relation to the elastic buckling load of the shell made of a single material with Young's modulus 200,000 MPa.

The analytical formula giving the critical pressure of the sandwich shell was perfectly satisfied by the multilayer shell calculation. However, the "QUA8" calculation, which is not based on the same shell assumptions, led to smaller critical pressures when the modulus of the core layer was less than, or equal to, one hundredth of the modulus of the skin. This overestimation became higher as the core layer became thicker and its stiffness became smaller.

Table 3 Sphere under external pressure: the calculated buckling loads for the different cases

Overall thickness (mm)	β	COQMULT $\frac{P_{cr}}{P_{euler}^{E=200000}}$	r_{sphere} solution	QUA8 $\frac{P_{cr}}{P_{euler}^{E=200000}}$	Ratio QUA8/ COQMULT
3	0.0001	0.80	0.80	0.022	0.30
3	0.001	0.80	0.80	0.0	0.56
3	0.01	0.80	0.80	0.256	0.94
3	0.1	0.82	0.82	0.82	0.99
3	1.	1.00	1.0	1.00	1.00
11	0.0001	0.286	0.286	0.022	0.078
11	0.001	0.286	0.286	0.073	0.25
11	0.01	0.294	0.294	0.257	0.87
11	0.1	0.364	0.364	0.369	0.98
11	1.	1.00	1.0	1.00	0.996
51	0.0001	0.065	0.065	0.0015	0.023
51	0.001	0.066	0.066	0.0092	0.13
51	0.01	0.075	0.075	0.066	0.85
51	0.1	0.160	0.160	0.165	0.97
51	1.	1.00	1.0	1.00	0.96

Figure 3 compares the critical loads obtained for all the cases with the two models. One can clearly see that both models led to the same buckling loads as long as the modulus of the core layer remained greater than one tenth that of the skins.

With the COQMULT model, once the modulus became smaller than one hundredth that of the skins, the core layer ceased to contribute to the buckling strength. Nevertheless, the critical load thus estimated still could be much greater than that obtained using a QUA8 model, which allows the two skins to have independent kinematics.

Figures (4A) and (4B) show two typical deformed shapes obtained using the COQMULT models.

In the case of the thicker shell with the smaller modulus, the critical load was overestimated by a factor of 50 when using the multilayer shell model, which is considerable. The reason was that in the case of the COQMULT model the predicted mode was such that the two skins were interdependent (Figure 4A), whereas with the QUA8 model the predicted buckling mode involved both skins when $\beta = 1$, but only the external skin when $\beta < 0.01$. This is clearly visible if one compares Figures (4C) and (4D).

The critical load of the skin alone can be obtained using Koiter's formula (Equation 32). In our case, one gets 0.242 MPa. One can observe that the lower the modulus of the core

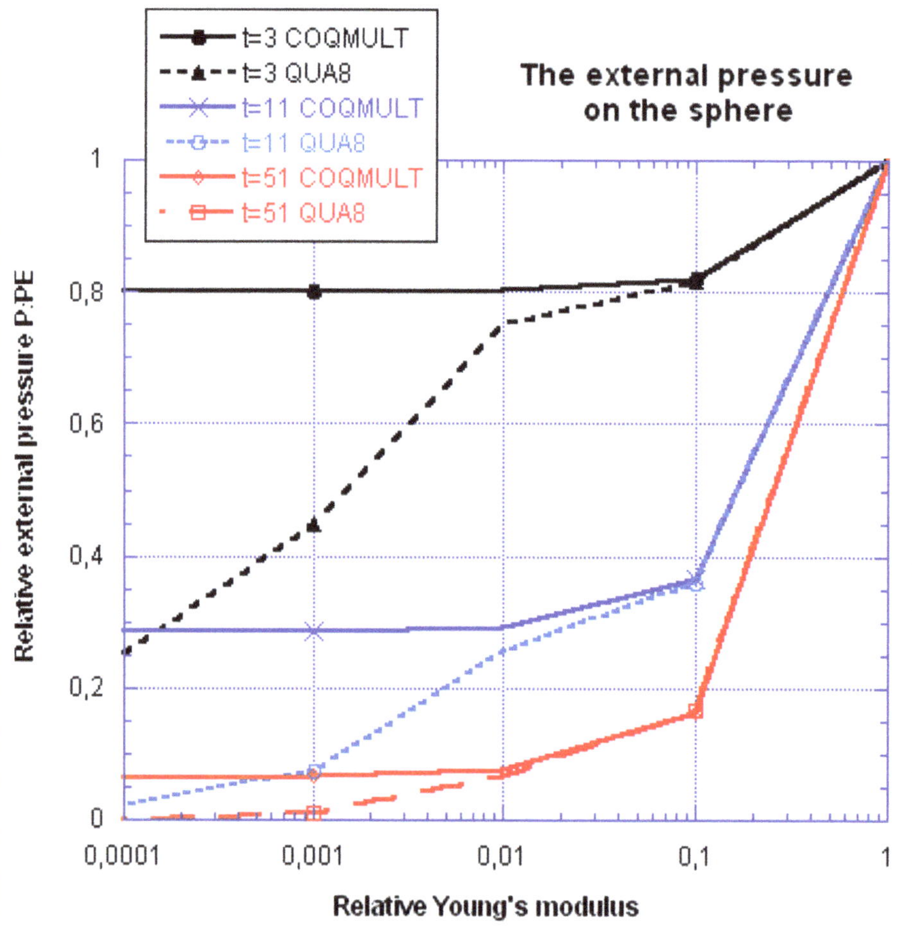

Figure 3 Comparison of the effects of the Young's modulus of the core layer on the sphere's buckling pressure using the two models (COQMULT and QUA8).

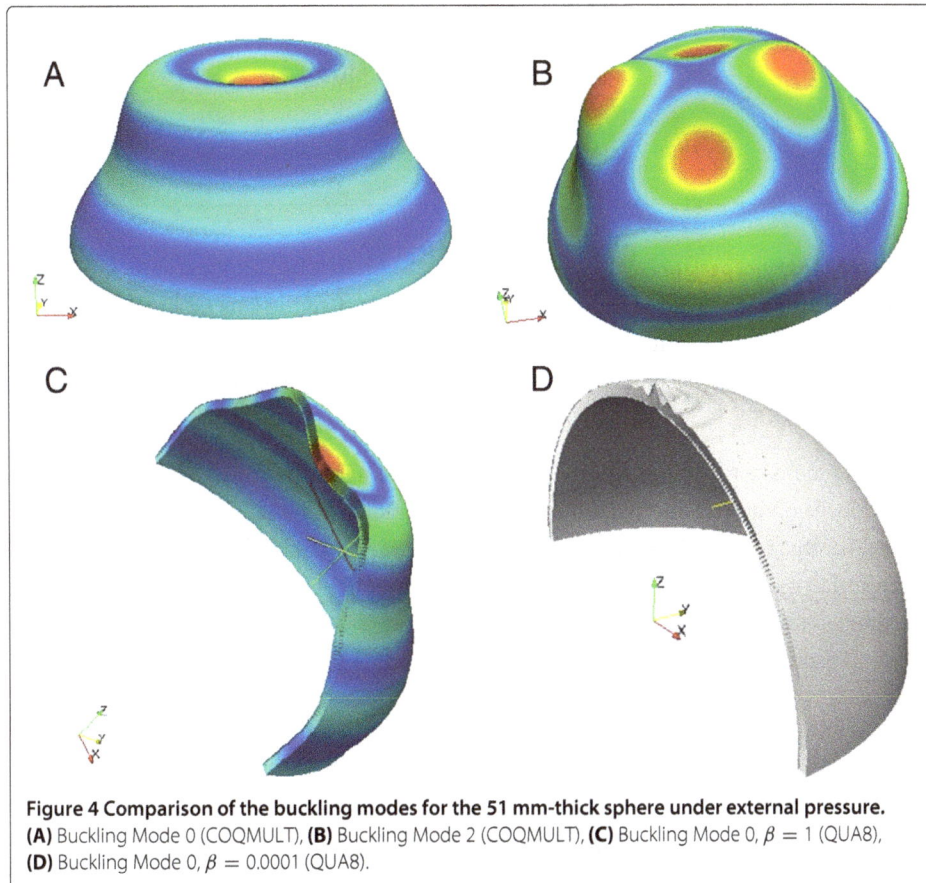

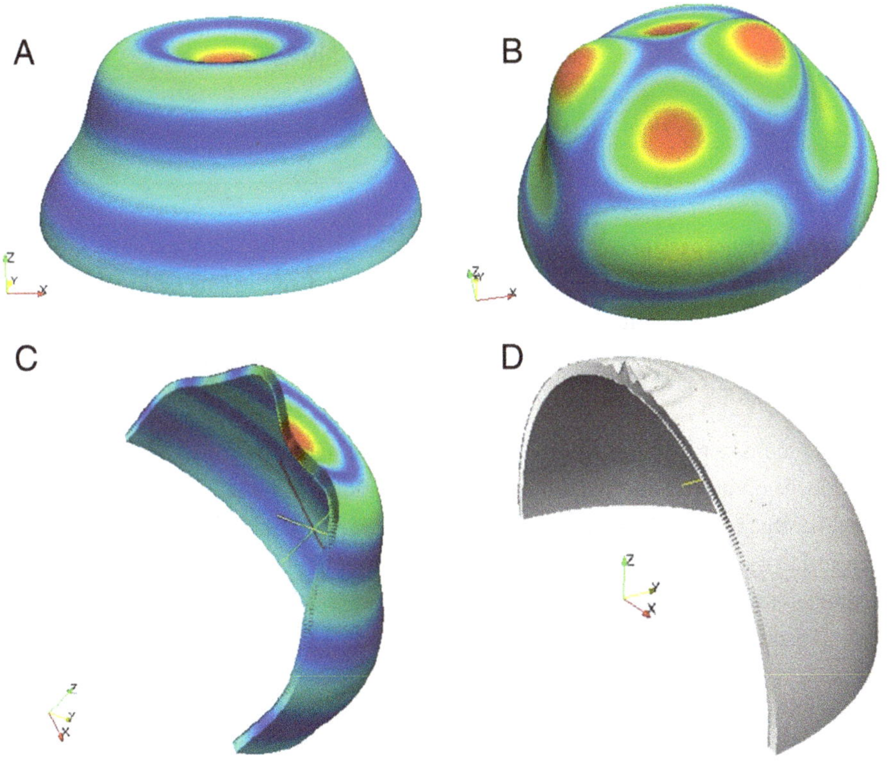

Figure 4 Comparison of the buckling modes for the 51 mm-thick sphere under external pressure.
(A) Buckling Mode 0 (COQMULT), **(B)** Buckling Mode 2 (COQMULT), **(C)** Buckling Mode 0, $\beta = 1$ (QUA8),
(D) Buckling Mode 0, $\beta = 0.0001$ (QUA8).

layer, the closer the critical load was to that value. Table 4 shows a comparison between the critical pressure of the skin alone and the critical pressure of the three stacked layers in the case where the modulus of the core layer was nearly zero (0.2 MPa): one can see that the critical pressure of the assembly was practically equal to that of the external skin alone in the case of the thickest shell and was 10% greater in the case of the thin shell.

The multilayer model overestimated the critical bucking load of the assembly by a factor of 100 in the case of the thickest shell, which is considerable.

The same comparison for the smallest Young's modulus of the parametric analysis ($\beta = 0.0001$) is shown in Table 5. The trend was the same, but one can see that the core layer still helped the external skin resist buckling. Nevertheless, the multilayer shell model overestimated the critical load by a factor of 3 in the case of the thin shell $\left(\frac{R}{t} = 333\right)$. This overestimation rose to a factor of 45 when the slenderness ratio was equal to $20 \left(\frac{R}{t} = 20\right)$.

Table 4 The elastic buckling loads for $E_{core} = 0.2$ MPa

Overall thickness mm	P_{lin}^{buck} (skin) (MPa)	COQMULT (MPa)	QUA8 (MPa)
3	0.242	1.74	0.316
11	0.242	8.34	0.270
51	0.242	39.7	0.246

Table 5 The elastic buckling loads for E_{core} = 20 MPa

Overall thickness mm	P^{buck}_{lin} (skin) (MPa)	COQMULT (MPa)	QUA8 (MPa)
3	0.242	1.74	0.55
11	0.242	8.34	0.65
51	0.242	39.7	0.9

The cones under external pressure

Usually, cones which are fixed along their base and whose fixed radius is smaller than the free radius buckle following a non-axisymmetric mode which will be called a "skirt mode". The associated displacement is maximal on the cone's free edge. A typical mode is shown in Figure (5).

- Let us first consider the cones meshed with QUA8 elements. In order to do that, the critical loads obtained with two different representations of the same mid-line will be compared. By meshing the cones with a horizontal free boundary ("HORI") instead of a free surface perpendicular to the the mean surface ("ORTHO") (Figure 6), the buckling pressure could be reduced significantly.
  A comparison of the two buckling modes corresponding to the two meshes of the 30° cone is shown in Figure 7. One can see that the smaller critical load obtained for the model with the horizontal free surface was due to a loss of stiffness at the top of the cone, leading to a smaller buckling load.
  Table 6 shows the results in the case of the thick cone (overall thickness 51 mm, single material), for which the difference was the greatest. One can note that for the 30° angle the model with the horizontal free surface led to a critical load which was six times smaller.
  From here on, the QUA8 mesh which conforms to the geometry of the COQMULT mesh will be retained.
- Now let us compare the results of the COQMULT models with the results of the QUA8 models. Table 7 shows the results of the calculations in the case of single-material cones. Again, in this case, one can see that the critical load obtained with the solid model was very close to that given by the shell model for the two thinnest walls, and was slightly smaller for the thickest wall. Once again, the problem lies near the limit of validity of the shell model.
  The typical buckling modes of the 60° and 30° cones are shown in Figure (5).

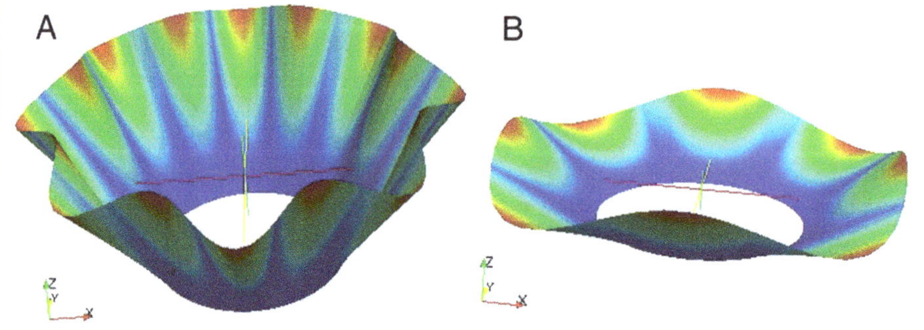

Figure 5 The typical buckling modes for two cones under external pressure. (A) "ORTHO" Mesh, **(B)** "HORI" Mesh.

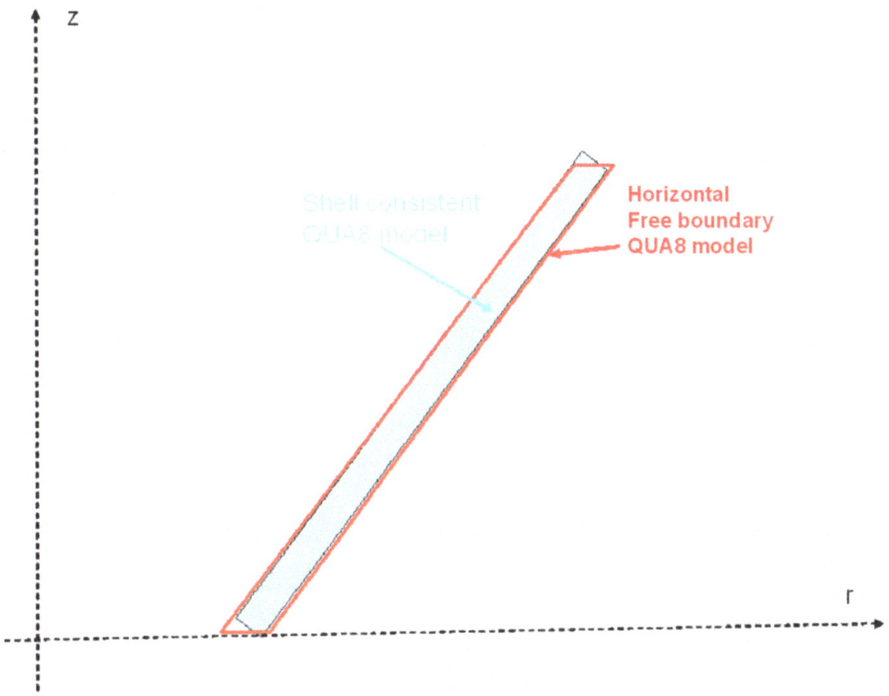

Figure 6 Schematic view of the two representations of a cone's geometry.

Tables 8, 9 and 10 show the normalized Euler loads calculated for the 9 cases and for each of the 3 cone angles studied. The last column also shows the ratio of the elastic buckling load predicted by the QUA8 model to that predicted by the COQMULT model. The agreement between the buckling pressures predicted by the COQMULT and QUA8 models was good for the two thinnest cones as long as $\beta \geq 0.01$. In the case when $\beta = 0.0001$, the critical load given by the shell model was overestimated by 30%. For the thickest cone, the critical load was overestimated by a factor of 3 to 5. Once again, the shell model precluded the buckling of the external skin alone. One can also note that when the Young's modulus of the core layer was very small the rank of the Fourier buckling mode found by the QUA8 model was always greater than that of the COQMULT model. For all the cone angles considered, the results in terms of relative pressure were quite similar, which indicates that the cone angle had very little influence on the relative loss of stiffness induced by a core layer with a smaller modulus. This can probably be explained by the facts that the mode

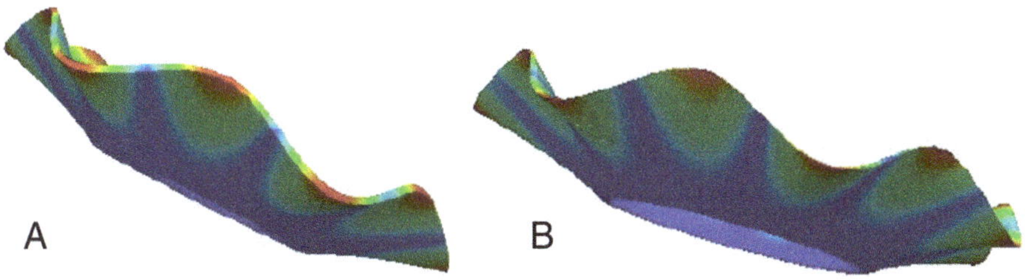

Figure 7 Comparison of the buckling modes for the 51 mm-thick 30° cone under external pressure. **(A)** 60° angle, **(B)** 30° angle.

Table 6 The calculated buckling pressures and modes for the 3 cone angles and the 3 models: Shell, QUA8 with the standard free surface ("ORTHO") and QUA8 with the horizontal free surface ("HORI")

Angle °	COQMULT (MPa)	Mode	Standard QUA8 (MPa)	Mode	Horizontal QUA8 (MPa)	Mode
30	7.34	4	7.28	4	1.15	5
45	9.21	4	9.06	4	3.607	5
60	7.83	4	8.08	4	5.529	4

developed in the cone's upper portion alone and that the external radius was the same for all the shells. Finally, let us note that the estimated reduction in the critical load proposed by Formula (40) was relatively accurate for the multilayer shell model, but it overestimated the critical pressure when the modulus of the core layer was very small because it is based on a pure shell kinematics.

Figure 8 compares the effects of the stiffness of the core layer on the critical buckling mode for the two models and the three types of cones (with overall thicknesses equal to 3, 11 and 51 mm). One can observe that the behavior was very similar for each of the cone geometries and was also similar to the case of the sphere. The estimate of the reduction in the critical load given by the analytical formula remained quite acceptable. The maximum error was less than 3%.

The cones under internal pressure

The same examples were calculated with an internal pressure load. In such a case, the cone buckles following an axisymmetric mode near the smaller radius. The results for the 60°, 45° and 30° cones are given in Tables 11, 12 and 13 respectively. Once again, one can observe that the multilayer shell model drastically overestimated the critical loads when the Young's modulus of the core layer was small. In this case, the overestimation could become considerable, reaching a factor of 50 in the case of the 45° angle and the thickest shell. The ratio of the critical pressure of the single-material shell to that of the three-layer shell still followed Equation 42 as was the case for the sphere under external pressure.

Table 7 The calculated linear external buckling pressure for the 3 cone angles and the 3 thicknesses

Angle °	Overall thickness mm	COQMULT (MPa)	Mode	QUA8 (MPa)	Mode
30	3	0.0052	9	0.0052	9
30	11	0.139	6	0.139	6
30	51	7.34	4	7.28	4
45	3	0.00709	9	0.00708	8
45	11	0.187	6	0.187	6
45	51	9.21	4	9.06	5
60	3	0.00677	8	0.00678	8
60	11	0.107	6	0.108	6
60	51	7.83	4	8.08	4

Table 8 The calculated normalized external buckling pressure for the 60° cone

Overall thickness (mm)	β	COQMULT $\frac{P_{cr}}{P^{E=200000}_{COQMULT}}$	$r_{cone-pext}$	Mode	QUA8 $\frac{P_{cr}}{P^{E=200000}_{QUA8}}$	Mode	Ratio QUA8/ COQMULT
3	0.0001	0.889	0.878	8	0.729	8	0.82
3	0.001	0.889	0.878	8	0.868	8	0.98
3	0.01	0.891	0.879	8	0.888	8	1.0
3	0.1	0.899	0.892	8	0.900	8	1.0
3	1.0	1.	1.	8	1.0	8	1.0
11	0.0001	0.366	0.360	5	0.176	7	0.49
11	0.001	0.366	0.361	5	0.325	5	0.91
11	0.01	0.376	0.367	5	0.358	5	0.98
11	0.1	0.443	0.411	5	0.429	5	0.99
11	1.0	1.	1.	6	1.	5	1.02
51	0.0001	0.097	0.087	3	0.022	6	0.24
51	0.001	0.098	0.088	3	0.072	4	0.76
51	0.01	0.11	0.097	3	0.101	4	0.95
51	0.1	0.196	0.183	4	0.187	4	0.99
51	1.0	1.	1.	4	1.	4	1.03

The evolutions of the relative critical pressures as functions of β, the ratio of the moduli, are shown in Figure 9. Again, the cone angle did not appear to have much influence.

The three buckling modes are shown in Figure 10. For all the cases, one can note the presence of an axisymmetric axial compression buckling mode at the base.

Table 9 The calculated normalized external buckling pressure for the 45° cone

Overall thickness (mm)	β	COQMULT $\frac{P_{cr}}{P^{E=200000}_{COQMULT}}$	$r_{cone-pext}$	Mode	QUA8 $\frac{P_{cr}}{P^{E=200000}_{QUA8}}$	Mode	Ratio QUA8/ COQMULT
3	0.0001	0.889	0.878	9	0.698	9	0.98
3	0.001	0.889	0.878	9	0.863	9	0.97
3	0.01	0.891	0.879	9	0.890	9	1.0
3	0.1	0.900	0.892	9	0.900	9	1.0
3	1.0	1.	1.	9	1.0	8	1.0
11	0.0001	0.360	0.360	6	0.152	8	0.42
11	0.001	0.361	0.361	6	0.316	6	0.87
11	0.01	0.367	0.367	6	0.361	6	0.98
11	0.1	0.412	0.430	6	0.425	6	1.03
11	1.0	1.	1.	6	1.	6	1.0
51	0.0001	0.089	0.087	4	0.017	4	0.19
51	0.001	0.090	0.088	4	0.066	4	0.72
51	0.01	0.099	0.097	4	0.095	4	0.94
51	0.1	0.143	0.183	4	0.151	4	1.03
51	1.0	1.	1.	4	1.	5	1.0

Table 10 The calculated normalized external buckling pressure for the 30° cone

Overall thickness (mm)	β	COQMULT $\frac{P_{cr}}{P^{E=200000}_{COQMULT}}$	$r_{cone-pext}$	Mode	QUA8 $\frac{P_{cr}}{P^{E=200000}_{QUA8}}$	Mode	Ratio QUA8/ COQMULT
3	0.0001	0.890	0.878	9	0.69	9	0.78
3	0.001	0.890	0.878	9	0.86	9	0.97
3	0.01	0.890	0.879	9	0.89	9	0.99
3	0.1	0.900	0.892	9	0.900	9	1.0
3	1.0	1.	1.	9	1.0	9	1.0
11	0.0001	0.360	0.360	6	0.150	8	0.41
11	0.001	0.364	0.361	6	0.314	6	0.86
11	0.01	0.370	0.367	6	0.362	6	0.97
11	0.1	0.430	0.43	6	0.423	6	1.0
11	1.0	1.	1.	6	1.	6	1.0
51	0.0001	0.096	0.087	4	0.016	9	0.16
51	0.001	0.096	0.088	4	0.067	4	0.69
51	0.01	0.104	0.097	4	0.098	4	0.93
51	0.1	0.185	0.183	4	0.184	4	0.98
51	1.0	1.	1.	4	1.	4	0.99

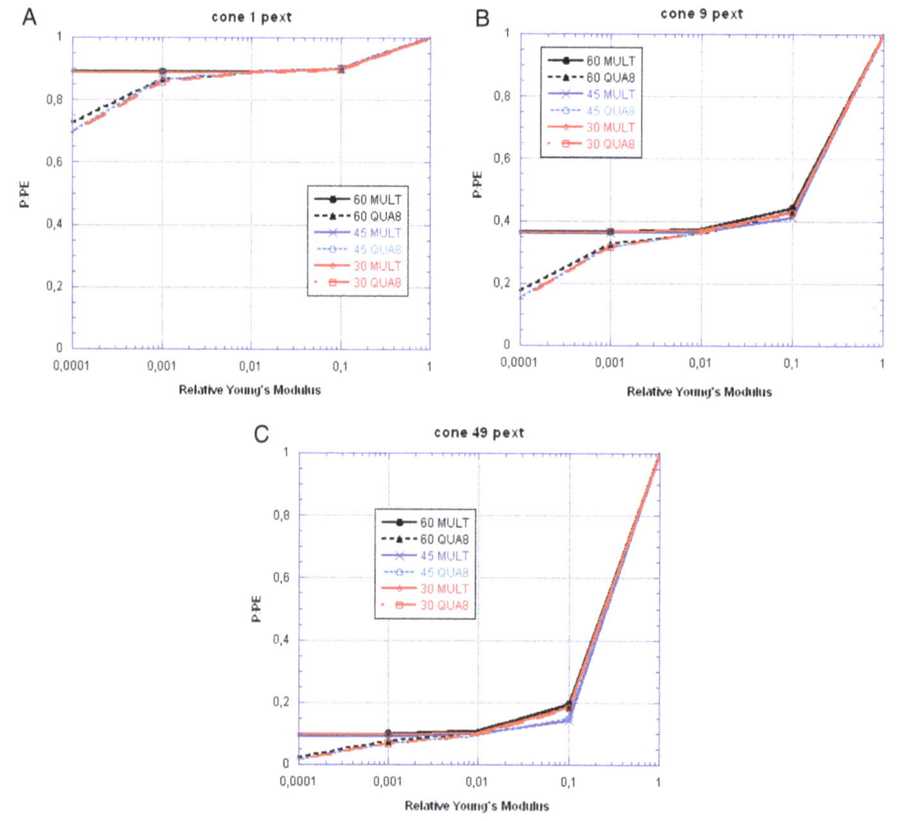

Figure 8 The effect of β (the relative stiffness of the internal layer) on the external buckling pressure of the cone. (A) Overall thickness 3 mm, **(B)** Overall thickness 11 mm, **(C)** Overall thickness 51 mm.

Table 11 The calculated relative internal buckling pressure for the 60° cone

Overall thickness (mm)	β	COQMULT $\frac{P_{cr}}{P^{E=200000}_{COQMULT}}$	$r_{cone-pint}$	QUA8 $\frac{P_{cr}}{P^{E=200000}_{QUA8}}$	Ratio QUA8/ COQMULT
3	0.0001	0.804	0.801	0.253	0.32
3	0.001	0.809	0.801	0.452	0.58
3	0.01	0.811	0.803	0.755	0.94
3	0.1	0.829	0.823	0.820	1.00
3	1.0	1.	1.0	1.	1.01
11	0.0001	0.300	0.287	0.021	0.07
11	0.001	0.300	0.287	0.072	0.24
11	0.01	0.307	0.295	0.264	0.86
11	0.1	0.377	0.365	0.371	0.98
11	1.0	1.	1.	1.0	1.0
51	0.0001	0.074	0.067	0.0015	0.02
51	0.001	0.075	0.068	0.0074	0.10
51	0.01	0.084	0.077	0.049	0.57
51	0.1	0.171	0.165	0.167	0.97
51	1.0	1.	1.	1.0	0.99

Geometrically non linear elastic-plastic buckling

The objective of this section is to answer the question: "Are the limitations of the multi-layer shell modeling of sandwich shells observed in the linear case still valid if one takes into account material and geometric nonlinearities?" In order to do that, elastic-plastic buckling was predicted by applying the method described previously to the calculation of the stability of the nonlinear elastic-plastic solution. A perfect elastic-plastic material which becomes plastic when the strain exceeds 0.001 is assumed.

Table 12 The calculated relative internal buckling pressure for the 45° cone

Overall thickness (mm)	β	COQMULT $\frac{P_{cr}}{P^{E=200000}_{COQMULT}}$	$r_{cone-pint}$	QUA8 $\frac{P_{cr}}{P^{E=200000}_{QUA8}}$	Ratio QUA8/ COQMULT
3	0.0001	0.813	0.801	0.25	0.31
3	0.001	0.813	0.801	0.49	0.60
3	0.01	0.815	0.803	0.77	0.95
3	0.1	0.833	0.823	0.83	1.00
3	1.0	1.	1.	1.0	1.00
11	0.0001	0.306	0.287	0.021	0.07
11	0.001	0.306	0.287	0.079	0.26
11	0.01	0.314	0.295	0.273	0.87
11	0.1	0.382	0.365	0.378	0.99
11	1.0	1.	1.	1.0	1.0
51	0.0001	0.078	0.067	0.0015	0.02
51	0.001	0.079	0.068	0.0067	0.08
51	0.01	0.089	0.077	0.0407	0.45
51	0.1	0.174	0.165	0.17	0.97
51	1.0	1.	1.	1.0	0.99

Table 13 The calculated relative internal buckling pressure for the 30° cone

Overall thickness (mm)	β	COQMULT $\frac{P_{cr}}{P^{E=200000}_{COQMULT}}$	$r_{cone-pint}$	QUA8 $\frac{P_{cr}}{P^{E=200000}_{QUA8}}$	Ratio QUA8/ COQMULT
3	0.0001	0.82	0.801	0.26	0.31
3	0.001	0.82	0.801	0.54	0.67
3	0.01	0.82	0.803	0.79	0.96
3	0.1	0.83	0.823	0.83	1.00
3	1.0	1.	1.0	1.	1.00
11	0.0001	0.313	0.287	0.022	0.07
11	0.001	0.314	0.287	0.097	0.31
11	0.01	0.321	0.295	0.287	0.89
11	0.1	0.389	0.365	0.385	0.99
11	1.0	1.	1.0	1.	1.00
51	0.0001	0.068	0.067	0.0017	0.026
51	0.001	0.069	0.068	0.0072	0.10
51	0.01	0.078	0.077	0.044	0.57
51	0.1	0.160	0.0165	0.162	1.00
51	1.0	1.	1.0	1.	1.01

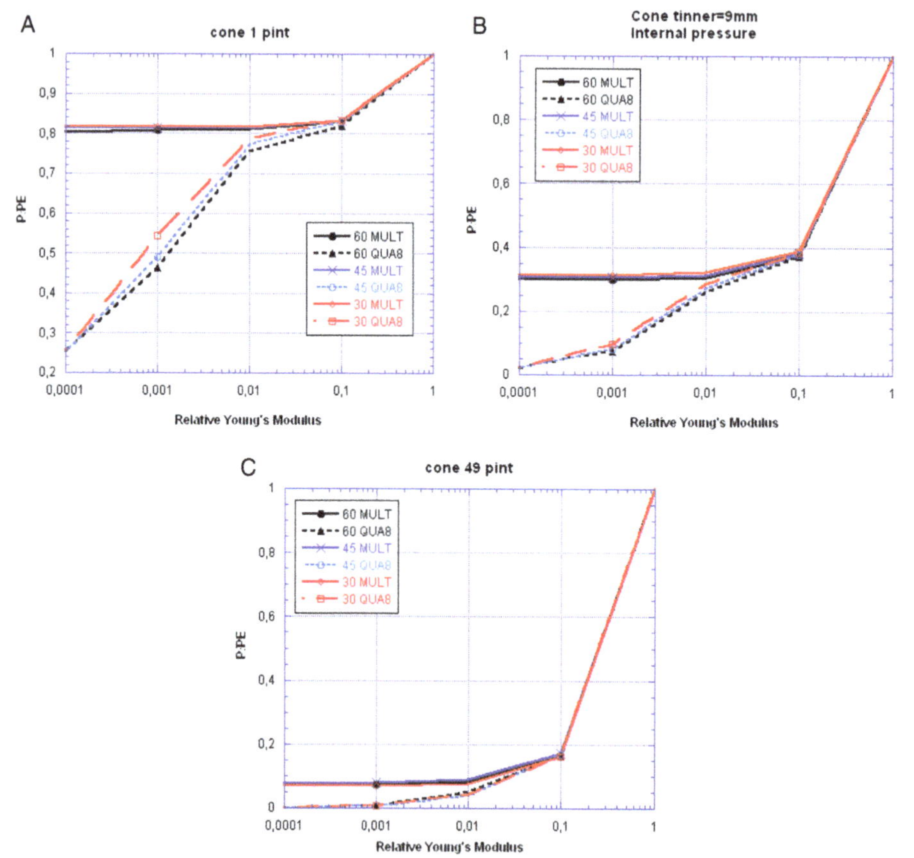

Figure 9 The effect of β (the relative stiffness of the core layer) on the cone's internal buckling pressure. (A) Overall thickness 3 mm, **(B)** Overall thickness 11 mm, **(C)** Overall thickness 51 mm.

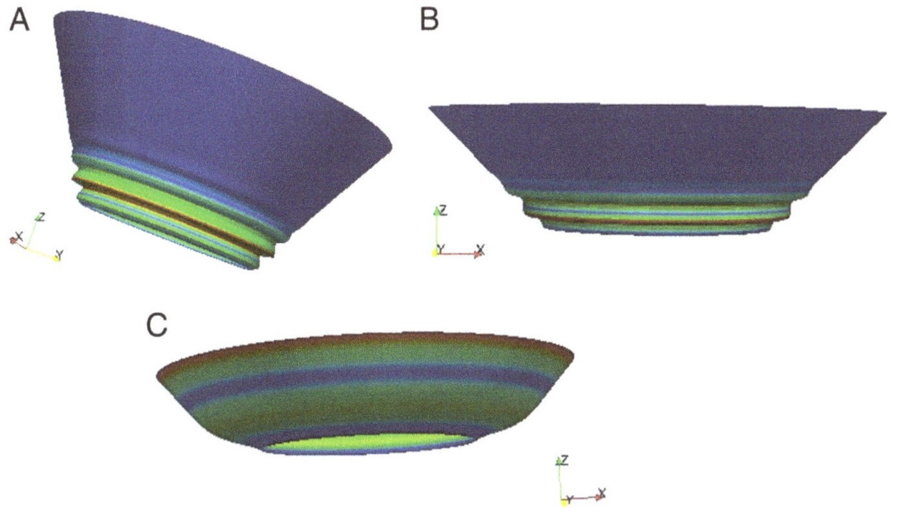

Figure 10 The buckling modes of the cones under internal pressure. **(A)** 60° angle, **(B)** 45° angle, **(C)** 30° angle.

The sphere under external pressure

- First, the nonlinear stability analysis was carried out for all the cases.
  For the sphere under external pressure, the results shown in Table 14 were obtained. One can note that plasticity reduced the buckling load by a factor of 2 for the thinnest spheres. The elastic-plastic critical load was 50 times smaller for the thickest spheres. One observes that the multilayer model is not dependable when the modulus of the core layer is very small. Once again, the local buckling mode of the external layer alone could not be predicted. The conclusion drawn in the case of elastic buckling still holds for nonlinear buckling.
- Now let us compare the critical loads and complete incremental responses of the COQMULT and QUA8 models in the case of more significant differences. This concerns the spheres with a 49 mm-thick core layer (overall thickness 51 mm) and a modulus equal to 20 MPa. This case can be calculated relatively easily because the

Table 14 The calculated elastic-plastic buckling loads for the sphere under external pressure

Overall thickness (mm)	β	COQMULT p_{crit}^{plast} (MPa)	$\frac{p_{crit}^{plast}}{P_E}$	QUA8 p_{crit}^{plast} (MPa)	$\frac{p_{crit}^{etan}}{P_E}$	Plasticity ratio QUA8/ COQMULT
3	0.0001	0.80	0.46	0.08	0.14	0.1
3	0.01	0.8	0.46	0.77	0.47	0.96
3	1.	1.17	0.54	1.16	0.53	0.99
11	0.0001	0.80	0.096	0.095	0.148	0.12
11	0.01	0.836	0.098	0.833	0.11	1.0
11	1.	4.4	0.15	4.39	0.15	1.
51	0.0001	0.802	0.02	0.018	0.021	0.023
51	0.01	1.0	0.022	0.98	0.025	0.98
51	1.	20.4	0.033	19.8	0.034	0.97

buckling modes are axisymmetric and, thus, one does not have to carry out coupled Fourier series calculations. Thus, one can use this case to compare the post-buckling behavior given by the two models. The results are summarized in Table 15. Several effects can be observed. For the COQMULT model with no defect, the two analyses (incremental and bifurcation prediction based upon the tangent modulus theory) led to approximately similar results. In the case of the QUA8 model, the predictions achieved with the three methods – buckling using the tangent modulus (etan), buckling using incremental plasticity theory (flow) and incremental analysis – led to different results. The incremental analysis without defect overestimated the critical load by 50%, except for the case when $\beta = 0.0001$. In the other cases, the estimates of the critical loads were identical for the COQMULT and QUA8 models regardless of the bifurcation analysis performed. In the case when $\beta = 0.0001$, the multilayer shell analysis overestimated the critical load by 50% when the buckling was not estimated using tangent modulus theory.

Figure 11 shows the comparison of the displacements at the top of the sphere obtained using the two incremental calculations COQMULT and QUA8 with no defect (the case when $\beta = 0.0001$). On this curve, one can clearly see that the thickness of the shell modeled with QUA8 elements diminishes at the top, a phenomenon which cannot be accounted for by the COQMULT model. The tangent modulus model appears to lead to an overly pessimistic estimate of the critical load in the case when $\beta = 0.0001$.

Figures 11 and 12 show a comparison of the displacements at the top of the sphere given by the COQMULT model respectively without and with a variable modal amplitude defect. The post-buckling behavior appears to be stable.

The cones under external pressure

The results of the elastic-plastic bifurcation calculations are given in Tables 16, 17 and 18 below. Let us note that all the bifurcations occurred for non-axisymmetric modes and that Fourier series analysis is very efficient in such cases. The lowest mode was always Mode 3. The thickest cone with the smallest core layer modulus meshed with QUA8 elements buckled following Mode 19. Of course, such an analysis could be carried out in 3D, but this would require significantly longer computation times: indeed, one would go from 300 2D Fourier elements to several tens of thousands of 3D elements.

For the cones under external pressure, one can note that the buckling was elastic for the two thinnest configurations. For the thickest shell, the buckling was elastic-plastic. No significant nonlinear geometric effect was observed for this type of buckling. Once

Table 15 Comparison of the estimated elastic-plastic buckling loads for the 51 mm-thick sphere under external pressure

β	COQMULT		QUA8		
	P^{plast}_{etan} MPa	P^{max}_{incr} MPa	P^{plas}_{etan} MPA	P^{plas}_{flow} MPa	P^{max}_{incr} MPa
0.0001	0.802	0.803	0.018	0.555	0.56
0.01	1.0	1.07	0.98	1.02	1.42
1.	20.4	20.5	19.8	20.44	33.7

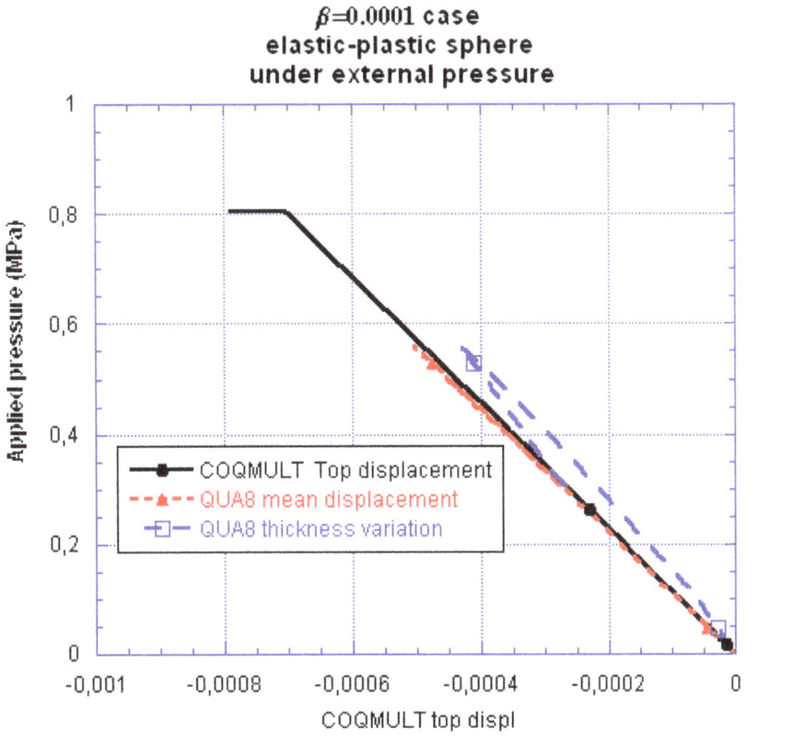

Figure 11 Comparison of the top inner displacements given by the COQMULT and QUA8 models.

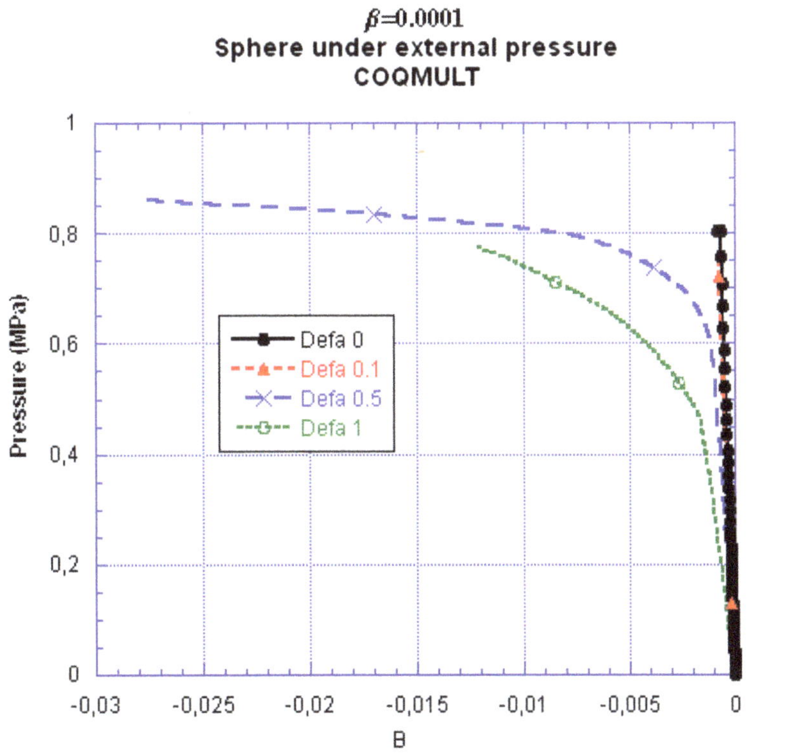

Figure 12 The effect of an initial imperfection (0, 0.01, 0.1) on the maximum pressure (COQMULT model).

Table 16 The calculated elastic-plastic buckling load under external pressure for the 60° cone

Overall thickness (mm)	β	COQMULT P^{plast}_{crit} (MPa)	$\frac{P^{plast}_{crit}}{P_E}$	QUA8 P^{plast}_{crit} (MPa)	$\frac{P^{plast}_{crit}}{P_E}$	Plasticity ratio QUA8/ COQMULT
3	0.0001	0.00605	1.0	0.00496	1.	0.82
3	0.01	0.00607	1.0	0.00605	1.	1.
3	1.	0.00859	1.27	0.00681	1.	0.79
11	0.0001	0.0627	1.0	0.028	0.9	0.44
11	0.01	0.0642	1.0	0.063	1.	0.98
11	1.	0.201	1.18	0.158	0.9	0.79
51	0.0001	0.175	0.23	0.032	0.18	0.18
51	0.01	0.219	0.25	0.196	0.24	0.89
51	1.	4.15	0.53	3.55	0.44	0.86

again, when the modulus of the core layer was very small, the buckling mode was a skin mode. This effect cannot be predicted using the COQMULT model. Because of elastic-plasticity, the thicker the cone, the smaller the ratio of elastic-plastic to elastic buckling pressure. These results confirm those obtained previously in the case of the sphere under external pressure. For the 30° case and a very low stiffness of the core layer, the critical load predicted by the elastic-plastic solid model was 10 times smaller.

The cones under internal pressure

Tables 19, 20 and 21 below show the complete results of the plastic bifurcation calculations. Let us note that all the bifurcations occurred for axisymmetric modes. These bifurcation analysis results can be compared to the results of nonlinear elastic-plastic analyses. In this case, the buckling was clearly plastic. Overall, the comparison between

Table 17 The calculated elastic-plastic buckling load under external pressure for the 45° cone

Overall thickness (mm)	β	COQMULT P^{plast}_{crit} (MPa)	$\frac{P^{plast}_{crit}}{P_E}$	QUA8 P^{plast}_{crit} (MPa)	$\frac{P^{plast}_{crit}}{P_E}$	Plasticity ratio QUA8 / COQMULT
3	0.0001	0.0063	1.0	0.00501	1.01	0.79
3	0.01	0.0063	1.0	0.00634	1.	1.0
3	1.	0.00898	1.27	0.00715	1.	0.8
11	0.0001	0.068	1.0	0.0134	0.47	0.20
11	0.01	0.069	1.0	0.068	1.	0.98
11	1.	0.212	1.13	0.188	1.	0.89
51	0.0001	0.144	0.175	0.0159	0.10	0.11
51	0.01	0.178	0.195	0.154	0.18	0.97
51	1.	9.21	0.382	2.66	0.294	0.76

Table 18 The calculated elastic-plastic buckling load under external pressure for the 30° cone

Overall thickness (mm)	β	COQMULT P^{plast}_{crit} (MPa)	$\frac{P^{plast}_{crit}}{P_E}$	QUA8 P^{plast}_{crit} (MPa)	$\frac{P^{plast}_{crit}}{P_E}$	Plasticity ratio QUA8/ COQMULT
3	0.0001	0.0046	1.0	0.0032	0.9	0.70
3	0.01	0.0046	1.0	0.0047	1.01	1.0
3	1.	0.0066	1.27	0.0054	1.04	0.82
11	0.0001	0.051	1.0	0.00708	0.37	0.15
11	0.01	0.052	1.0	0.051	1.02	0.98
11	1.	0.159	1.14	0.14	1.01	0.89
51	0.0001	0.097	0.14	0.01	0.08	0.10
51	0.01	0.12	0.16	0.104	0.15	0.87
51	0.1	2.58	0.35	1.75	0.24	0.68

the COQMULT and QUA8 models leads to the same conclusions as for elastic buckling: the COQMULT model drastically overestimated the critical load when the modulus of the core layer of the sandwich shell was very small. Again, in this case, one had a plastic buckling mode of the skin alone, which the COQMULT model is unable to predict. Depending on the case, the nonlinear elastic-plastic critical load varied between one hundredth and 90% of the elastic buckling load.

In this case, one finds again, although to a lesser degree than for the sphere, that tangent modulus theory led to a safe estimate of the critical load for $\beta = 0.0001$, the QUA8 model and the thickest shell. For the other cases, the COQMULT became more unsafe as the shell's thickness increased. The critical plastic buckling pressure predicted using incremental plasticity theory along with the QUA8 model was

Table 19 The calculated elastic-plastic buckling load under internal pressure for the 60° cone

Overall thickness (mm)	β	COQMULT P^{plast}_{crit} (MPa)	$\frac{P^{plast}_{crit}}{P_E}$	QUA8 P^{plast}_{crit} (MPa)	$\frac{P^{plast}_{crit}}{P_E}$	Plasticity ratio QUA8 / COQMULT
3	0.0001	0.156	0.3	0.0186	0.11	0.12
3	0.01	0.158	0.3	0.15	0.31	0.95
3	1.	0.307	0.48	0.224	0.35	0.73
11	0.0001	0.175	0.062	0.0306	0.15	0.18
11	0.01	0.183	0.064	0.142	0.06	0.78
11	1.	1.04	0.11	0.892	0.09	0.86
51	0.0001	0.183	0.01	0.0039	0.01	0.02
51	0.01	0.253	0.012	0.183	0.016	0.72
51	1.	4.64	0.019	3.3	0.014	0.71

Table 20 The calculated elastic-plastic buckling load under internal pressure for the 45° cone

Overall thickness (mm)	β	COQMULT P^{plast}_{crit} (MPa)	$\frac{P^{plast}_{crit}}{P_E}$	QUA8 P^{plast}_{crit} (MPa)	$\frac{P^{plast}_{crit}}{P_E}$	Plasticity ratio QUA8/ COQMULT
3	0.0001	0.127	0.35	0.04	0.37	0.33
3	0.01	0.128	0.35	0.11	0.31	0.85
3	1.	0.248	0.55	0.18	0.39	0.71
11	0.0001	0.145	0.07	0.01	0.07	0.07
11	0.01	0.152	0.07	0.12	0.06	0.79
11	1.	0.860	0.125	0.59	0.09	0.69
51	0.0001	0.145	0.01	0.006	0.02	0.04
51	0.01	0.184	0.01	0.048	0.006	0.26
51	1.	3.80	0.02	0.77	0.004	0.2

up to 50% greater than using tangent modulus theory. However, both these critical loads were always smaller than, or equal to, that obtained using the COQMULT model.

Conclusions

This paper presented the analytical formulae for the elastic buckling loads of sandwich cones and spheres under pressure loading. These formulae were successfully compared to multilayer shell calculations. Then, one showed the limitations of that formulation which relies on a shell kinematics which does not allow the skins to wrinkle independently of each other. The calculations were compared to continuum mechanics calculations, which are the most relevant calculations in these cases.

All these calculations showed that when the Young's modulus of the sandwich shell's core layer is more than one tenth that of the skins the multilayer model suffices. When the ratio of the moduli is less than one hundredth (which is often the case with foam

Table 21 The calculated elastic-plastic buckling load under internal pressure for the 30° cone

Overall thickness (mm)	β	COQMULT P^{plast}_{crit} (MPa)	$\frac{P^{plast}_{crit}}{P_E}$	QUA8 P^{plast}_{crit} (MPa)	$\frac{P^{plast}_{crit}}{P_E}$	Plasticity ratio QUA8 / COQMULT
3	0.0001	0.086	0.44	0.004	0.06	0.05
3	0.01	0.087	0.44	0.06	0.34	0.75
3	1.	0.165	0.69	0.11	0.48	0.70
11	0.0001	0.10	0.09	0.01	0.09	0.09
11	0.01	0.11	0.09	0.08	0.75	0.75
11	1.	0.61	0.16	0.39	0.64	0.64
51	0.0001	0.11	0.015	0.011	0.06	0.10
51	0.01	0.13	0.016	0.095	0.02	0.71
51	0.1	2.70	0.025	1.19	0.01	0.44

materials) skin modes are likely to occur at pressures which are much lower than those predicted by the multilayer model. These conclusions are valid both in elasticity and in nonlinear elastic-plasticity. The analysis which is proposed here for the axisymmetric case is particularly efficient since a linear calculation such as those which were carried out in this study takes no more than a few seconds on a laptop PC and a nonlinear solid calculation takes only a few minutes.

Competing interests

The authors declare that they have no competing interests.

Authors' contributions

The author has programmed the software made all the computations and developements presented in this work. The article was first written in French, then translated in English by a professionnal translator.

References

1. Budiansky B (1999) On the minimum weight of compression structures. Int J Solids Struct 36:3677–3708
2. Hutchinson JW, He MY (2000) Buckling of cylindrical sandwich shells with metal foam cores. Int J Solids Struct 37(46-47):6777–6794
3. Bazant Z, Beghini A (2004) Sandwich buckling formulas and applicability of standard computational algorithms for finite strains. Composites Part B Eng 35(6-8):573–581
4. Anon (1965) Buckling of thin circular cylinders. NASA Space Vehicle Design Criteria, NASA SP-8007
5. Ji W, Waas AM (2007) Global and local buckling of sandwich beams. J Engineering Mech. 230–237
6. Zhang Z, Liu S, Tang Z (2010) Crashworthiness investigation of kagome honeycomb sandwich cylindrical colums under axial crushing loads. Thin Walled Struct 48:9–18
7. Walker M, Smith M (2006) A procedure to select the best material combination and optimally design sandwich composite cylindrical shells for minimum mass. Mater Des 27(2):160–165
8. Leotoing L, Drapier S, Vautrin A (2002) First application of a novel unified model for local and global buckling of sandwich columns. Eur J Mech A/Solids 21:683–701
9. Golterman P, Molmann H (1989) Interactive buckling in thin walled beams: II applications. Int J Solids Struct 25(7):729–749
10. Molmann H, Golterman P (1989) Interactive buckling in thin walled beams: I theory. Int J Solids Struct 25(7):715–728
11. Szyniszewski S, Smith BH, Hajjar JF, Arwade SR, Schafer BW (2012) Local buckling strength of steel foam sandwich panels. Thin-Walled Struct 59:11–19
12. Koiter WT (1970) On the stability of elastic equilibrium. AFFDL-TR-70-25 Technical report
13. Bushnell D (1985) Computerized buckling analysis of shells. Springer, Series: Mech Elastic Stability 9 ISBN 978-90-247-3099-5
14. Bushnell D (1974) Bifurcation buckling of shells of revolution including large deflections, plasticity and creep. Int J Solids Struct 10:1287–1305
15. Ahmad S, Irons BM, Zienkiewicz OC (2005) Analysis of thick and thin shell structures by curved finite elements. Int J Numerical Methods Eng 2(3):419–451
16. Combescure A (1986) Static and dynamic buckling of large thin shells. Nucl Eng Des 92(3):339–354
17. Ricks E (1979) An incremental approach to the solution of snapping and buckling problems. Int J Solids Struct 15:524–551
18. Crisfield M (1980) A fast incremental-iterative solution procedure that handles snap through. Comput Struct 13:57-62
19. Gusic G, Combescure A, Jullien JF (2000) The influence of circumferential thickness variations on the buckling of cylindrical shells under external pressure. Comput Struct 74(4):461–477
20. Volmir AS (1963) Stability of elastic systems. Gos. Izd-vo Fiz.-Mat. Lit., Moscow

2

Adjoint-consistent formulations of slip models for coupled electroosmotic flow systems

Vikram V Garg[1*], Serge Prudhomme[2], Kris G van der Zee[3] and Graham F Carey[4^]

*Correspondence: vikram@ices.utexas.edu
^Deceased
[1] Massachusetts Institute of Technology, 77 Massachusetts Avenue, Cambridge, MA 02139, USA
Full list of author information is available at the end of the article

Abstract

Background: Models based on the Helmholtz 'slip' approximation are often used for the simulation of electroosmotic flows. The objectives of this paper are to construct adjoint-consistent formulations of such models, and to develop adjoint-based numerical tools for adaptive mesh refinement and parameter sensitivity analysis.

Methods: We show that the direct formulation of the 'slip' model is adjoint inconsistent, and leads to an ill-posed adjoint problem. We propose a modified formulation of the coupled 'slip' model, which is shown to be well-posed, and therefore automatically adjoint-consistent.

Results: Numerical examples are presented to illustrate the computation and use of the adjoint solution in two-dimensional microfluidics problems.

Conclusions: An adjoint-consistent formulation for Helmholtz 'slip' models of electroosmotic flows has been proposed. This formulation provides adjoint solutions that can be reliably used for mesh refinement and sensitivity analysis.

Keywords: Adjoint techniques; Electroosmosis; Microfluidics; Slip boundary conditions

Background

Introduction

Emerging micro- and nano-electromechanical systems (MEMS/NEMS) have a growing number of applications, ranging from lab-on-a-chip DNA analysis to micro-actuators [1]. By scaling down processes, these systems offer savings in space, cost, and energy for scientific and technological advancement [2]. However, the high manufacturing costs and complex architectures of these systems necessitate the use of numerical simulation tools for optimal design and precise control of their operation [3]. Microfluidic devices operate over various length scales and are best described using multiphysics modeling that involves hydrodynamics, electroosmosis, and chemical species transport models. The development of accurate and efficient computational simulators of these devices is therefore challenging and resource intensive.

Numerical simulations of such complex engineering systems are typically targeted towards the calculation of specific Quantities of Interest (QoI) associated with the systems. Accurate estimation of local QoIs can be achieved using goal-oriented error estimation and adaptive techniques based on the use of adjoint methods [4,5]. Adjoint methods can also be used to improve the computational performance of parameter sensitivity analyses [6], especially for systems with a large number of parameters. However, the application of adjoint methods to such coupled flow systems is an open area of study.

The objective of the current research work is to apply an adjoint-based Adaptive Finite Element Method (AFEM) to microfluidics problems for mesh refinement and parameter sensitivity analysis. There has been a growing interest in the modeling and numerical simulation of microfluidic systems [7-12]. A key issue that one faces while modeling and simulating microfluidic systems is the application of 'slip' boundary conditions. Prachittham et al. [13] presented a space-time adaptive finite element method applied to an electroosmotic flow using large aspect ratio elements. However, their work did not consider adjoint techniques and did not use the 'slip' boundary coupling condition. On the other hand, van Brummelen et al. [14] and Estep et al. [15] have shown the importance of the treatment of boundary flux coupling for the use of adjoint-based techniques. To our best knowledge, no advances in the application of adjoint-based techniques to microfluidics applications, particularly those involving 'slip' boundary coupling, have yet been published in the literature.

In this work, we investigate coupled systems arising in electroosmotic flow (EOF). We show that a naive formulation of the 'slip' model leads to an ill-posed adjoint problem and adjoint inconsistency [14]. We provide numerical evidence that the corresponding adjoint solution exhibits spurious oscillations on a simple example dealing with a straight channel flow. Accordingly, we propose a modified variational formulation of the 'slip' model, for which the adjoint problem is well-posed and can be computed and used in adaptive mesh refinement and parameter sensitivity analysis.

The paper is organized as follows: Section "Microfluidics modeling" describes the 'slip' electroosmotic flow (EOF) model, with a brief discussion of the relevant physics and the applicability of such a model. Then, a variational formulation for a modified version of the 'slip' model is presented and its adjoint is derived in Section "Variational formulation of the slip BC EOF model". An analysis of the ill-posed adjoint problem for the naive formulation is presented in Section "Ill-posedness of the adjoint problem for the naive formulation". Next, a variational formulation using penalty boundary conditions (henceforth called the penalty formulation) is proposed in Section "Penalty formulation of the slip BC EOF model". This new formulation is also employed to derive the corresponding adjoint problem and it is shown that the adjoint problem thus obtained is asymptotically consistent with the one derived in Section "Variational formulation of the slip BC EOF model". Numerical experiments are presented in Section "Results and discussion" that support the fact that the adjoint problem obtained using the penalty formulation is well-posed and that the adjoint solution is free of spurious artifacts. The adjoint obtained using the penalty formulation is used for goal-oriented adaptive mesh refinement and sensitivity analyses for a T-channel problem. Section "Conclusions" provides some concluding remarks, followed by a discussion of further work and future applications related to this class of coupled flow models. Finally, the appendix elaborates on the well-posedness of the modified formulation and presents relevant theoretical developments.

Microfluidics modeling

Microfluidics is the branch of fluid mechanics concerned with the understanding, modeling, and control of flows that occur on the micron scale, i.e. where the characteristic length (L) is of the order 10^{-6} m. Squires and Quake [16], and later Whitesides [17], have presented reviews of the physics and applications of microfluidics. Prominent among them are microflows driven by applied electric fields, through electroosmosis, electrophoresis, or both. In this paper, we consider electroosmotic flow devices, which find wide use in commercial and industrial applications [8,18]. These devices utilize the properties of the electric double layer (also called the Debye layer) to drive a bulk fluid flow. More detailed descriptions of the electric double layer and electroosmotically driven microfluidic devices can be found in the microfluidics literature [1,19,20].

Consider a rectangular open domain $\Omega \subset \mathbb{R}^d$, with d = 2, and boundary $\partial\Omega$. The boundary $\partial\Omega$ is composed of the channel wall Γ_w and its inlet/outlet Γ_{io}, such that $\partial\Omega = \overline{\Gamma_w \cup \Gamma_{io}}$. For simplicity, we consider a single species flow through the channel. The one-way coupled, steady-state EOF in a straight rectangular channel can be modeled with the following system of equations,

$$-\Delta\Psi = K^2 \sinh(\Psi) \quad \text{in } \Omega, \tag{1a}$$

$$-\nabla \cdot (\sigma_c \nabla\phi) = 0 \quad \text{in } \Omega, \tag{1b}$$

$$-\mu\Delta\mathbf{u} + \nabla p = -\rho_e(\Psi)\nabla\phi \quad \text{in } \Omega, \tag{1c}$$

$$\nabla \cdot \mathbf{u} = 0 \quad \text{in } \Omega, \tag{1d}$$

where Ψ is the non-dimensional electric potential associated with the double layer, K is a non-dimensional constant, called the Debye-Huckel parameter, ϕ is the applied potential, σ_c is the fluid conductivity, and $\mathbf{u}$ and p are the flow velocity and pressure, respectively. The charge density is given by $\rho_e = -2z_v n_\infty e \sinh(\Psi)$ and μ is the viscosity of the fluid. The parameters that the charge density depends on are the ion valence z_v, electron charge e, and the bulk concentration of ions n_∞. The above equations are supplemented with the boundary conditions,

$$\Psi = \Psi_0 \quad \text{on } \Gamma_w, \tag{2a}$$

$$\mathbf{n} \cdot \nabla\Psi = 0 \quad \text{on } \Gamma_{io}, \tag{2b}$$

$$\mathbf{n} \cdot (\sigma_c \nabla\phi) = 0 \quad \text{on } \Gamma_w, \tag{2c}$$

$$\phi = \phi_{io} \quad \text{on } \Gamma_{io}, \tag{2d}$$

$$\mathbf{u} = \mathbf{0} \quad \text{on } \Gamma_w, \tag{2e}$$

$$\mathbf{u} \cdot \mathbf{t} = 0 \quad \text{on } \Gamma_{io}, \tag{2f}$$

$$\mathbf{n} \cdot (\boldsymbol{\sigma} \cdot \mathbf{n}) = 0 \quad \text{on } \Gamma_{io}, \tag{2g}$$

where Ψ_0 is the electric zeta potential on the walls Γ_w of the channel, ϕ_{io} is the external potential applied at the inlet and outlet of the channel Γ_{io}, $\mathbf{n}$ is the unit outward normal vector to $\partial\Omega$, $\mathbf{t}$ is a unit tangential vector along $\partial\Omega$, and $\boldsymbol{\sigma}$ is the stress tensor:

$$\boldsymbol{\sigma} = -p\mathbf{I} + \mu\left(\nabla\mathbf{u} + (\nabla\mathbf{u})^T\right). \tag{3}$$

The no-slip condition (Eq. (2e)) is applied at the channel walls for the flow velocity. The no-slip conditions, in combination with the body force term in Eq. (1c), which is significant only near the boundary, lead to sharp velocity boundary layer near the wall. This makes the system given by Eq. (1) and Eq. (2) multi-scale. Note that the last two boundary (Eq. (2f) and (2g)) conditions imply that the velocity is normal to Γ_{io} and that the pressure vanishes on Γ_{io} (this is in the case of planar boundaries, see [21]), i.e.

$$p = 0 \quad \text{on } \Gamma_{io}. \tag{4}$$

Eqs. 1 and (2) define the complete EOF model and constitute a challenging system of coupled multiscale equations. They are computationally expensive, especially for complex geometries, due to the presence of extremely sharp layers near the channel walls because of the electric double layer. Therefore, to reduce the complexity and the computational cost associated with the full model, the Helmholtz-Smoluchowski velocity approximation is introduced into the model. The approximation states that the body-force term in the Stokes equations can be replaced by an effective 'slip velocity' on Γ_w as,

$$\mathbf{u}_{\text{wall}} = \frac{\epsilon \Psi_0}{\mu}(-\nabla \phi) = \lambda \mathbf{E}, \tag{5}$$

where $\mathbf{E}$ denotes the applied electric field, and a new parameter $\lambda = \epsilon \Psi_0/\mu$ has been introduced. Here, ϵ is the permitivity (or dielectric constant) of the fluid medium. A detailed derivation of this approximation can be found in a standard reference [19]. The validity of this approximation has been verified for several examples through both experiments and numerical simulations [10].

The associated slip model of EOF can thus be written as:

$$-\nabla \cdot (\sigma_c \nabla \phi) = 0 \quad \text{in } \Omega, \tag{6a}$$

$$-\mu \Delta \mathbf{u} + \nabla p = \mathbf{0} \quad \text{in } \Omega, \tag{6b}$$

$$\nabla \cdot \mathbf{u} = 0 \quad \text{in } \Omega, \tag{6c}$$

complemented by the boundary conditions,

$$\sigma_c \partial_n \phi = 0 \quad \text{on } \Gamma_w, \tag{7a}$$

$$\phi = \phi_{io} \quad \text{on } \Gamma_{io}, \tag{7b}$$

$$\mathbf{u} + \lambda \nabla \phi = \mathbf{0} \quad \text{on } \Gamma_w, \tag{7c}$$

$$\mathbf{u} \cdot \mathbf{t} = 0 \quad \text{on } \Gamma_{io}, \tag{7d}$$

$$\mathbf{n} \cdot (\boldsymbol{\sigma} \cdot \mathbf{n}) = 0 \quad \text{on } \Gamma_{io}. \tag{7e}$$

We see that the slip model spares one from solving the Poisson-Boltzmann equation, whose solution exhibits a thin layer near the wall. As a remark, the slip boundary approximation model given by Eq. (6) and (7) is widely used throughout the microfluidics research and development community for modeling and simulation [1]. The model is even included in the commercial Finite Element software package COMSOL Multiphysics [12].

The slip condition/coupling makes sense only in the direction tangent to the wall. In Eq. (7c), the no-flux boundary condition on the potential (Eq. (7a)) automatically enforces a no penetration boundary condition on the velocity. Thus, expressing the coupling condition as Eq. (5) is a matter of convenience from a notational standpoint. However, as shown in Section "Ill-posedness of the adjoint problem for the naive formulation", using Eq. (5) as such in the formulation of the coupled problem leads to an ill-posed adjoint problem. This ill-posedness will also be illustrated on numerical examples in Section "Results and discussion". We decouple one of the velocity components from the potential as follows,

$$\begin{cases} \mathbf{u}\cdot\mathbf{t}+\lambda\partial_t\phi=0 \\ \mathbf{u}\cdot\mathbf{n}+\lambda\partial_n\phi=0 \end{cases} \quad\Rightarrow\quad \begin{cases} \mathbf{u}\cdot\mathbf{t}+\lambda\partial_t\phi=0, \\ \mathbf{u}\cdot\mathbf{n}=0, \end{cases} \tag{8}$$

where we have denoted the normal ($\nabla\phi\cdot\mathbf{n}$) and tangential ($\nabla\phi\cdot\mathbf{t}$) derivatives of ϕ by $\partial_n\phi$ and $\partial_t\phi$, respectively. This coupling is equivalent to the one given by Eq. (7c). We now proceed to derive the weak formulation for Eq. (6) using the modified coupling given by Eq. (8).

Methods

Variational formulation of the slip BC EOF model

Variational formulation of primal problem

We derive here the weak formulation of the equations in Eq. (6) and then combine them to give the coupled problem. The potential ϕ satisfies,

$$-\nabla\cdot(\sigma_c\nabla\phi)=0 \quad \text{in } \Omega, \tag{9a}$$

$$\phi=\phi_{io} \quad \text{on } \Gamma_{io}, \tag{9b}$$

$$\sigma_c\partial_n\phi=0 \quad \text{on } \Gamma_w, \tag{9c}$$

We define $C^+(\Omega)$ as the space of continuous, strictly positive functions on Ω, and assume that the conductivity $\sigma_c \in C^+(\Omega)$. This assumption allows us to simplify the analysis later on and can be easily relaxed. We now introduce the spaces of admissible trial and test functions:

$$Z=H^1(\Omega), \tag{10}$$

$$Z_0=\{v\in Z;\ v=0 \text{ on } \Gamma_{io}\}, \tag{11}$$

$$Z_{io}=\{v\in Z;\ v=\phi_{io} \text{ on } \Gamma_{io}\}. \tag{12}$$

We can use a lift function $\tilde{\phi}\in Z_{io}$ such that $\phi=\varphi+\tilde{\phi}$ with $\varphi\in Z_0$. In this case, the weak form of the problem is given by:

Given $\sigma_c\in C^+(\Omega)$ and $\tilde{\phi}\in Z_{io}$, find $\varphi\in Z_0$ such that

$$\int_\Omega \sigma_c\nabla\varphi\cdot\nabla\psi\, dx=-\int_\Omega \sigma_c\nabla\tilde{\phi}\cdot\nabla\psi\, dx, \quad \forall\psi\in Z_0. \tag{13}$$

We now consider the non-dimensionalized stationary Stokes equations (with $\mu=1$) with slip boundaries,

$$-\Delta\mathbf{u}+\nabla p=\mathbf{0} \quad \text{in } \Omega, \tag{14a}$$

$$\nabla\cdot\mathbf{u}=0 \quad \text{in } \Omega, \tag{14b}$$

$$\mathbf{u}\cdot\mathbf{t}+\lambda\partial_t\phi=0 \quad \text{on } \Gamma_w, \tag{14c}$$

$$\mathbf{u}\cdot\mathbf{n}=0 \quad \text{on } \Gamma_w, \tag{14d}$$

$$\mathbf{u}\cdot\mathbf{t}=0 \quad \text{on } \Gamma_{io}, \tag{14e}$$

$$\mathbf{n}\cdot(\boldsymbol{\sigma}\cdot\mathbf{n})=0 \quad \text{on } \Gamma_{io}. \tag{14f}$$

We look for the velocity and pressure fields in the function spaces,

$$X = \left[H^1(\Omega)\right]^2, \tag{15a}$$

$$X_0 = \left\{\mathbf{v} \in X;\ \mathbf{v} = \mathbf{0} \text{ on } \Gamma_w,\ \mathbf{v}\cdot\mathbf{t} = 0 \text{ on } \Gamma_{io}\right\}, \tag{15b}$$

$$M = \left\{p \in L^2(\Omega);\ \int_\Omega p\, dx = 0\right\}. \tag{15c}$$

Eq. (14c) requires that ϕ lie in the space $H^2(\Omega)$. Note that if $\partial\Omega$ is Lipschitz and convex, or C^1-continuous, then the elliptic regularity theorem guarantees that $\phi \in H^2(\Omega)$. However, to derive a well-posed adjoint problem for the slip EOF model, we need to enforce the coupling boundary conditions in a weak sense. In fact, if we enforce the condition in a weak sense, the curl theorem shows that it is sufficient to have $\phi \in Z$. Consider the function space $H_0^{\frac{1}{2}}(\partial\Omega)$,

$$H_0^{\frac{1}{2}}(\partial\Omega) = \{v \in H^{\frac{1}{2}}(\partial\Omega);\ v = 0 \text{ on } \Gamma_{io}\}. \tag{16}$$

We have the following proposition.

Proposition 1. *Let Ω be Lipschitz. Further, let $\Phi \in Z$, $\lambda \in \mathbb{R}$ and $w, g \in H_0^{\frac{1}{2}}(\partial\Omega)$. Define $b(g,w) = \langle g, w\rangle_{\left(H^{\frac{1}{2}}, H^{\frac{1}{2}}\right)}$ and $f(w) = \langle -\lambda\, \partial_t\Phi, w\rangle_{\left(H^{-\frac{1}{2}}, H^{\frac{1}{2}}\right)}$. Then, the variational problem,*

Find $g \in H_0^{\frac{1}{2}}(\partial\Omega)$, such that

$$b(g,w) = f(w) \quad \forall w \in H^{\frac{1}{2}}(\partial\Omega), \tag{17}$$

is well-posed, and

$$\|g\|_{H^{\frac{1}{2}}(\partial\Omega)} \le c(\Omega)\, |\lambda|\, \|\Phi\|_{H^1(\Omega)}, \tag{18}$$

where $c(\Omega)$ is a positive constant that depends only on the domain Ω.

Proof. Since the bilinear form $b(g,w)$ is defined using the $H^{\frac{1}{2}}(\partial\Omega)$ inner product, it is bounded and coercive. We can use the curl theorem and a special extension of w to show the boundedness of $f(w)$,

$$f(w) = \langle -\lambda\partial_t\Phi, w\rangle_{\left(H^{-\frac{1}{2}}, H^{\frac{1}{2}}\right)} = \int_{\partial\Omega} -\lambda\partial_t\Phi\, w\, ds.$$

Consider an $H^1(\Omega)$ extension of the function w, denoted by $\tilde{w}$. From corollary B. 53, on page 488 in [22], we know that for each $w \in H^{\frac{1}{2}}(\partial\Omega)$, there exists an extension $w^* \in H^1(\Omega)$, such that,

$$\|w^*\|_{H^1(\Omega)} \le c(\Omega)\, \|w\|_{H^{\frac{1}{2}}(\partial\Omega)}. \tag{19}$$

From Proposition 12 in (F.J. Sayas: Weak normal derivatives, normal and tangential traces, and tangential differential operators on Lipschitz boundaries, unpublished), we have,

$$f(w) = \langle -\lambda\partial_t\Phi, w\rangle_{\left(H^{-\frac{1}{2}}, H^{\frac{1}{2}}\right)} = \lambda \int_\Omega \left(\nabla\Phi\cdot(\nabla\times(w^*\mathbf{k})) - (\nabla\times\nabla\Phi)\cdot(w^*\mathbf{k})\right) dx.$$

Using the fact that the gradient of the curl is the zero vector, and the vector cross product identity, we have,

$$f(w) = -\lambda \int_{\Omega} \left(\nabla\Phi \times \nabla w^{*}\right) \cdot \mathbf{k}. \tag{20}$$

We then easily have the following bound on $|f(w)|$,

$$\begin{aligned} |f(w)| &\leq |\lambda| \, \|\Phi\|_{H^{1}(\Omega)} \, \|w^{*}\|_{H^{1}(\Omega)}, \\ &\leq |\lambda| \, \|\Phi\|_{H^{1}(\Omega)} \, \mathrm{c}(\Omega) \, \|w\|_{H^{\frac{1}{2}}(\partial\Omega)}. \end{aligned} \tag{21}$$

Therefore, $\|f\|_{H^{-\frac{1}{2}}} \leq \mathrm{c}(\Omega) \, |\lambda| \, \|\Phi\|_{H^{1}(\Omega)}$, and an application of the Lax-Milgram theorem gives the well-posedness of the variational problem, and the bound $\|g\|_{H^{\frac{1}{2}}(\partial\Omega)} \leq \mathrm{c}(\Omega) \, |\lambda| \, \|\Phi\|_{H^{1}(\Omega)}$. □

We can now define the operator $\boldsymbol{\ell} : Z \to X$,

$$\boldsymbol{\ell}(\Phi) = (g, 0) \text{ on } \partial\Omega, \text{where } g \text{ satisfies Eq. (17)}, \tag{22a}$$

$$\|\boldsymbol{\ell}(\Phi)\|_{X} \leq \mathrm{c}\|g\|_{H^{\frac{1}{2}}(\partial\Omega)}. \tag{22b}$$

The bounded extension corollary B. 53, on page 488 in [22] implies that there is at least one possible lift for any given Φ. For the purposes of deriving the variational formulation, we can pick any admissible lift. It is trivial to show that the operator $\boldsymbol{\ell}$ is linear. Proposition 1 further gives,

$$\|\boldsymbol{\ell}(\Phi)\|_{X} \leq \mathrm{c}\|g\|_{H^{\frac{1}{2}}(\partial\Omega)} \leq \mathrm{c}(\Omega) \, |\lambda| \, \|\Phi\|_{H^{1}(\Omega)}. \tag{23}$$

Therefore, $\boldsymbol{\ell}$ is a bounded operator. Using this lift operator, we can write $\mathbf{u} = \mathbf{w} + \boldsymbol{\ell}(\phi) = \mathbf{w} + \boldsymbol{\ell}(\varphi) + \boldsymbol{\ell}(\tilde{\phi})$, where $\mathbf{w} \in X_0$, and write a weak form for Eq. (14),

Given $\phi \in Z$, find $(\mathbf{w}, p) \in X_0 \times M$ such that,

$$\begin{aligned} &\int_{\Omega} \left[\nabla\mathbf{w} \cdot \nabla\mathbf{v} - p\,\nabla \cdot \mathbf{v} - q\,\nabla \cdot \mathbf{w}\right] dx \\ &\quad = -\int_{\Omega} \left[\nabla\boldsymbol{\ell}(\phi) \cdot \nabla\mathbf{v} - q\,\nabla \cdot \boldsymbol{\ell}(\phi)\right] dx \quad \forall(\mathbf{v}, q) \in X_0 \times M. \end{aligned} \tag{24}$$

Combining Eq. (13) and Eq. (24) together, we get the coupled variational statement:

Given $\sigma_c \in \mathrm{C}^{+}(\Omega)$ and $\tilde{\phi} \in Z_b$, find $(\varphi, \mathbf{w}, p) \in Z_0 \times X_0 \times M$ such that,

$$\begin{aligned} &\int_{\Omega} \sigma_c \nabla\varphi \cdot \nabla\psi \, dx \\ &\quad + \int_{\Omega} \left[\nabla[\mathbf{w} + \boldsymbol{\ell}(\varphi)] \cdot \nabla\mathbf{v} - p\,\nabla \cdot \mathbf{v} - q\,\nabla \cdot [\mathbf{w} + \boldsymbol{\ell}(\varphi)]\right] dx \\ &\quad = -\int_{\Omega} \sigma_c \nabla\tilde{\phi} \cdot \nabla\psi \, dx \\ &\quad - \int_{\Omega} \left[\nabla\boldsymbol{\ell}(\tilde{\phi}) \cdot \nabla\mathbf{v} - q\,\nabla \cdot \boldsymbol{\ell}(\tilde{\phi})\right] dx \quad \forall(\psi, \mathbf{v}, q) \in Z_0 \times X_0 \times M, \end{aligned} \tag{25}$$

where we emphasize that the integrals involving the lift velocity, which depend on the solution φ, will be part of the bilinear form for this coupled problem. We can recast the bilinear form above in more compact notation as,

Given $\sigma_c \in C^+(\Omega)$ and $\tilde{\phi} \in Z_b$, find $\mathbf{U} \in Z_0 \times X_0 \times M$ such that,

$$A(\mathbf{U}, \mathbf{V}) = F(\mathbf{V}) \quad \forall \mathbf{V} \in Z_0 \times X_0 \times M, \tag{26}$$

where $\mathbf{U} = (\varphi, \mathbf{w}, p)$ and $\mathbf{V} = (\psi, \mathbf{v}, q)$.

We have thus incorporated the coupling condition within our bilinear form and can prove that the coupled problem (26) is well-posed. We refer the interested reader to Appendix A and proceed with the derivation of the corresponding adjoint problem.

Adjoint problem

Given the primal weak form Eq. (26), we have the corresponding weak form for the adjoint problem associated with the quantity of interest $Q : Z_0 \times X_0 \times M \to \mathbb{R}$

Given $\sigma_c \in C^+(\Omega)$, find $\mathbf{U}^* \in Z_0 \times X_0 \times M$ such that ,

$$A(\mathbf{V}, \mathbf{U}^*) = Q(\mathbf{V}) \quad \forall \mathbf{V} \in Z_0 \times X_0 \times M, \tag{27}$$

where $\mathbf{U}^* = (\varphi^*, \mathbf{w}^*, p^*)$ is the adjoint solution and $Q(\mathbf{U})$ is a linear functional that corresponds to the quantity of interest (QoI). The full weak form for the adjoint problem reads,

$$\begin{aligned} &\int_\Omega \sigma_c \nabla\psi \cdot \nabla\varphi^* \, dx + \int_\Omega (\nabla\mathbf{v} \cdot \nabla\mathbf{w}^* - q\, \nabla \cdot \mathbf{w}^* - p^*\, \nabla \cdot \mathbf{v})\, dx \\ &\quad + \int_\Omega (\nabla \boldsymbol{\ell}(\psi) \cdot \nabla\mathbf{w}^* - \nabla \cdot \boldsymbol{\ell}(\psi)\, p^*)\, dx \\ &= Q(\mathbf{V}) \quad \forall\, \mathbf{V} = (\psi, \mathbf{v}, q) \in Z_0 \times X_0 \times M. \end{aligned} \tag{28}$$

Note that the adjoint Stokes problem solution $\mathbf{w}^*$ and p^* are coupled to the adjoint potential φ^* through the lift operator $\boldsymbol{\ell}(\cdot)$ acting on the test function ψ. The adjoint problem (28) is also well-posed, see Corollary 2 in Appendix A.

We recall that applying the forward coupling condition Eq. (14c) in a strong sense required that the forward potential $\phi \in H^2(\Omega)$. Correspondingly, to obtain a strong form corresponding to Eq (28), we require that the adjoint velocity $\mathbf{w}^* \in \left[H^2(\Omega)\right]^2$ and the adjoint pressure $p^* \in H^1(\Omega)$. We also have the adjoint stress tensor,

$$\boldsymbol{\sigma}^* = -p^*\mathbf{I} + (\nabla\mathbf{u}^* + (\nabla\mathbf{u}^*)^T). \tag{29}$$

Now integrating by parts we obtain,

$$\begin{aligned} &\int_\Omega -\nabla \cdot (\sigma_c \nabla\varphi^*)\, \psi \, dx + \int_{\partial\Omega} \sigma_c \partial_n \varphi^*\, \psi \, ds \\ &\quad + \int_\Omega -\nabla \cdot \boldsymbol{\sigma}^* \cdot \left(\mathbf{v} + \boldsymbol{\ell}(\psi)\right) + \int_{\partial\Omega} \left(\mathbf{v} + \boldsymbol{\ell}(\psi)\right) \cdot (\sigma^* \cdot \mathbf{n})\, ds \\ &\quad + \int_\Omega -q\, \nabla \cdot \mathbf{w}^* \, dx = Q(\mathbf{V}). \end{aligned} \tag{30}$$

The boundary term in the formulation of the adjoint problem becomes

$$
\begin{aligned}
&\int_{\partial\Omega} \left[\mathbf{v} + \boldsymbol{\ell}(\psi)\right] \cdot (\sigma^* \cdot \mathbf{n})\, ds \\
&= \int_{\Gamma_w} \left[\mathbf{v} + \boldsymbol{\ell}(\psi)\right] \cdot (\sigma^* \cdot \mathbf{n})\, ds + \int_{\Gamma_{io}} \left[\mathbf{v} + \boldsymbol{\ell}(\psi)\right] \cdot (\sigma^* \cdot \mathbf{n})\, ds \\
&= \int_{\Gamma_w} \underbrace{\left(\mathbf{n} \cdot \left[\mathbf{v} + \boldsymbol{\ell}(\psi)\right]\right)}_{0} \left(\mathbf{n} \cdot (\sigma^* \cdot \mathbf{n})\right) + \underbrace{\left(\mathbf{t} \cdot \left[\mathbf{v} + \boldsymbol{\ell}(\psi)\right]\right)\left(\mathbf{t} \cdot (\sigma^* \cdot \mathbf{n})\right)}_{-\lambda \partial_t \psi \left(\mathbf{t} \cdot (\sigma^* \cdot \mathbf{n})\right)} ds \\
&\quad + \int_{\Gamma_{io}} \left(\mathbf{n} \cdot \left[\mathbf{v} + \boldsymbol{\ell}(\psi)\right]\right)\left(\mathbf{n} \cdot (\sigma^* \cdot \mathbf{n})\right) + \underbrace{\left(\mathbf{t} \cdot \left[\mathbf{v} + \boldsymbol{\ell}(\psi)\right]\right)}_{0} \left(\mathbf{t} \cdot (\sigma^* \cdot \mathbf{n})\right) ds \\
&= \int_{\Gamma_w} \left[\nabla_{\Gamma_w} \cdot \left(\lambda \mathbf{t} \cdot (\sigma^* \cdot \mathbf{n})\right)\mathbf{t}\right] \psi\, ds + \int_{\Gamma_{io}} (\mathbf{n} \cdot \mathbf{v})\left(\mathbf{n} \cdot (\sigma^* \cdot \mathbf{n})\right) ds,
\end{aligned}
\tag{31}
$$

where we have used integration by parts for the tangential derivative term along Γ_w [23],

$$
\int_{\Gamma_w} -(\lambda \partial_t \psi)(\mathbf{t} \cdot \mathbf{z})\, ds = \int_{\Gamma_w} \left[\nabla_{\Gamma_w} \cdot (\lambda \mathbf{t} \cdot \mathbf{z})\mathbf{t}\right] \psi\, ds, \tag{32}
$$

where $\nabla_{\Gamma_w} \cdot \mathbf{v}$ denotes the surface divergence of the vector $\mathbf{v}$. Replacing these terms in the adjoint formulation and setting,

$$
Q(\mathbf{U}) = \int_{\Omega} k(\mathbf{x})\, \mathbf{u} \cdot \boldsymbol{\alpha}\, dx + \int_{\Gamma_{io}} k_s(s)\, \mathbf{u} \cdot \mathbf{n}\, ds. \tag{33}
$$

We can write Eq. (31) as,

$$
\begin{aligned}
&\int_{\Omega} \psi \left[-\nabla \cdot (\sigma_c \nabla \varphi^*)\right] dx + \int_{\Gamma_w} \psi \left[\mathbf{n} \cdot (\sigma_c \nabla \varphi^*) + \nabla_{\Gamma_w} \cdot \left(\left(\lambda \mathbf{t} \cdot (\sigma^* \cdot \mathbf{n})\right)\mathbf{t}\right)\right] ds \\
&\quad - \int_{\Omega} \left[\mathbf{v} + \boldsymbol{\ell}(\psi)\right] \cdot \left(\nabla \cdot \boldsymbol{\sigma}^* + k(\mathbf{x})\alpha\right) + \int_{\Gamma_{io}} \left(\mathbf{n} \cdot \left[\mathbf{v} + \boldsymbol{\ell}(\psi)\right]\right)\left(\mathbf{n} \cdot (\sigma^* \cdot \mathbf{n} - k_s(s))\right) ds \\
&\quad - \int_{\Omega} q\left[\nabla \cdot \mathbf{w}^*\right] dx = 0 \qquad \forall (\psi, \mathbf{v}, q) \in Z_0 \times X_0 \times M.
\end{aligned}
\tag{34}
$$

The strong form of the adjoint system then reads:

$$
\nabla \cdot (\sigma_c \nabla \varphi^*) = 0 \quad \text{in } \Omega, \tag{35a}
$$

$$
-\Delta \mathbf{w}^* + \nabla p^* = k\boldsymbol{\alpha} \quad \text{in } \Omega, \tag{35b}
$$

$$
\mathbf{w}^* = 0 \quad \text{in } \Omega, \tag{35c}
$$

with three boundary conditions on Γ_{io}:

$$
\varphi^* = 0 \quad \text{on } \Gamma_{io}, \tag{36a}
$$

$$
\mathbf{w}^* \cdot \mathbf{t} = 0 \quad \text{on } \Gamma_{io}, \tag{36b}
$$

$$
\mathbf{n} \cdot (\sigma^* \cdot \mathbf{n}) = k_s \quad \text{on } \Gamma_{io}, \tag{36c}
$$

and three boundary conditions on Γ_w:

$$
\sigma_c \partial_n \varphi^* + \nabla_{\Gamma_w} \cdot \left(\left(\lambda \mathbf{t} \cdot (\sigma^* \cdot \mathbf{n})\right)\mathbf{t}\right) = 0 \qquad \text{on } \Gamma_w, \tag{37a}
$$

$$
\mathbf{w}^* = \mathbf{0} \qquad \text{on } \Gamma_w. \tag{37b}
$$

We readily observe that the adjoint Stokes problem can be solved first, independently of the adjoint potential problem, but that the latter does depend on the former through the Neumann coupling condition Eq. (37a). We also note that this coupling condition involves the tangential derivatives of the adjoint stress tensor on the boundary.

Ill-posedness of the adjoint problem for the naive formulation

If both the normal and tangential coupling components are considered in the formulation of the primal problem, the corresponding adjoint boundary integrals (see Eq. (31)) would read,

$$
\begin{aligned}
&\int_{\partial\Omega} \left[\mathbf{v}+\boldsymbol{\ell}(\psi)\right]\cdot(\sigma^*\cdot\mathbf{n})\,ds \\
&= \int_{\Gamma_w} \left[\mathbf{v}+\boldsymbol{\ell}(\psi)\right]\cdot(\sigma^*\cdot\mathbf{n})\,ds + \int_{\Gamma_{io}} \left[\mathbf{v}+\boldsymbol{\ell}(\psi)\right]\cdot(\sigma^*\cdot\mathbf{n})\,ds \\
&= \int_{\Gamma_w} \left(\mathbf{n}\cdot\left[\mathbf{v}+\boldsymbol{\ell}(\psi)\right]\right)\left(\mathbf{n}\cdot(\sigma^*\cdot\mathbf{n})\right) + \underbrace{\left(\mathbf{t}\cdot\left[\mathbf{v}+\boldsymbol{\ell}(\psi)\right]\right)\left(\mathbf{t}\cdot(\sigma^*\cdot\mathbf{n})\right)}_{-\lambda\partial_t\psi\left(\mathbf{t}\cdot(\sigma^*\cdot\mathbf{n})\right)}\,ds \\
&\quad + \int_{\Gamma_{io}} \left(\mathbf{n}\cdot\left[\mathbf{v}+\boldsymbol{\ell}(\psi)\right]\right)\left(\mathbf{n}\cdot(\sigma^*\cdot\mathbf{n})\right) + \underbrace{\left(\mathbf{t}\cdot\left[\mathbf{v}+\boldsymbol{\ell}(\psi)\right]\right)}_{0}\left(\mathbf{t}\cdot(\sigma^*\cdot\mathbf{n})\right)\,ds \\
&= \int_{\Gamma_w} -\lambda\,\partial_n\psi\left(\mathbf{n}\cdot(\sigma^*\cdot\mathbf{n})\right) + \int_{\Gamma_w} \left[\nabla_{\Gamma_w}\cdot\left(\lambda\mathbf{t}\cdot(\sigma^*\cdot\mathbf{n})\right)\mathbf{t}\right]\psi\,ds \\
&\quad + \int_{\Gamma_{io}} \left(\mathbf{n}\cdot\mathbf{v}\right)\left(\mathbf{n}\cdot(\sigma^*\cdot\mathbf{n})\right)\,ds.
\end{aligned}
$$

The boundary conditions on Γ_w for the strong form of the adjoint system would then read:

$$\mathbf{n}\cdot(\sigma_c\nabla\varphi^*)+\nabla_{\Gamma_w}\cdot\left((\lambda\mathbf{t}\cdot(\sigma^*\cdot\mathbf{n}))\mathbf{t}\right)=0 \qquad \text{on } \Gamma_w, \tag{38a}$$

$$\mathbf{n}\cdot(\sigma^*\cdot\mathbf{n})=0 \qquad \text{on } \Gamma_w, \tag{38b}$$

$$\mathbf{w}^*=\mathbf{0} \qquad \text{on } \Gamma_w. \tag{38c}$$

Thus the adjoint problem for the ill-posed primal problem contains four boundary conditions, despite there being only three variables. Therefore, the ill-posed primal problem leads to an adjoint inconsistent formulation, specifically in the boundary terms. Such an adjoint inconsistency can lead to oscillations in the discrete solutions of the adjoint problem [24].

Penalty formulation of the slip BC EOF model

Penalty formulation of the primal problem

The imposition of the boundary conditions for the primal and adjoint problems derived in the previous section can be extremely challenging, mainly due to the regularity requirements for the corresponding spaces in the interior and the difficulty of constructing appropriate Finite Element spaces. Therefore, we seek to impose the coupling constraint using a penalty method and relax the regularity requirements on the spaces containing the primal solution and the adjoint. The penalty method is an easy and robust approach for applying Dirichlet boundary conditions. See [25] and [26] for analysis of the method. As we shall see, the penalty method is a natural method for weak enforcement of the coupling. In addition, the penalty formulation gives us an adjoint that is asymptotically consistent with the one obtained using the lift technique. The variational formulation Eq. (25) with equivalent penalty boundary conditions can be given as:

Given $\sigma_c \in C^+(\Omega)$ and $\phi_{io} \in H^{\frac{1}{2}}(\Gamma_{io})$, find $(\phi_\epsilon, \mathbf{u}_\epsilon, p_\epsilon) \in Z \times X \times M$ such that,

$$\int_\Omega \sigma_c \nabla\phi_\epsilon \cdot \nabla\psi \, dx + \frac{1}{\epsilon}\int_{\Gamma_{io}} \phi_\epsilon \, \psi \, ds + \int_\Omega \left[\nabla\mathbf{u}_\epsilon \cdot \nabla\mathbf{v} - p_\epsilon \nabla\cdot\mathbf{v} - q\nabla\cdot\mathbf{u}_\epsilon\right] dx$$
$$+ \frac{1}{\epsilon}\int_{\Gamma_w} (\mathbf{u}_\epsilon \cdot \mathbf{n})\,(\mathbf{v}\cdot\mathbf{n})\, ds + \frac{1}{\epsilon}\int_{\Gamma_w} (\mathbf{u}_\epsilon \cdot \mathbf{t})\,(\mathbf{v}\cdot\mathbf{t})\, ds + \frac{1}{\epsilon}\int_{\Gamma_{io}} (\mathbf{u}_\epsilon \cdot \mathbf{t})\,(\mathbf{v}\cdot\mathbf{t})\, ds$$
$$+ \frac{1}{\epsilon}\int_{\Gamma_w} \lambda(\partial_t\phi_\epsilon)\,(\mathbf{v}\cdot\mathbf{t})\, ds = \frac{1}{\epsilon}\int_{\Gamma_{io}} \phi_{io}\,\psi \, ds \quad \forall(\psi,\mathbf{v},q) \in Z\times X\times M. \tag{39}$$

As was done in Eq. (26), we can recast the bilinear form above in more compact notation as,

Given $\sigma_c \in C^+(\Omega)$ and $\phi_{io} \in H^{\frac{1}{2}}(\Gamma_{io})$, find $\mathbf{U}_\epsilon \in Z \times X \times M$ such that,

$$A_\epsilon(\mathbf{U}_\epsilon, \mathbf{V}) = F_\epsilon(\mathbf{V}) \quad \forall \mathbf{V} \in Z\times X\times M, \tag{40}$$

where $\mathbf{U}_\epsilon = (\phi_\epsilon, \mathbf{u}_\epsilon, p_\epsilon)$ and $\mathbf{V} = (\psi, \mathbf{v}, q)$. We now verify that the weak form in Eq. (39) is indeed consistent and converges to the BVP Eq. (9) and Eq. (14) in the limit as the penalty parameter ϵ tends to zero. Integrating Eq. (39) by parts, we obtain,

$$\int_\Omega -\nabla\cdot(\sigma_c\nabla\phi_\epsilon)\,\psi\, dx + \int_{\Gamma_w} \sigma_c\partial_n\phi_\epsilon\,\psi\, ds + \int_{\Gamma_{io}} \left(\sigma_c\partial_n\phi_\epsilon + \frac{1}{\epsilon}(\phi_\epsilon - \phi_{io})\right)\psi\, ds$$
$$+ \int_\Omega \left((-\Delta\mathbf{u}_\epsilon + \nabla p_\epsilon)\cdot\mathbf{v} - q\nabla\cdot\mathbf{u}_\epsilon\right) ds$$
$$+ \int_{\Gamma_w} \left(\mathbf{n}\cdot(\boldsymbol{\sigma}_\epsilon\cdot\mathbf{n}) + \frac{1}{\epsilon}(\mathbf{u}_\epsilon\cdot\mathbf{n})\right)(\mathbf{v}\cdot\mathbf{n})\, ds$$
$$+ \int_{\Gamma_w} \left(\mathbf{t}\cdot(\boldsymbol{\sigma}_\epsilon\cdot\mathbf{n}) + \frac{1}{\epsilon}(\lambda\partial_t\phi_\epsilon) + \mathbf{u}_\epsilon\cdot\mathbf{t})\right)(\mathbf{v}\cdot\mathbf{t})\, ds + \int_{\Gamma_{io}} (\mathbf{n}\cdot(\boldsymbol{\sigma}_\epsilon\cdot\mathbf{n}))\,(\mathbf{v}\cdot\mathbf{n})\, ds$$
$$+ \int_{\Gamma_{io}} \left(\mathbf{t}\cdot(\boldsymbol{\sigma}\cdot\mathbf{n}) + \frac{1}{\epsilon}(\mathbf{u}_\epsilon\cdot\mathbf{t})\right)\mathbf{v}\cdot\mathbf{t}\, ds = 0 \quad \forall(\psi,\mathbf{v},q)\in Z\times X\times M, \tag{41}$$

where $\boldsymbol{\sigma}_\epsilon$ is the stress tensor corresponding to the penalized solution,

$$\boldsymbol{\sigma}_\epsilon = -p_\epsilon\mathbf{I} + (\nabla\mathbf{u}_\epsilon + (\nabla\mathbf{u}_\epsilon)^T). \tag{42}$$

The equivalent strong form for finite non-zero ϵ is,

$$-\nabla \cdot (\sigma_c \nabla \phi_\epsilon) = 0 \quad \text{in } \Omega, \tag{43a}$$
$$-\Delta \mathbf{u}_\epsilon + \nabla p_\epsilon = \mathbf{0} \quad \text{in } \Omega, \tag{43b}$$
$$\nabla \cdot \mathbf{u}_\epsilon = 0 \quad \text{in } \Omega, \tag{43c}$$

with the boundary conditions on Γ_{io}:

$$\sigma_c \partial_n \phi_\epsilon + \frac{1}{\epsilon}(\phi_\epsilon - \phi_{io}) = 0 \quad \text{on } \Gamma_{io}, \tag{44a}$$
$$\mathbf{n} \cdot (\boldsymbol{\sigma}_\epsilon \cdot \mathbf{n}) = 0 \quad \text{on } \Gamma_{io}, \tag{44b}$$
$$\mathbf{t} \cdot (\boldsymbol{\sigma}_\epsilon \cdot \mathbf{n}) + \frac{1}{\epsilon}(\mathbf{u}_\epsilon \cdot \mathbf{t}) = 0 \quad \text{on } \Gamma_{io}, \tag{44c}$$

and the boundary conditions on Γ_w:

$$\sigma_c \partial_n \phi_\epsilon = 0 \quad \text{on } \Gamma_w, \tag{45a}$$
$$\mathbf{n} \cdot (\boldsymbol{\sigma}_\epsilon \cdot \mathbf{n}) + \frac{1}{\epsilon}(\mathbf{u}_\epsilon \cdot \mathbf{n}) = 0 \quad \text{on } \Gamma_w, \tag{45b}$$
$$\mathbf{t} \cdot (\boldsymbol{\sigma}_\epsilon \cdot \mathbf{n}) + \frac{1}{\epsilon}(\lambda \partial_t \phi_\epsilon + \mathbf{u}_\epsilon \cdot \mathbf{t}) = 0 \quad \text{on } \Gamma_w. \tag{45c}$$

One can observe that the penalty method replaces the Dirichlet boundary conditions with a Robin condition. However, upon taking the limit $\epsilon \to 0$ one formally recovers the original problems Eq. (9) and Eq. (14).

Adjoint problem associated with the penalty formulation

We now derive the adjoint problem corresponding to the weak form Eq. (39),

Find $(\phi_\epsilon^*, \mathbf{u}_\epsilon^*, p_\epsilon^*) \in Z \times X \times M$ such that,

$$\begin{aligned}
&\int_\Omega \sigma_c \nabla \phi_\epsilon^* \cdot \nabla \psi \, dx + \frac{1}{\epsilon} \int_{\Gamma_{io}} \phi_\epsilon^* \, \psi \, ds + \frac{1}{\epsilon} \int_{\Gamma_w} \lambda \partial_t \psi_\epsilon \, (\mathbf{u}_\epsilon^* \cdot \mathbf{t}) \, ds \\
&\quad + \frac{1}{\epsilon} \int_{\Gamma_w} (\mathbf{u}_\epsilon^* \cdot \mathbf{n}) \, (\mathbf{v} \cdot \mathbf{n}) \, ds + \frac{1}{\epsilon} \int_{\Gamma_w} (\mathbf{u}_\epsilon^* \cdot \mathbf{t}) \, (\mathbf{v} \cdot \mathbf{t}) \, ds + \frac{1}{\epsilon} \int_{\Gamma_{io}} (\mathbf{u}_\epsilon^* \cdot \mathbf{t}) \, (\mathbf{v} \cdot \mathbf{t}) \, ds \\
&\quad + \int_\Omega (\nabla \mathbf{u}_\epsilon^* \cdot \nabla \mathbf{v} - p_\epsilon^* \nabla \cdot \mathbf{v} - q \nabla \cdot \mathbf{u}_\epsilon^*) \, dx \\
&= \int_\Omega k(\mathbf{x}) \, \mathbf{v} \cdot \boldsymbol{\alpha} \, dx + \int_{\partial\Omega} k_s(s) \, \mathbf{v} \cdot \mathbf{n} \, ds \quad \forall (\psi, \mathbf{v}, q) \in Z \times X \times M.
\end{aligned} \tag{46}$$

Using integration by parts for the term involving the tangential derivative along Γ_w [23], i.e.

$$\int_{\Gamma_w} \lambda \partial_t \psi_\epsilon \, (\mathbf{u}_\epsilon^* \cdot \mathbf{t}) \, ds = -\int_{\Gamma_w} \nabla_{\Gamma_w} \cdot \big((\lambda \mathbf{u}_\epsilon^* \cdot \mathbf{t}) \mathbf{t} \big) \, \psi_\epsilon \, ds, \tag{47}$$

and upon integrating by parts the higher-order terms and combining integrals with same test functions, one obtains:

$$\begin{aligned}
&\int_{\Omega}\Big(-\nabla\cdot(\sigma_c\nabla\phi_\epsilon^*)\Big)\,\psi\,dx+\int_{\Gamma_w}\Big(\sigma_c\partial_n\phi_\epsilon^*-\frac{1}{\epsilon}\nabla_{\Gamma_w}\cdot\big((\lambda\mathbf{u}_\epsilon^*\cdot\mathbf{t})\mathbf{t}\big)\Big)\,\psi\,ds\\
&+\int_{\Gamma_{io}}\Big(\sigma_c\partial_n\phi_\epsilon^*+\frac{1}{\epsilon}\phi_\epsilon^*\Big)\,\psi\,ds+\int_{\Omega}\Big(-\Delta\mathbf{u}_\epsilon^*+\nabla p_\epsilon^*-k(\mathbf{x})\boldsymbol{\alpha}\Big)\cdot\mathbf{v}\,dx\\
&+\int_{\Omega}\Big(-\nabla\cdot\mathbf{u}_\epsilon^*\Big)\,q\,dx+\int_{\Gamma_w}\Big(\mathbf{n}\cdot(\boldsymbol{\sigma}_\epsilon^*\cdot\mathbf{n})+\frac{1}{\epsilon}(\mathbf{u}_\epsilon^*\cdot\mathbf{n})\Big)\,(\mathbf{v}\cdot\mathbf{n})\,ds\\
&+\int_{\Gamma_w}\Big(\mathbf{t}\cdot(\boldsymbol{\sigma}_\epsilon^*\cdot\mathbf{n})+\frac{1}{\epsilon}(\mathbf{u}_\epsilon^*\cdot\mathbf{t})\Big)\,(\mathbf{v}\cdot\mathbf{t})\,ds+\int_{\Gamma_{io}}\Big(\mathbf{n}\cdot(\boldsymbol{\sigma}_\epsilon^*\cdot\mathbf{n})-k_s(s)\Big)\,(\mathbf{v}\cdot\mathbf{n})\,ds\\
&+\int_{\Gamma_{io}}\Big(\mathbf{t}\cdot(\boldsymbol{\sigma}_\epsilon^*\cdot\mathbf{n})+\frac{1}{\epsilon}(\mathbf{u}_\epsilon^*\cdot\mathbf{t})\Big)\,(\mathbf{v}\cdot\mathbf{t})\,ds=0\quad\forall(\psi,\mathbf{v},q)\in Z\times X\times M,
\end{aligned}\tag{48}$$

where $\boldsymbol{\sigma}_\epsilon^*$ is the penalized adjoint stress tensor. The equivalent strong form for finite non-zero ϵ is,

$$-\Delta\mathbf{u}_\epsilon^*+\nabla p_\epsilon^*=k\quad\text{in }\Omega,\tag{49a}$$

$$\nabla\cdot\mathbf{u}_\epsilon^*=0\quad\text{in }\Omega,\tag{49b}$$

$$\nabla\cdot(\sigma_c\nabla\phi_\epsilon^*)=0\quad\text{in }\Omega,\tag{49c}$$

with the three boundary conditions on Γ_{io}:

$$\mathbf{n}\cdot(\boldsymbol{\sigma}_\epsilon^*\cdot\mathbf{n})=k_s(s)\quad\text{on }\Gamma_{io},\tag{50a}$$

$$\mathbf{t}\cdot(\boldsymbol{\sigma}_\epsilon^*\cdot\mathbf{n})+\frac{1}{\epsilon}(\mathbf{u}_\epsilon^*\cdot\mathbf{t})=0\quad\text{on }\Gamma_{io},\tag{50b}$$

$$\sigma_c\partial_n\phi_\epsilon^*+\frac{1}{\epsilon}\phi_\epsilon^*=0\quad\text{on }\Gamma_{io},\tag{50c}$$

and the three boundary conditions on Γ_w:

$$\mathbf{n}\cdot(\boldsymbol{\sigma}_\epsilon^*\cdot\mathbf{n})+\frac{1}{\epsilon}(\mathbf{u}_\epsilon^*\cdot\mathbf{n})=0\quad\text{on }\Gamma_w,\tag{51a}$$

$$\mathbf{t}\cdot(\boldsymbol{\sigma}_\epsilon^*\cdot\mathbf{n})+\frac{1}{\epsilon}(\mathbf{u}_\epsilon^*\cdot\mathbf{t})=0\quad\text{on }\Gamma_w,\tag{51b}$$

$$\sigma_c\partial_n\phi_\epsilon^*-\frac{1}{\epsilon}\nabla_{\Gamma_w}\cdot\big((\lambda\mathbf{u}_\epsilon^*\cdot\mathbf{t})\mathbf{t}\big)=0\quad\text{on }\Gamma_w.\tag{51c}$$

In the next section, we show that above problem is asymptotically consistent with the previous formulation of the adjoint problem, in the sense that we recover the adjoint corresponding to the strong problem, i.e. Eq. (35), Eq. (36), and Eq. (37) as ϵ tends to zero.

Consistency of the adjoint penalty problem

The main issue is to ensure that the adjoint solution u_ϵ^* to the adjoint problem obtained from the penalized formulation does in fact converge to the adjoint solution u^* obtained from the primal formulation as the penalty parameter ϵ tends to zero, as illustrated in Figure 1. In this case, one has to show that the resulting boundary conditions associated with the penalized and non-penalized formulations of the adjoint problems are consistent.

$$
\begin{array}{ccc}
A(u,v) & \xrightarrow{\epsilon} & A_\varepsilon(u_\varepsilon, v) \\
\downarrow & & \downarrow \\
A(v,u^*) & \overset{?}{\Longleftarrow} & A_\varepsilon(v, u_\varepsilon^*)
\end{array}
$$

Figure 1 Consistency of the adjoint problems associated with the original and penalty formulations. The question here is whether the adjoint problem obtained from the penalty formulation converges to the adjoint problem derived from the original formulation in the limit when the penalty parameter ϵ tends to zero.

Recall that the non-penalized adjoint solution $(\phi^*, \mathbf{u}^*)$ for the problem of interest satisfies the following boundary conditions

$$\phi^* = 0 \qquad \text{on } \Gamma_{io}, \tag{52a}$$

$$\mathbf{n} \cdot (\sigma^* \cdot \mathbf{n}) = k \qquad \text{on } \Gamma_{io}, \tag{52b}$$

$$\mathbf{u}^* \cdot \mathbf{t} = 0 \qquad \text{on } \Gamma_{io}, \tag{52c}$$

$$\mathbf{u}^* = \mathbf{0} \qquad \text{on } \Gamma_{w}, \tag{52d}$$

$$\sigma_c \partial_n \phi^* + \nabla_{\Gamma_w} \cdot \left(\lambda \mathbf{t} \cdot (\sigma^* \cdot \mathbf{n})\mathbf{t}\right) = 0 \qquad \text{on } \Gamma_{w}, \tag{52e}$$

while the penalized adjoint solution $(\phi_\epsilon^*, \mathbf{u}_\epsilon^*)$ satisfies,

$$\sigma_c \partial_n \phi_\epsilon^* - \frac{1}{\epsilon}\phi^* = 0 \qquad \text{on } \Gamma_{io}, \tag{53a}$$

$$\mathbf{n} \cdot (\boldsymbol{\sigma}_\epsilon^* \cdot \mathbf{n}) = k \qquad \text{on } \Gamma_{io}, \tag{53b}$$

$$\mathbf{t} \cdot (\boldsymbol{\sigma}_\epsilon^* \cdot \mathbf{n}) + \frac{1}{\epsilon}(\mathbf{u}_\epsilon^* \cdot \mathbf{t}) = 0 \qquad \text{on } \Gamma_{io}, \tag{53c}$$

$$\mathbf{t} \cdot (\boldsymbol{\sigma}_\epsilon^* \cdot \mathbf{n}) + \frac{1}{\epsilon}(\mathbf{u}_\epsilon^* \cdot \mathbf{t}) = 0 \qquad \text{on } \Gamma_{w}, \tag{53d}$$

$$\mathbf{n} \cdot (\boldsymbol{\sigma}_\epsilon^* \cdot \mathbf{n}) + \frac{1}{\epsilon}(\mathbf{u}_\epsilon^* \cdot \mathbf{n}) = 0 \qquad \text{on } \Gamma_{w}, \tag{53e}$$

$$\sigma_c \partial_n \phi_\epsilon^* - \frac{1}{\epsilon}\nabla_{\Gamma_w} \cdot \left((\lambda \mathbf{u}_\epsilon^* \cdot \mathbf{t})\mathbf{t}\right) = 0 \qquad \text{on } \Gamma_{w}. \tag{53f}$$

To formally interpret Eq. (53f), we can substitute Eq. (53d) into Eq. (53f) as follows,

$$
\begin{aligned}
&\mathbf{t} \cdot (\boldsymbol{\sigma}_\epsilon^* \cdot \mathbf{n}) + \frac{1}{\epsilon}(\mathbf{u}_\epsilon^* \cdot \mathbf{t}) = 0 \Rightarrow \frac{\lambda}{\epsilon}(\mathbf{u}_\epsilon^* \cdot \mathbf{t}) = -\lambda \mathbf{t} \cdot (\boldsymbol{\sigma}_\epsilon^* \cdot \mathbf{n}), \\
&\sigma_c \partial_n \phi_\epsilon^* = \nabla_{\Gamma_w} \cdot \left(\left(\frac{\lambda \mathbf{u}_\epsilon^* \cdot \mathbf{t}}{\epsilon} \right) \mathbf{t} \right) \Rightarrow \sigma_c \partial_n \phi_\epsilon^* = -\nabla_{\Gamma_w} \cdot \left(\lambda \mathbf{t} \cdot (\boldsymbol{\sigma}_\epsilon^* \cdot \mathbf{n})\mathbf{t}\right).
\end{aligned} \tag{54}
$$

We can thus derive the following boundary conditions for the adjoint potential

$$\sigma_c \partial_n \phi_\epsilon^* + \nabla_{\Gamma_w} \cdot \left(\lambda \mathbf{t} \cdot (\boldsymbol{\sigma}_\epsilon^* \cdot \mathbf{n})\mathbf{t}\right) = 0 \qquad \text{on } \Gamma_{w}, \tag{55a}$$

$$\phi_\epsilon^* + \epsilon\, \sigma_c \partial_n \phi_\epsilon^* = 0 \qquad \text{on } \Gamma_{io}, \tag{55b}$$

which are the same as those for the non-penalized adjoint in the limit $\epsilon \to 0$. Eq. 53d corresponds to a penalty representation of the tangential boundary flux. Further discussion of this representation is presented in [27]. We thus see that the penalized formulation of the electroosmotic flow problem is adjoint consistent in the limit as $\epsilon \to 0$, and the use of the discrete penalized adjoint solution $((\phi_\epsilon^*)_h, (\mathbf{u}_\epsilon^*)_h, (p_\epsilon^*)_h)$ in adjoint-based error estimation and sensitivity analysis is justified. Thus, if the forward problem is computed numerically using a discrete representation of Eq. (39), the adjoint solution can also be easily computed by taking the transpose of the stiffness matrix associated with the forward problem.

Results and discussion

We now consider the application of the new EOF formulation on specific microfluidic examples. First, we simulate a flow in a straight microchannel driven purely by electroosmosis. The objective here is to highlight the convergence and stability properties of the adjoint solution. We then showcase an adjoint-based adaptive strategy for mesh refinement on a T-shaped microchannel flow and adjoint-based parameter sensitivity analyses. We discuss the improvement of the convergence rates with respect to quantities of interest and their sensitivities when using adjoint-based techniques. Simulations are performed using the adjoint capabilities added to the `libMesh` Finite Element library [28]. For both applications, second-order Lagrange elements are employed for the potential and velocity approximations. Linear Lagrange elements are selected to approximate the pressure field in order to satisfy the inf-sup condition. Initial meshes in all the experiments dealing with the straight and T-channel domains consist of structured meshes of bi-quadratic quadrilateral elements. Numerical errors to generate the convergence plots are estimated in this work using so-called overkilled reference solutions of the two problems. These are obtained on a uniform mesh of 428,676 degrees of freedom for the straight channel problem and a combined adaptive-uniform mesh with 288,160 degrees of freedom for the T-channel problems. Numerical solutions are calculated using an ILU preconditioned GMRES iterative method for both problems. The linear algebra library PETSc is accessed through `libMesh` to obtain these solutions. The penalty parameter ϵ was set to the constant value of 10^{-8} for all the numerical experiments.

Electroosmotic flow in a straight channel

Numerical experiments are performed here in the case of an electroosmotic flow in a straight channel. The channel has unit width and the length is five times the width. Since the objective of these simulations is to illustrate the numerical properties of the adjoint solution obtained by using Eq. (39), we set arbitrary values of the model parameters rather than choosing values representative of an actual flow. The fluid viscosity μ, electroosmotic slip parameter κ, and fluid density ρ are all taken to be unity. Constant potentials $\phi_i = 8$ and $\phi_o = 0$ are prescribed at the inlet Γ_{in} and outlet Γ_{out} boundaries, respectively. The electric conductivity of the fluid is chosen as $\sigma_c = 1 + x$ (note that this particular form of the conductivity is chosen for no other reason than better illustrate the properties of the computed adjoint).

The quantity of interest is defined here in terms of the bounded linear functional:

$$Q(U) = \int_{\Omega} \mathbf{u} \cdot \boldsymbol{\alpha} \, dx, \qquad (56)$$

where $\boldsymbol{\alpha} = (1, 1)$. Such a bounded functional ensures that any oscillations observed in the numerical results solely arise from the definition of the bilinear form in the adjoint problem. We consider the formulation of the adjoint problem as given in Eq. (46). After computing the forward solution using the numerical set-up as above, we obtain the adjoint potential ϕ_ϵ^* as seen in Figure 2. By observing the norm of the computed adjoint

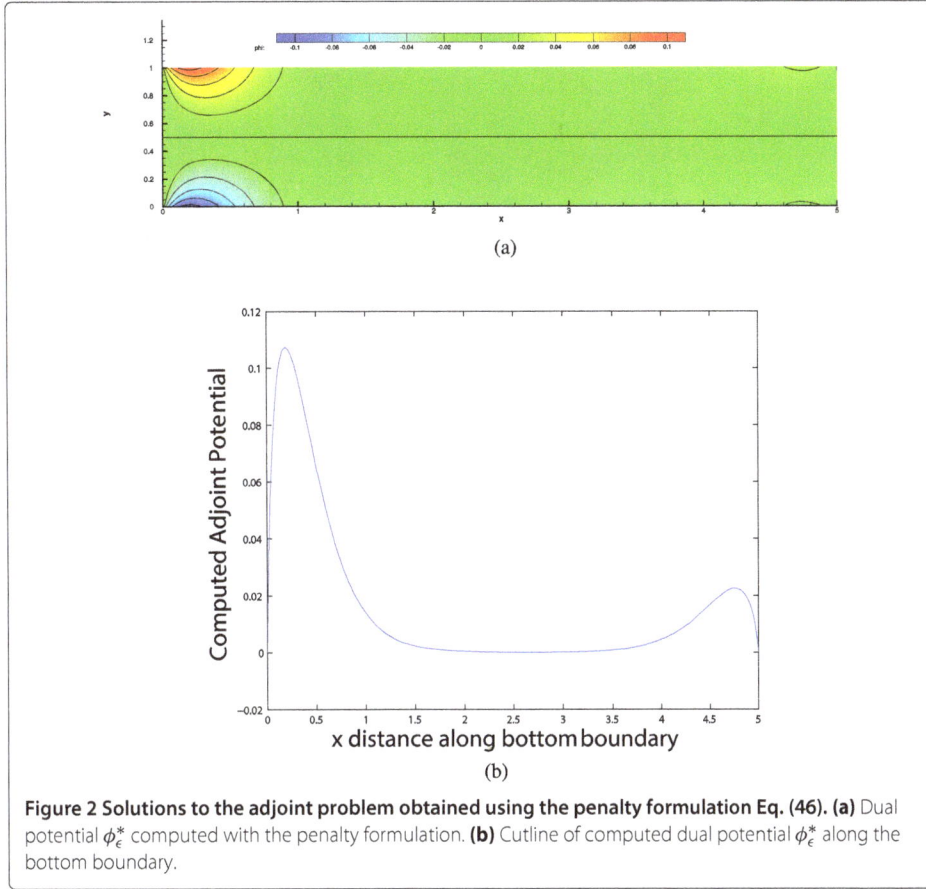

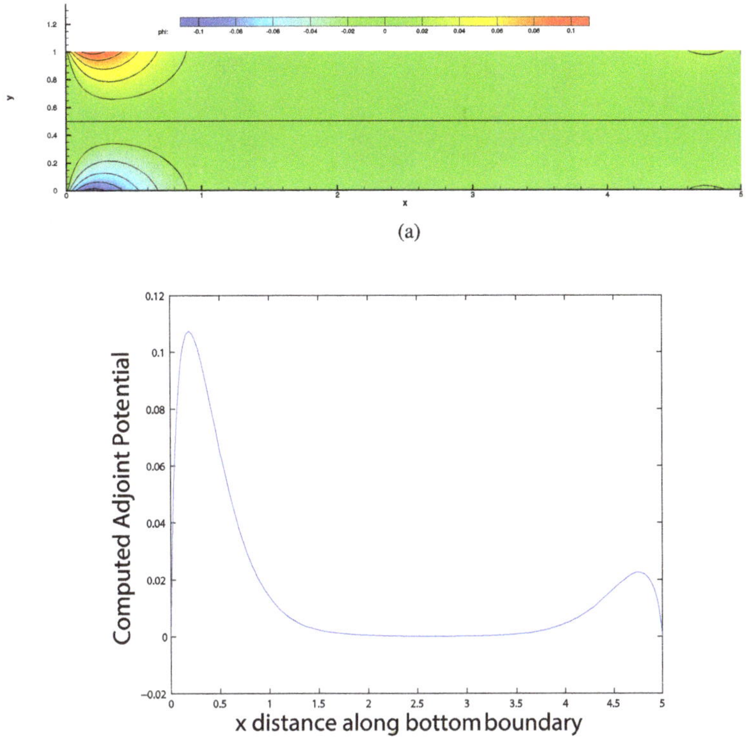

(a)

(b)

Figure 2 Solutions to the adjoint problem obtained using the penalty formulation Eq. (46). (a) Dual potential ϕ_ϵ^* computed with the penalty formulation. **(b)** Cutline of computed dual potential ϕ_ϵ^* along the bottom boundary.

solutions with successive mesh refinements, it was numerically verified that the adjoint potential and velocities were all in $H^1(\Omega)$.

We also studied the convergence rates for the approximate primal and adjoint potential and x-component of the velocity. Recall that the potential and velocity fields are both approximated using second-order Lagrange elements so that one would expect first-order convergence rates with respect to the number of degrees of freedom in the H^1 norm. However, one can observe in Figure 3 that the primal velocity and the adjoint potential converge at a slower than optimal rate while the primal potential and adjoint velocity converge at the optimal rate. This is due to the tangential 'slip' coupling given by Eq. (8) for the forward problem and the Neumann conditions Eq. (55a) for the adjoint problem. Essentially, we can say that the forward Stokes problem and the adjoint potential problem have non-accurate data, leading to higher errors in the computation of their solutions.

Consequences of coupling both normal and tangential components As we saw in Section "Ill-posedness of the adjoint problem for the naive formulation", coupling both the normal and tangential components of the velocity to the potential leads to an ill-posed adjoint problem. This ill-posedness is inherited by the penalized formulation as well. On directly enforcing the constraint Eq. (7c) on the wall boundary rather than splitting the two velocity components as in Eq. (8), one observes spurious oscillations in the numerical adjoint potential field ϕ_ϵ^*, as shown in Figure 4. One clearly observes in Figure 4(a) the

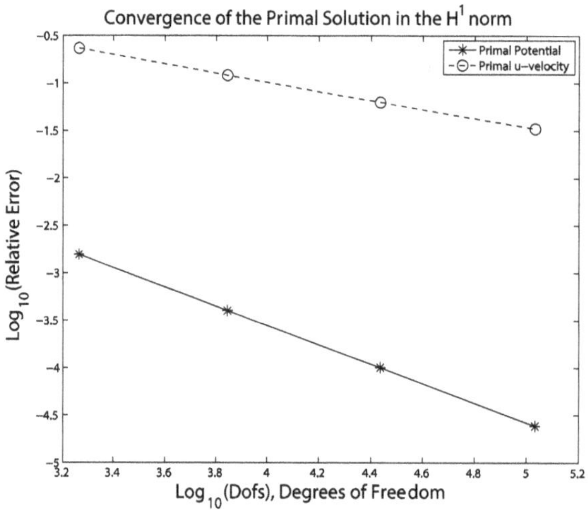

(a)

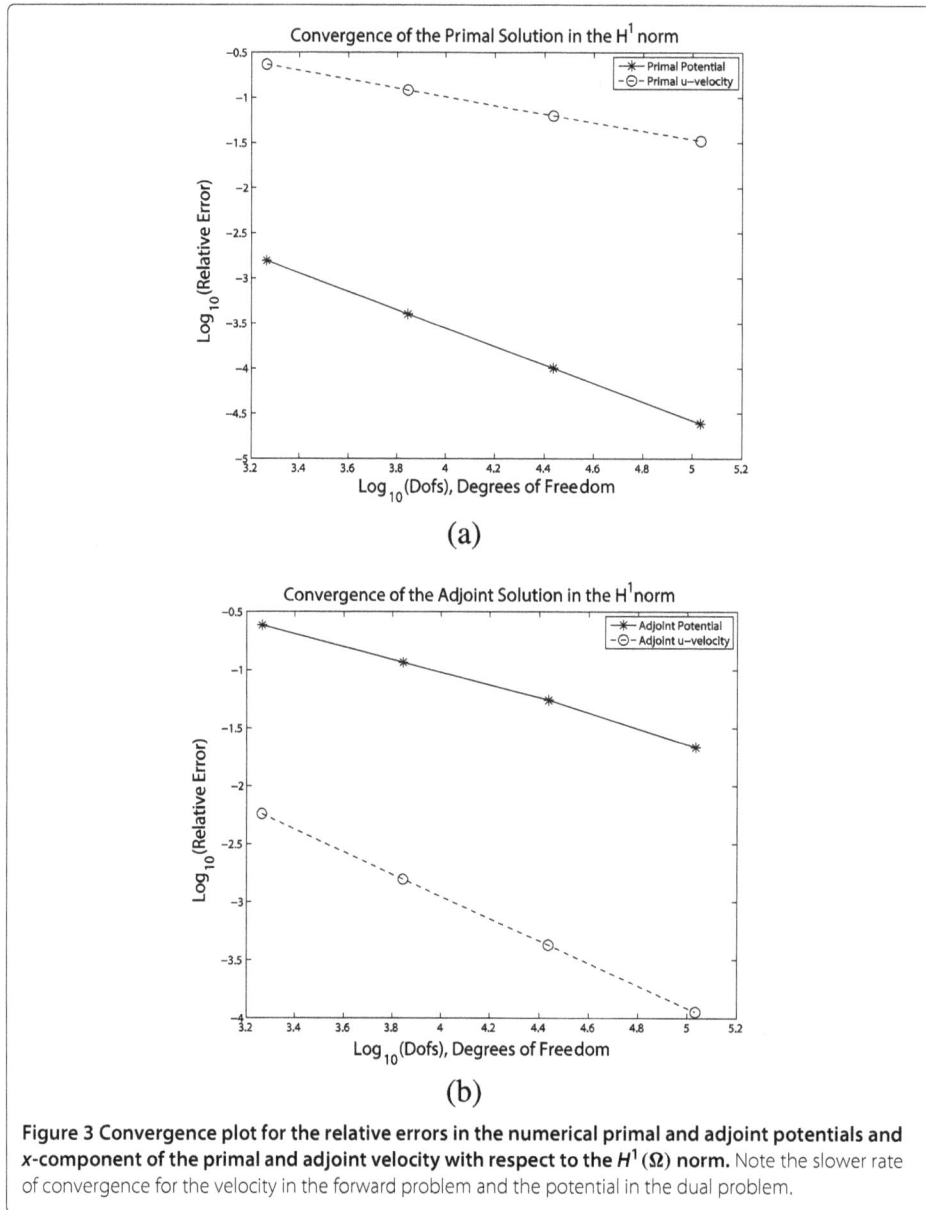

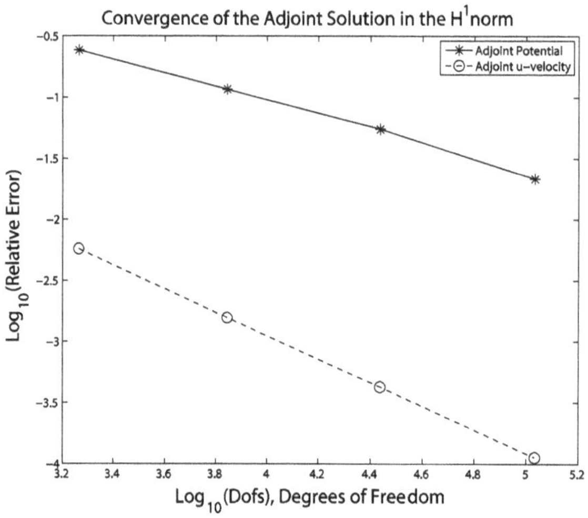

(b)

Figure 3 Convergence plot for the relative errors in the numerical primal and adjoint potentials and *x*-component of the primal and adjoint velocity with respect to the $H^1(\Omega)$ norm. Note the slower rate of convergence for the velocity in the forward problem and the potential in the dual problem.

presence of closed contour lines along the top and bottom wall boundaries. This result is confirmed in Figure 4(b), which shows the solution ϕ_ϵ^* along the top boundary. For uniform meshes, the instabilities manifest not as oscillations, but a blow-up of the adjoint potential solution in the $H^1(\Omega)$ norm.

Electroosmotic flow in a T-channel

Crossing T- and H-channels are commonly utilized in microfluidics. Applications typically involve mixing of two chemical species [8], purification [16], or fluid identification [12]. However, numerical modeling of electroosmotic flows with slip boundary conditions in such geometrical configurations poses distinctive challenges due to the presence of corner singularities [11]. One immediate consequence is the observation of reduced convergence rates in the approximation of the global solution. A possible remedy

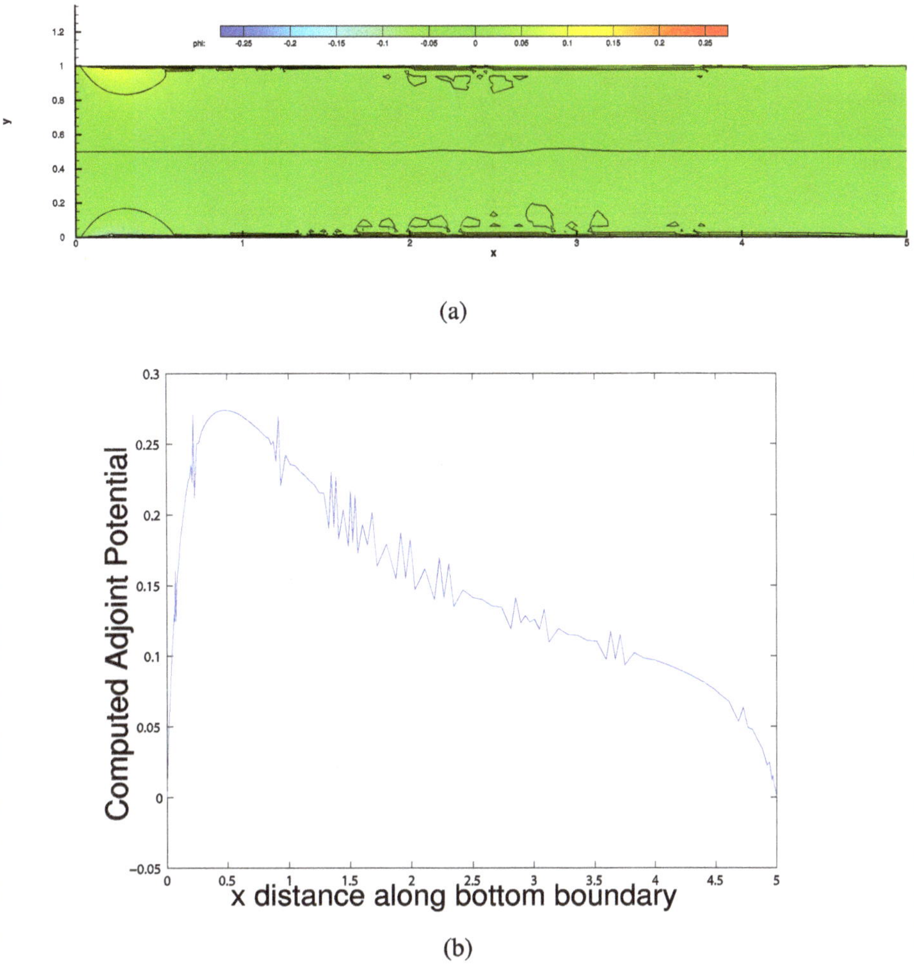

Figure 4 The solutions to the adjoint problems obtained using an ill-posed penalty formulation. (a) Dual potential ϕ_ϵ^* computed with the ill-posed formulation. **(b)** Cutline of computed dual potential ϕ_ϵ^* along bottom boundary.

is to use adaptive finite element methods to help restore the optimal convergence properties of such singular problems [29]. Likewise, adaptive methods can also improve the convergence behavior of the adjoint solution and potentially restore the optimal rates that one may expect when estimating linear QoIs.

We consider below a T-channel geometry. The two upper ends of the T-channel, $\Gamma_{i,l}$ and $\Gamma_{i,r}$, correspond to the left and right inlets, respectively, at which a high potential ϕ_i is prescribed, while the bottom end of the channel Γ_o, the flow outlet, is set to the ground potential $\phi_o = 0$. The flow is assumed here to be purely electrically driven, in which case Dirichlet pressure boundary conditions $p = 0$ are considered at the inlet and outlet boundaries. The flow parameters used for the numerical experiments are provided in Table 1.

In the numerical experiments below, we consider the following quantity of interest:

$$Q(U) = \int_{\Gamma_o} \mathbf{u} \cdot \mathbf{n} \, ds. \tag{57}$$

Table 1 Values of the input parameters in the case of the T-channel flow

Parameter	Symbol	Value
Conductivity	σ_c	1.0
Inlet potentials	ϕ_i	8.0
Outlet potential	ϕ_o	0.0
Fluid viscosity	μ	1.0
Slip parameter	λ	1.0

We also estimate the sensitivity of the QoI with respect to the parameters ϕ_i, ϕ_o, and λ, evaluated in terms of the first derivatives $dQ/d\phi_i$, $dQ/d\phi_o$ and $dQ/d\lambda$. We used ten adaptive refinement steps followed by two uniform refinements (for a total of 288,160 dofs) to calculate the reference values of these quantities. These values are reported in Table 2 and were used as exact values to compute numerical errors.

The adaptive strategy for mesh refinement with respect to the QoI is described in Algorithm 1, which has been implemented in `libMesh`. We show in Figure 5 the horizontal and vertical components of the primal velocity **u**. We note that the vertical component of the velocity, shown in Figure 5(b), is close to zero near the inlets, but then undergoes a stiff acceleration around the corners. Likewise, we observe in Figure 5(a) the rapid deceleration of the horizontal velocity near the corners. This clearly induces a singular behavior of the solution at the two corners. Note also that the solution is symmetric about the centerline of the vertical channel, as expected, given that the inlet potentials at stations $\Gamma_{i,r}$ and $\Gamma_{i,l}$ are equal.

Algorithm 1 Compute the finite element solution to Eq. (39) that either reaches a prescribed mesh size h_{min} or is obtained after a given number of adaptive steps n_{max} using an adaptive meshing strategy based on the dual approach with respect to the QoI Eq. (57).

1: Start step counter n_{step}
2: Compute the finite element solution u_h to the problem using a uniform mesh M_{start} of resolution $h_{\text{elem}} = h_{\text{start}}$
3: Compute an a posteriori error indicator $\tilde{e}_h$ for the QoI based on the adjoint residual error indicator (see [27]) and flag elements to be refined
4: **if** $h_{\text{elem}} \leq h_{\text{min}}$ OR $n_{\text{step}} > n_{\text{max}}$ **then**
5: Go to step 11
6: **else**
7: Refine the top 30 percent of the flagged elements to obtain an adaptive mesh M_{adaptive}
8: Increment n_{step} by 1
9: Repeat steps 2, 3, and 4 using the adapted mesh M_{adaptive}
10: **end if**
11: Postprocess results.

Table 2 Estimated reference values of QoI and of its sensitivity to ϕ_i, ϕ_o, and κ

$Q(U)$	$dQ/d\phi_o$	$dQ/d\phi_i$	$dQ/d\lambda$
1.0205649	-0.1275705	0.0637853	1.0203276

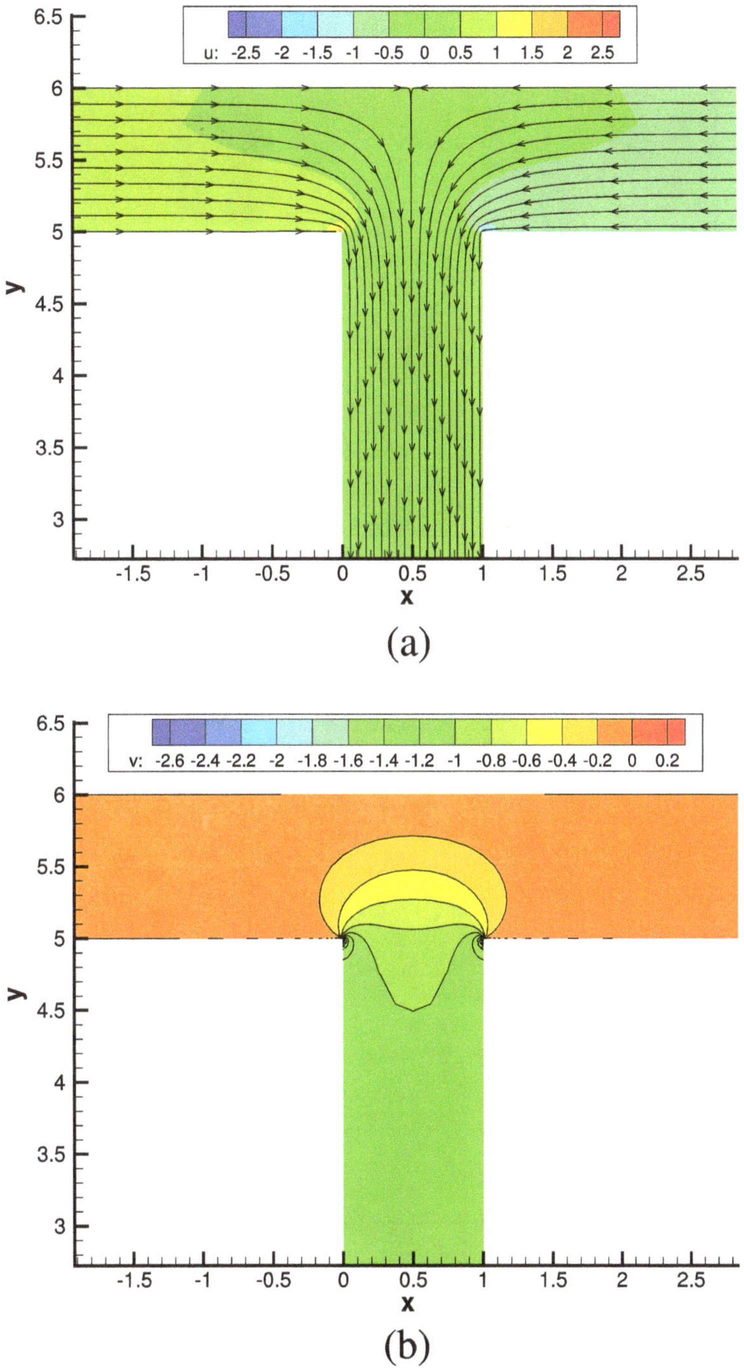

Figure 5 Contour plot of the primal solution obtained using the penalty formulation. The singularities are clearly visible in the vicinity of the corners. The solution is smooth away from the corners. **(a)** x-component u_1 of velocity **u**. **(b)** y-component u_2 of velocity **u**.

Next, we show the adjoint solutions computed using the adaptive procedure described in Algorithm 1. The vertical velocity, displayed in Figure 6(a), exhibits a parabolic profile that reaches the maximum value along the centerline of the vertical channel and vanishes on its boundaries. Therefore, the presence of corners is solely responsible for the singular behavior in the velocity field. The adjoint potential solution is shown in Figure 6(a). The

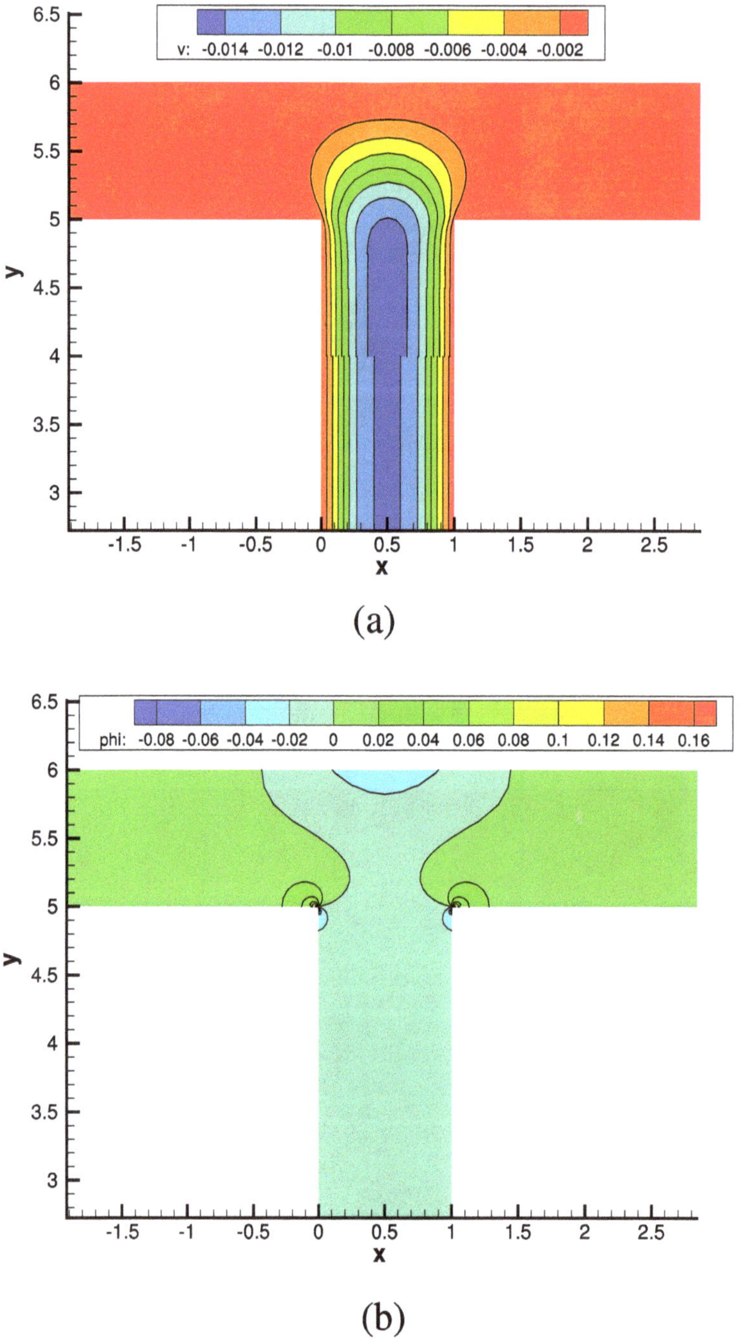

Figure 6 Contour plot of the *y*-component of the adjoint velocity $\mathbf{u}^*$ and of the adjoint potential ϕ^*. **(a)** y-component u_2^* of adjoint velocity $\mathbf{u}^*$. It is mainly different from zero inside the vertical channel indicating that the primal solution needs to be accurate in that region. **(b)** Adjoint potential ϕ^*. Note that ϕ^* almost vanishes everywhere except at the corners and along the middle section of the top wall.

potential is of course singular at the corners due to the geometrical discontinuity and to the fact that the coupling boundary condition, although almost zero everywhere along the boundaries, becomes non-zero near the corners, since $\nabla_{\Gamma_w} \cdot \left(\lambda \mathbf{t} \cdot (\boldsymbol{\sigma}_\epsilon^* \cdot \mathbf{n})\mathbf{t}\right)$ may not be zero there. This should imply extensive refinement near the corners, as confirmed by the adapted mesh shown in Figure 7.

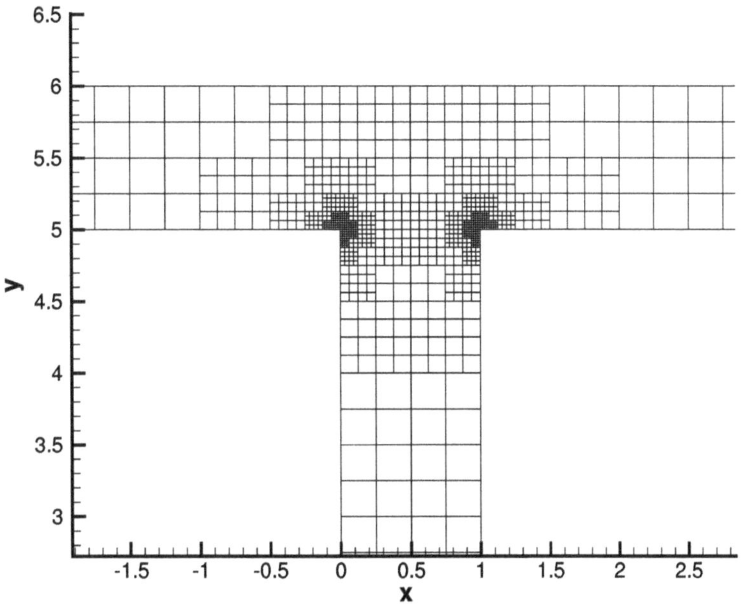

Figure 7 Adaptive mesh obtained using adjoint-based error estimates. Note that the elements get refined almost exclusively near the corners due to the singularities in the primal velocity and adjoint potential.

We also used an adjoint method to compute parameter sensitivities for the given QoI to the parameters ϕ_i, ϕ_o, and λ. The advantage of using an adjoint method for sensitivity analysis is that the sensitivity to all three parameters could be found with a single adjoint solve. This is considerably more efficient than using a finite difference or a forward sensitivity method. In addition, we can also combine the adjoint-based mesh refinement and sensitivity analysis for further improvements in the convergence of the sensitivities.

Convergence plots are shown in Figure 8. In particular, the relative error in the quantity of interest estimated using uniform refinement and adjoint-based adaptive refinement is shown in Figure 8(a) against the total number of degrees of freedom (dofs). Relative errors in the estimated sensitivities of the QoI with respect to parameters are displayed in Figure 8(b). We note that the adaptive refinement strategy offers much improved error reduction than uniform refinement for both the estimation of the quantity of interest and its sensitivity derivatives.

In fact, on account of the geometric corner singularities present in the problem, we obtain an inferior convergence rate on using uniform refinement. However, with the adaptive method we obtain a rate of 1.5 (vs dofs) for the QoI, which can be said to be semi-optimal. We had observed earlier that there is a loss of one order in the convergence rate for the forward velocity and adjoint potential for the straight channel problem where there are no corner singularities. We recall that with second-order Lagrange Finite Elements this would result in a convergence rate of 1.5 ($N^1 \times N^{\frac{1}{2}}$) for a linear QoI.

Conclusions

We have presented an analysis of an electroosmotic flow model with slip boundary conditions and its adjoint. The slip boundary conditions require the evaluation of potential

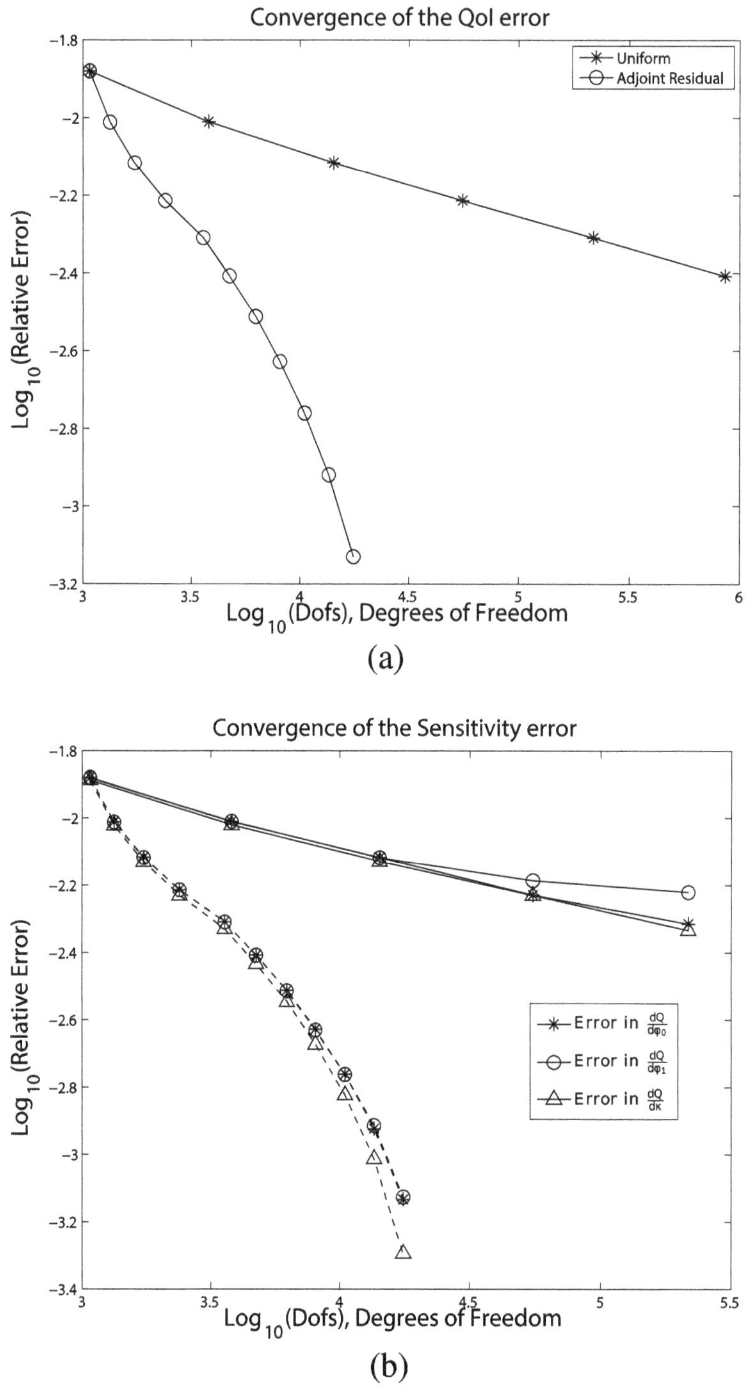

Figure 8 Convergence plots for the approximation of the quantity of interest and its sensitivity to the parameters ϕ_i, ϕ_o, and λ. (a) Convergence plots for the relative error in QoI Eq. (57) using uniform and adjoint-based refinements. **(b)** Convergence plots for the relative errors in the sensitivities of the QoI Eq. (57) using uniform and adjoint-based refinements.

derivatives on the boundary, which increases the regularity requirements on the potential. We emphasize that a naive enforcement of the standard slip boundary condition leads to an ill-posed adjoint problem (see Section "Ill-posedness of the adjoint problem for the naive formulation"). This leads to instabilities in the computed adjoint, illustrated by numerical experiments in Section "Results and discussion". A well-posed adjoint problem can be obtained by modifying the slip boundary condition ($\mathbf{u} + \lambda\nabla\phi = \mathbf{0}$), i.e. specifying the normal velocity at the wall independently of the potential ($\mathbf{u}\cdot\mathbf{n} = 0, \mathbf{u}\cdot\mathbf{t}+\lambda\nabla\phi\cdot\mathbf{t} = 0$).

We further proposed a penalty formulation of the forward problem that requires no extra regularity for the potential, and leads to a well-posed, asymptotically consistent adjoint formulation as well. The penalty boundary conditions lead to a weak enforcement of the boundary coupling, allowing us to easily compute the adjoint problem using the adjoint capabilities of `libMesh`.

Finally, we presented numerical experiments for a simple straight channel microflow and a more challenging T-channel flow. The convergence results for the straight channel problem indicate that the primal velocity and the adjoint potential converge at sub-optimal rates due to the nature of the coupling between the potential and the velocity. For the T-channel, we presented QoI computation and QoI adjoint sensitivity results for a practical engineering QoI. We observed a loss of convergence order due to the singularities in the T-channel geometry, and substantial improvements in the rate on using an adjoint-based adaptive method. However, the fully optimal convergence rate for the QoI could not be achieved, possibly due to the convergence properties of the adjoint potential.

Future work will involve the application of more sophisticated adjoint-based error estimators to further improve the convergence properties of the approximate solutions and obtain reliable a posteriori error estimates for complex applications.

Appendix A: Well-posedness of coupled formulation

Let Z and X denote two reflexive Banach spaces. Consider the following abstract linear one-way coupled problem:

$$\text{Find } (\phi, u) \in Z \times X :$$

$$\mathcal{A}(\phi, \psi) = \mathcal{F}(\psi) \qquad \forall \psi \in Z, \tag{58a}$$

$$\mathcal{B}(u, v) = \mathcal{G}(v) - \mathcal{C}(\phi, v) \qquad \forall v \in X, \tag{58b}$$

where $\mathcal{A}(\cdot,\cdot)$, $\mathcal{B}(\cdot,\cdot)$, and $\mathcal{C}(\cdot,\cdot)$ and continuous bilinear forms, and $\mathcal{F}(\cdot)$ and $\mathcal{G}(\cdot)$ are continuous linear forms.

If $\mathcal{A}(\cdot,\cdot)$ and $\mathcal{B}(\cdot,\cdot)$ satisfy the inf-sup conditions on $Z \times Z$ and on $X \times X$, respectively, then the above problem can be solved sequentially: First solve (58a) for $\phi \in Z$. Then, since $\mathcal{C}(\phi,\cdot)$ is a continuous linear form on X, Eq. (58b) can be solved for $u \in X$.

The fact that one-way coupled problems (with bounded coupling) are well-posed is probably well known. We now provide an extension of this result for the aggregated bilinear form $\mathcal{A}(\phi,\psi) + \mathcal{B}(u,v) + \mathcal{C}(\phi,v)$.

Theorem. Let $\mathscr{B}\big((\cdot,\cdot),(\cdot,\cdot)\big)$ denote the aggregated bilinear form:

$$\mathscr{B}\big((\phi,u),(\psi,v)\big) := \mathcal{A}(\phi,\psi) + \mathcal{B}(u,v) + \mathcal{C}(\phi,v)\,.$$

If $\mathcal{A}(\cdot,\cdot)$ and $\mathcal{B}(\cdot,\cdot)$ satisfy the inf–sup conditions, i.e., there exist constants $\gamma_{\mathcal{A}} > 0$ and $\gamma_{\mathcal{B}} > 0$ such that

$$\inf_{\phi\in Z\setminus\{0\}} \sup_{\psi\in Z\setminus\{0\}} \frac{\mathcal{A}(\phi,\psi)}{\|\phi\|_Z\|\psi\|_Z} \geq \gamma_{\mathcal{A}}, \tag{59a}$$

$$\forall\psi \in Z,\ \big(\forall\phi \in Z,\ \mathcal{A}(\phi,\psi) = 0\big) \Longrightarrow \big(\psi = 0\big), \tag{59b}$$

$$\inf_{u\in X\setminus\{0\}} \sup_{v\in X\setminus\{0\}} \frac{\mathcal{B}(u,v)}{\|u\|_X\|v\|_X} \geq \gamma_{\mathcal{B}}, \tag{59c}$$

$$\forall v \in X,\ \big(\forall u \in X,\ \mathcal{B}(u,v) = 0\big) \Longrightarrow \big(v = 0\big), \tag{59d}$$

and if $\mathcal{C}(\cdot,\cdot)$ is a continuous bilinear form, i.e., there is a constant $c_{\mathcal{C}} > 0$ such that

$$\big|\mathcal{C}(\phi,v)\big| \leq c_{\mathcal{C}}\|\phi\|_Z\|v\|_X \qquad \forall\phi \in Z, \forall v \in X,$$

then $\mathscr{B}\big((\cdot,\cdot),(\cdot,\cdot)\big)$ satisfies the inf–sup conditions:

$$\inf_{(\phi,u)\in Z\times X\setminus\{0\}} \sup_{(\psi,v)\in Z\times X\setminus\{0\}} \frac{\mathscr{B}\big((\phi,u),(\psi,v)\big)}{\big(\|\phi\|_Z + \|u\|_X\big)\big(\|\psi\|_Z + \|v\|_X\big)} \geq \gamma, \tag{60a}$$

$$\forall(\psi,v) \in Z\times X,\ \Big(\forall(\phi,u) \in Z\times X,\ \mathscr{B}\big((\phi,u),(\psi,v)\big) = 0\Big) \Longrightarrow \Big((\psi,v) = (0,0)\Big), \tag{60b}$$

with

$$1/\gamma = \frac{1}{\gamma_{\mathcal{A}}} + \frac{1}{\gamma_{\mathcal{B}}}\Big(1 + \frac{c_{\mathcal{C}}}{\gamma_{\mathcal{A}}}\Big).$$

In other words, problem (58) is well-posed. Moreover, the following a priori estimates hold:

$$\|\phi\|_Z \leq \frac{1}{\gamma_{\mathcal{A}}}\|\mathcal{F}\|_{Z^*},$$
$$\|u\|_X \leq \frac{1}{\gamma_{\mathcal{B}}}\Big(\|\mathcal{G}\|_{X^*} + \frac{c_{\mathcal{C}}}{\gamma_{\mathcal{A}}}\|\mathcal{F}\|_{Z^*}\Big).$$

□

Proof. The proof is similar to the Brezzi–Babuška equivalence theorem; see e.g. [22, Proposition 2.36].

We first prove (60a). Using inf-sup stability of $\mathcal{A}(\cdot,\cdot)$ we obtain:

$$\begin{aligned}
\|\phi\|_Z &\leq \frac{1}{\gamma_{\mathcal{A}}} \sup_{\psi\in Z\setminus\{0\}} \frac{\mathcal{A}(\phi,\psi)}{\|\psi\|_Z} \\
&= \frac{1}{\gamma_{\mathcal{A}}} \sup_{\psi\in Z\setminus\{0\}} \frac{\mathscr{B}\big((\phi,u),(\psi,0)\big)}{\|\psi\|_Z} \\
&\leq \frac{1}{\gamma_{\mathcal{A}}} \sup_{(\psi,v)\in Z\times X\setminus\{0\}} \frac{\mathscr{B}\big((\phi,u),(\psi,v)\big)}{\|\psi\|_Z + \|v\|_X}.
\end{aligned}$$

From (58b), inf-sup stability of $\mathcal{B}(\cdot,\cdot)$, and continuity of $\mathcal{C}(\cdot,\cdot)$ we obtain:

$$\begin{aligned}
\|u\|_X &\le \frac{1}{\gamma_\mathcal{B}} \sup_{v\in X\setminus\{0\}} \frac{\mathcal{B}(u,v)}{\|v\|_X} \\
&= \frac{1}{\gamma_\mathcal{B}} \sup_{v\in X\setminus\{0\}} \frac{\mathcal{A}(\phi,0)+\mathcal{B}(u,v)+\mathcal{C}(\phi,v)-\mathcal{C}(\phi,v)}{\|v\|_X} \\
&\le \frac{1}{\gamma_\mathcal{B}} \sup_{v\in X\setminus\{0\}} \frac{\mathcal{A}(\phi,0)+\mathcal{B}(u,v)+\mathcal{C}(\phi,v)}{\|v\|_X} + \frac{1}{\gamma_\mathcal{B}} \sup_{v\in X\setminus\{0\}} \frac{\mathcal{C}(\phi,v)}{\|v\|_X} \\
&= \frac{1}{\gamma_\mathcal{B}} \sup_{v\in X\setminus\{0\}} \frac{\mathscr{B}\big((\phi,u),(0,v)\big)}{\|v\|_X} + \frac{1}{\gamma_\mathcal{B}} \sup_{v\in X\setminus\{0\}} \frac{\mathcal{C}(\phi,v)}{\|v\|_X} \\
&\le \frac{1}{\gamma_\mathcal{B}} \sup_{v\in X\setminus\{0\}} \frac{\mathscr{B}\big((\phi,u),(0,v)\big)}{\|v\|_X} + \frac{c_\mathcal{C}}{\gamma_\mathcal{B}}\|\phi\|_Z \\
&\le \frac{1}{\gamma_\mathcal{B}} \sup_{(\psi,v)\in Z\times X\setminus\{0\}} \frac{\mathscr{B}\big((\phi,u),(\psi,v)\big)}{\|\psi\|_Z+\|v\|_X} + \frac{c_\mathcal{C}}{\gamma_\mathcal{B}}\|\phi\|_Z \\
&\le \frac{1}{\gamma_\mathcal{B}}\Big(1+\frac{c_\mathcal{C}}{\gamma_\mathcal{A}}\Big) \sup_{(\psi,v)\in Z\times X\setminus\{0\}} \frac{\mathscr{B}\big((\phi,u),(\psi,v)\big)}{\|\psi\|_Z+\|v\|_X}.
\end{aligned}$$

Finally, summing both contributions,

$$\|\phi\|_Z+\|u\|_X \le \left(\frac{1}{\gamma_\mathcal{A}}+\frac{1}{\gamma_\mathcal{B}}\Big(1+\frac{c_\mathcal{C}}{\gamma_\mathcal{A}}\Big)\right) \sup_{(\psi,v)\in Z\times X\setminus\{0\}} \frac{\mathscr{B}\big((\phi,u),(\psi,v)\big)}{\|\psi\|_Z+\|v\|_X}.$$

To prove (60b), let $(\psi,v)\in Z\times X$ such that

$$0=\mathscr{B}\big((\phi,u),(\psi,v)\big) \quad \forall(\phi,u)\in Z\times X.$$

Choosing $(\phi,u)=(0,u)$, we obtain

$$0=\mathscr{B}\big((0,u),(\psi,v)\big)=\mathcal{B}(u,v) \quad \forall u\in X,$$

which upon invoking (59d) yields $v=0$. Next, choosing $(\phi,u)=(\phi,0)$, we obtain

$$0=\mathscr{B}\big((\phi,0),(\psi,v)\big)=\mathcal{A}(\phi,\psi)+\mathcal{C}(\phi,v)=\mathcal{A}(\phi,\psi) \quad \forall\phi\in Z,$$

which upon invoking (59b) yields $\phi=0$.

Next, we derive the a priori estimates. From inf-sup stability of $\mathcal{A}(\cdot,\cdot)$ and (58a) we obtain:

$$\|\phi\|_Z \le \frac{1}{\gamma_\mathcal{A}} \sup_{\psi\in Z\setminus\{0\}} \frac{\mathcal{A}(\phi,\psi)}{\|\psi\|_Z} = \frac{1}{\gamma_\mathcal{A}}\|\mathcal{F}\|_{Z^*}.$$

Similarly, using inf-sup stability of $\mathcal{B}(\cdot,\cdot)$, (58b), and continuity of $\mathcal{C}(\cdot,\cdot)$ we obtain:

$$\begin{aligned}
\|u\|_X &\le \frac{1}{\gamma_\mathcal{B}} \sup_{v\in X\setminus\{0\}} \frac{\mathcal{B}(u,v)}{\|v\|_X} \\
&= \frac{1}{\gamma_\mathcal{B}} \sup_{v\in X\setminus\{0\}} \frac{\mathcal{G}(v)-\mathcal{C}(\phi,v)}{\|v\|_X} \\
&\le \frac{1}{\gamma_\mathcal{B}}\Big(\|\mathcal{G}\|_{X^*}+c_\mathcal{C}\|\phi\|_Z\Big) \\
&\le \frac{1}{\gamma_\mathcal{B}}\Big(\|\mathcal{G}\|_{X^*}+\frac{c_\mathcal{C}}{\gamma_\mathcal{A}}\|\mathcal{F}\|_{Z^*}\Big).
\end{aligned}$$

□

Corollary. The weak formulation of the slip-BC EOF model, see (25), is well-posed. We have,

$$\mathcal{A}(\varphi,\psi) = \int_\Omega \sigma_c \nabla\varphi \cdot \nabla\psi \, dx, \tag{61}$$

$$\mathcal{B}((\mathbf{w},p),(\mathbf{v},q)) = \int_\Omega \nabla\mathbf{w}\cdot\nabla\mathbf{v} - p\nabla\cdot\mathbf{v} - q\nabla\cdot\mathbf{w} \, dx, \tag{62}$$

$$\mathcal{C}(\varphi,\mathbf{v}) = \int_\Omega \nabla\boldsymbol{\ell}(\varphi)\cdot\nabla\mathbf{v} - q\nabla\cdot\boldsymbol{\ell}(\varphi) \, dx. \tag{63}$$

Moreover, the following a priori estimates hold:

$$\|\varphi\|_Z \leq \frac{\max(\sigma_c)}{\min(\sigma_c)} |\tilde{\phi}|_{H^1(\Omega)}, \tag{64}$$

$$\|w\|_X + \|p\|_M \leq \frac{|\lambda| \mathrm{c}(\Omega)}{\gamma_{\mathcal{B}}} \frac{\max(\sigma_c)}{\min(\sigma_c)} |\tilde{\phi}|_{H^1(\Omega)}. \tag{65}$$

where $\gamma_{\mathcal{B}}$ is the inf-sup constant for the bilinear form $\mathcal{B}((\mathbf{w},p),(\mathbf{v},q))$, λ as required by Proposition 1 and $\mathrm{c}(\Omega)$ is as given by Eq. (19). □

Proof. It is well known that the variational forms $\mathcal{A}(\varphi,\psi)$ and $\mathcal{B}((\mathbf{w},p),(\mathbf{v},q))$ satisfy the inf-sup condition [22]. We can use the definition of the operator $\boldsymbol{\ell}$ in Eq. (22) to easily show that $\mathcal{C}(\varphi,\mathbf{v})$ is bounded,

$$\begin{aligned}
\mathcal{C}(\varphi,\mathbf{v}) &= \int_\Omega \nabla\boldsymbol{\ell}(\varphi)\cdot\nabla\mathbf{v} - q\nabla\cdot\boldsymbol{\ell}(\varphi) \, dx \\
&\leq |\int_\Omega \nabla\boldsymbol{\ell}(\varphi)\cdot\nabla\mathbf{v} \, dx| + |\int_\Omega q\nabla\cdot\boldsymbol{\ell}(\varphi) \, dx| \\
&\leq |\boldsymbol{\ell}|(\boldsymbol{\varphi})_{H^1(\Omega)}|\mathbf{v}|_{H^1(\Omega)} + \|q\|_{L^2(\Omega)}\|\boldsymbol{\ell}(\varphi)\|_{H^1(\Omega)} \\
&\leq \|\boldsymbol{\ell}(\varphi)\|_{H^1(\Omega)} \left(\|\mathbf{v}\|_{H^1(\Omega)} + \|q\|_{L^2(\Omega)}\right) \\
&\leq \mathrm{c}(\Omega)\,|\lambda|\,\|\varphi\|_{H^1(\Omega)} \left(\|\mathbf{v}\|_{H^1(\Omega)} + \|q\|_{L^2(\Omega)}\right).
\end{aligned} \tag{66}$$

We thus satisfy all the conditions of Theorem 4. We also identfy $\mathcal{F}$ as,

$$\mathcal{F}(\psi) = -\int_\Omega \sigma_c \nabla\tilde{\phi}\cdot\nabla\psi \, dx, \tag{67}$$

while $\mathcal{G}$ is simply the null map. This easily gives the a-priori bounds on φ and $(\mathbf{w},p)$. □

Corollary. The weak adjoint problem given by Eq. (28) is well-posed. Moreover, the following a priori estimates hold:

$$\|w^*\|_X + \|p^*\|_M \leq \frac{1}{\gamma_{\mathcal{B}}} \|Q\|_{X^*} \tag{68}$$

$$\|\varphi\|_Z \leq \frac{\mathrm{c}(\Omega)\,|\lambda|}{\gamma_{\mathcal{B}}\gamma_{\mathcal{A}}} \|Q\|_{X^*}. \tag{69}$$

□

Proof. The well-posedness of the adjoint problem follows directly from the well-posedness of the primal problem (see proposition A.9 in [30]). The bilinear forms for the adjoint variational formulation are,

$$\mathcal{A}(\varphi, \psi) = \int_{\Omega} \sigma_c \nabla \varphi^* \cdot \nabla \psi \, dx, \tag{70}$$

$$\mathcal{B}((\mathbf{w}^*, p^*), (\mathbf{v}, q)) = \int_{\Omega} \nabla \mathbf{w}^* \cdot \nabla \mathbf{v} - q \nabla \cdot \mathbf{w}^* - p^* \nabla \cdot \mathbf{v} \, dx, \tag{71}$$

$$\mathcal{C}((\mathbf{w}^*, p^*), \psi) = \int_{\Omega} \nabla \boldsymbol{\ell}(\psi) \cdot \nabla \mathbf{w}^* - p^* \nabla \cdot \boldsymbol{\ell}(\psi) \, dx. \tag{72}$$

In an analogous fashion to the one-way coupled formulation of the primal problem (see Eq. (58)), a one-way coupled formulation of the adjoint problem would read,

Find $(u, \varphi) \in Z \times X$:

$$\mathcal{B}(u, v) = Q(v) \qquad \forall v \in X, \tag{73a}$$

$$\mathcal{A}(\varphi, \psi) = -\mathcal{C}(u, \psi) \qquad \forall \psi \in Z. \tag{73b}$$

The inf-sup stability of $\mathcal{B}(\cdot,\cdot)$, $\mathcal{A}(\cdot,\cdot)$ and Eq. (73a) and (73b) then easily give the a priori error estimates. □

Competing interests

The authors declare that they have no competing interests.

Authors' contributions

The author's have presented an adjoint-consistency analysis of the widely used Helmholtz 'slip' electroosmotic flow models. It was shown that a direct formulation of such models leads to an ill-posed adjoint problem. A modified formulation is proposed and it is shown that this formulation is well-posed and adjoint-consistent. Numerical experiments show that the adjoint obtained from the modified formulation can be used for goal-oriented adaptive mesh refinement and adjoint-based sensitivity analysis. All authors read and approved the final manuscript.

Acknowledgements

Vikram Garg is grateful for the support of the Bruton fellowship and the University Continuing fellowship from The University of Texas at Austin. Kris van der Zee is grateful for the support of this work by the 2010 NWO Innovational Research Incentives Scheme (IRIS) Grant 639.031.033. Serge Prudhomme is sponsored by a Discovery Grant from the Natural Sciences and Engineering Research Council of Canada. Serge Prudhomme is also a participant of the KAUST SRI center for Uncertainty Quantification in Computational Science and Engineering.

Author details

[1]Massachusetts Institute of Technology, 77 Massachusetts Avenue, Cambridge, MA 02139, USA. [2]Ecole Polytechnique de Montréal, C.P. 6079, succ. Centre-Ville, Montréal, Canada. [3]The University of Nottingham, Nottingham, UK. [4]The University of Texas at Austin, 1 University Station, C0200 Austin, USA.

References

1. Karniadakis G, Beskok A, Aluru NR (2005) Microflows and Nanoflows: Fundamentals and Simulation. Springer, New York
2. Feynman RP (1960) There's plenty of room at the bottom. Eng Sci 23(5):22–36
3. Knio OM, Ghanem RG, Matta A, Najm HN, Debusschere B, Le Maître OP (2005) Quantitative Uncertainty Assessment and Numerical Simulation of Micro-Fluid Systems. Technical report, Johns Hopkins University, Baltimore, MD
4. Becker R, Rannacher R (2003) An optimal control approach to a posteriori error estimation in finite element methods. Acta Numerica 10:1–102
5. Estep D, Carey V, Ginting V, Tavener S, Wildey T (2008) A posteriori error analysis of multiscale operator decomposition methods for multiphysics models. J Phys: Conf Ser 125:012075. IOP Publishing
6. Ionescu-Bujor M, Cacuci DG (2004) A comparative review of sensitivity and uncertainty analysis of large-scale systems. i: Deterministic methods. Nuclear Sci Eng 147(3):189–203
7. Zhang Y, Wong TN, Yang C, Ooi KT (2005) Electroosmotic flow in irregular shape microchannels. Int J Eng Sci 43(19-20):1450–1463

8. Ren L, Sinton D, Li D (2003) Numerical simulation of microfluidic injection processes in crossing microchannels. J Micromech Microeng 13:739
9. MacInnes JM, Du X, Allen RWK (2003) Prediction of electrokinetic and pressure flow in a microchannel T-junction. Phys Fluids 15:1992
10. Hahm J, Balasubramanian A, Beskok A (2007) Flow and species transport control in grooved microchannels using local electrokinetic forces. Phys Fluids 19:013601
11. Craven TJ, Rees JM, Zimmerman WB (2008) On slip velocity boundary conditions for electroosmotic flow near sharp corners. Phys Fluids 20:043603
12. Zimmerman WB, Rees JM, Craven TJ (2006) Rheometry of non-newtonian electrokinetic flow in a microchannel T-junction. Microfluidics Nanofluidics 2(6):481–492
13. Prachittham V, Picasso M, Gijs MAM (2010) Adaptive finite elements with large aspect ratio for mass transport in electroosmosis and pressure-driven microflows. Int J Numerical Methods Fluids 63(9):1005–1030
14. van Brummelen EH, van der Zee KG, Garg VV, Prudhomme S (2012) Flux evaluation in primal and dual boundary-coupled problems. J Appl Mech 79(1):010904. American Society of Mechanical Engineers
15. Estep D, Tavener S, Wildey T (2010) A posteriori error estimation and adaptive mesh refinement for a multi-discretization operator decomposition approach to fluid-solid heat transfer. J Comput Phys 229:4143–4158
16. Squires TM, Quake SR (2005) Microfluidics: Fluid physics at the nanoliter scale. Rev Modern Phys 77(3):977–1026
17. Whitesides GM (2006) The origins and the future of microfluidics. Nature 442(7101):368–373
18. Chen C, Lin H, Lele S, Santiago J (2005) Convective and absolute electrokinetic instability with conductivity gradients. J Fluid Mech 524:263–303
19. Masliyah JH, Bhattacharjee S (2006) Electrokinetic and Colloid Transport Phenomena. Wiley-Interscience, New York
20. Barz D (2009) Comprehensive model of electrokinetic flow and migration in microchannels with conductivity gradients. Microfluidics and Nanofluidics 7(2):249–265. Springer
21. Barth WL, Carey GF (2007) On a boundary condition for pressure-driven laminar flow of incompressible fluids. Int J Numerical Methods Fluids 54(11):1313–1325
22. Ern A, Guermond JL (2004) Theory and Practice of Finite Elements, volume 159 of Applied Mathematical Sciences. Springer, New York
23. Delfour M, Zolésio J-P (2001) Shapes and Geometries: Analysis, Differential Calculus, and Optimization. SIAM, New York Vol. 4 of SIAM Series on Advances in Design and Control
24. Hartmann R (2007) Adjoint consistency analysis of discontinuous galerkin discretizations. SIAM J Numer Anal 45(6):2671–2696
25. Babuška I (1973) The finite element method with penalty. Math Comp 27(122):221–228
26. Utku M, Carey GF (1982) Boundary penalty techniques. Comput Methods Appl Mech Eng 30(1):103–118
27. Garg VV (2012) Coupled flow systems, adjoint techniques and uncertainty quantification. PhD Thesis, The University of Texas at Austin
28. Kirk B, Peterson J, Stogner R, Carey G (2006) libMesh: a C++ library for parallel adaptive mesh refinement/coarsening simulations. Eng Comput 22(3):237–254
29. Demkowicz L (2006) Computing with Hp-adaptive Finite Elements: One and Two Dimensional Elliptic and Maxwell Problems. CRC Press, New York
30. Van der Zee K (2009) Goal-adaptive discretization of fluid-structure interaction Dissertation. Delft Institute of Technology, ISBN: 9789079488544 TU Delft institutional repository

3

Separated representations of 3D elastic solutions in shell geometries

Brice Bognet, Adrien Leygue and Francisco Chinesta*

*Correspondence:
Francisco.Chinesta@ec-nantes.fr
GeM UMR CNRS-Centrale Nantes, 1 rue de la Noe, F-44300 Nantes, France

Abstract

Background: The solution of 3D models in degenerated geometries in which some characteristic dimensions are much lower than the other ones -e.g. beams, plates, shells, ...- is a tricky issue when using standard mesh-based discretization techniques.

Methods: Separated representations allow decoupling the meshes used for approximating the solution along each coordinate. Thus, in plate or shell geometries 3D solutions can be obtained from a sequence of 2D and 1D problems allowing fine and accurate representation of the solution evolution along the thickness coordinate while keeping the computational complexity characteristic of 2D simulations. In a former work this technique was considered for addressing the 3D solution of thermoelastic problems defined in plate geometries. In this work, the technique is extended for addressing the solution of 3D elastic problems defined in shell geometries.

Results: The capabilities of the proposed approach are illustrated by considering some numerical examples involving different degrees of complexity, from simple shells to composite laminates involving stiffeners.

Conclusions: The analyzed examples prove the potentiality and efficiency of the proposed strategy, where the computational complexity was found evolving as reported in our former works, proving that 3D solutions can be computed at a 2D cost.

Keywords: PGD; Separated representations; Model reduction; Shell geometries

Background

Plates and shells are very common in nature and thus they inspired engineers that used both from the very beginning of structural mechanics. Shells offer a diversity of possible shapes and geometries, some of them with simple curvature and most of them with double curvature. Many times they are assembled in complex structural systems, in many applications they contain many stiffeners as in the case of aircraft fuselages.

In general the design of such structural elements requires the calculation of stresses, strains and displacements for the design loads. Strains and stresses are related by the so-called constitutive law. The simplest one consists of the linear elasticity. Despite its simplicity many structures are designed for working precisely within the elastic domain. Other designs require considering more complex behaviors (e.g. non-linear elasticity due to material or geometrical non linearities, elastoplastic behaviors usually encountered in material forming – forging, bending, ... –, or complex multiphysics behaviors as the ones encountered in composites manufacturing processes implying change of phases, crystallization, polymerization, ... coupled with rich thermomechanical mechanisms).

Design problems always involve the solution of a set of partial differential equations in the degenerate domain of the plate or the shell with appropriate initial and boundary conditions. These domains are degenerated because one of its characteristic dimensions (the thickness in the present case) is much lower that the other characteristic dimensions. We will understand the consequences of such degeneracy later. When analytical solutions are neither available nor possible because the geometrical or behavior complexities, the solution must be calculated by invoking any of the available numerical techniques (finite elements, finite differences, finite volumes, methods of particles, ...).

In the numerical framework the solution will be only obtained in a discrete number of points, usually called nodes, distributed in the domain. From the solution at those points, it can be interpolated at any other point in the domain. In general regular nodal distributions are preferred because they offer the best accuracy. In the case of degenerated plate or shell domains one could expect that if the solution evolves significantly in the thickness direction, a large enough number of nodes must be distributed along the thickness direction to ensure the accurate representation of the field evolution in that direction. In that case, a regular nodal distribution in the whole domain will imply the use of an extremely large number of nodes with the consequent impact on the numerical solution efficiently.

When simple behaviors and domains were considered semi-analytical models can be considered [1]. For addressing more complex scenarios plate and shell theories were developed allowing, through the introduction of some hypotheses, reducing the 3D complexity to the 2D related to the problem now formulated by considering the in-plane coordinates. The use of these theories have been extended gradually for addressing larger and more complex geometries (anisotropic laminates, ...) and behaviors.

There are thousand of papers concerning the proposal and application of plate and shell models (the interested reader can refer to the recent reviews [2,3] and the references therein). Some models are based on the introduction of kinematic hypotheses in the thickness (e.g. [4] among many others). Transverse shear can be also taken into account [5]. Recent zig-zag representations [6,7], layer-wise models [8-10] and solid-shell approaches [11,12], allow addressing accurately more complex scenarios, by increasing the computational complexity slightly. Stiffeners require an appropriate coupling of beam and shell models in order to perform calculations at a moderate computational cost [13].

However, as soon as richer physics are involved in the models and the considered geometries differ of those ensuring the validity of the different reduction hypotheses, efficient simulations are compromised. For example in composites manufacturing processes of large parts many reactions and thermal processes inducing significant evolutions on the thermomechanical fields in the thickness occur. These inhomogeneities are at the origin of residual stresses and the associated distortion of the formed parts [14].

In these circumstances as just indicated the reduction from the 3D model to a 2D simplified one is not obvious, and 3D simulations appear many times as the only valid route for addressing such models, that despite the fact of being defined in degenerated geometries (plate or shell) they seem requiring a fully 3D solution. However in order to integrate such calculations (fully 3D and implying an impressive number of degrees of freedom) in usual design procedures, a new efficient (fast and accurate) solution procedure is needed.

The Saint Venant's principle was extensively used in the Ladeveze's works for defining elegant and efficient 3D simplified models [15]. This technique was then generalized to dynamics [16]. This technique allowed significant reduction of computational complexity.

Later, a new discretization technique based on the use of separated representations was proposed for addressing space-time nonlinear models [17] and then it was generalized for defining general separated representations of solutions involving conformational coordinates [18], space and time and even parameters considered as extra-coordinates. The interested reader can refer to the recent reviews [19-22] and the references therein.

A direct consequence was the separated representations involving the space coordinates. Thus in plate domains an in-plane-out-of-plane decomposition was proposed for solving flow problems in laminates [20], then for solving thermal problems in extruded geometries [23], elasticity problems [24] and coupled multiphycisc problems [25]. In those cases the 3D solution was obtained from the solution of a sequence of 2D problems (the ones involving the in-plane coordinates) and 1D problems (the ones involving the coordinate related to the plate thickness).

It is important emphasizing the fact that these approaches are radically different to standard plate and shell approaches. We proposed a 3D solver able to compute the different unknown fields without the necessity of introducing any hypothesis. The most outstanding advantage is that 3D solutions can be obtained with a computational cost characteristic of standard 2D solutions. Moreover, as noticed in [24] no locking effects were found, possibly because the fully 3D solution accomplished.

In this work we will generalize the just referred approach considered in the case of plate domains for calculating the fully 3D solution of the elastic problem in shell domains. The 3D solution will be calculated again from the solution of a sequence of 2D and 1D problems thanks to the in-plane-out-of-plane separated representation.

It is important to note that in this paper we are not addressing a new shell modeling, and by this reason we do not need neither establishing a precise state of the art on shell theories nor comparing our approach with the solutions obtained by using shell models. As we are proposing a new procedure for calculating 3D solutions (keeping a computational complexity characteristic of 2D solution procedures) we will compare our solutions with the ones obtained by considering the fully 3D elastic solution in the shell geometries computed with standard 3D solvers (e.g. finite elements).

Before generalizing the technique proposed in [24] for treating elastic problems defined in shell domains we are summarizing it.

In-plane-out-of-plane separated representation of elastic problems defined in plate domains

We proposed in [24] and original in-plane-out-of-plane decomposition of the 3D elastic solution in a plate geometry. The elastic problem was defined in a plate domain $\Xi = \Omega \times \mathcal{I}$ with $(x_1, x_2) \in \Omega$, $\Omega \subset \mathcal{R}^2$ and $x_3 \in \mathcal{I}, \mathcal{I} = [0, H] \subset \mathcal{R}$, being H the plate thickness. The separated representation of the displacement field $\mathbf{u} = (u_1, u_2, u_3)$ reads:

$$\mathbf{u}(x_1, x_2, x_3) = \begin{pmatrix} u_1(x_1, x_2, x_3) \\ u_2(x_1, x_2, x_3) \\ u_3(x_1, x_2, x_3) \end{pmatrix} \approx \sum_{i=1}^{N} \begin{pmatrix} P_1^i(x_1, x_2) \cdot T_1^i(x_3) \\ P_2^i(x_1, x_2) \cdot T_2^i(x_3) \\ P_3^i(x_1, x_2) \cdot T_3^i(x_3) \end{pmatrix} \tag{1}$$

where P_k^i, $k = 1,2,3$, are functions of the in-plane coordinates (x_1, x_2) whereas T_k^i, $k = 1,2,3$, are functions involving the thickness coordinate x_3. In [24] we compared the first modes of such separated representations with the kinematic hypotheses usually considered in plate theories. Similar behavior was noticed in the case of elastic solutions in shell domains with respect to classical shell theories.

Expression (1) can be written in a more compact form by using the Hadamard product:

$$\mathbf{u}(x_1, x_2, x_3) \approx \sum_{i=1}^{N} \mathbf{P}^i(x_1, x_2) \circ \mathbf{T}^i(x_3) \tag{2}$$

where vectors $\mathbf{P}^i$ and $\mathbf{T}^i$ contains functions P_k^i and T_k^i respectively.

Because neither the number of terms in the separated representation of the displacement field nor the dependence on x_3 of functions T_k^i are assumed *a priori*, the approximation is flexible enough for representing the fully 3D solution, being obviously more general than theories assuming particular *a priori* evolutions in the thickness direction x_3.

Let's consider a linear elasticity problem on a plate domain $\Xi = \Omega \times \mathcal{I}$. The weak formulation reads:

$$\int_{\Xi} \boldsymbol{\epsilon}(\mathbf{u}^*)^T \cdot \mathbf{K} \cdot \boldsymbol{\epsilon}(\mathbf{u})\, d\mathbf{x} = \int_{\Xi} \mathbf{u}^* \cdot \mathbf{f}_d \, d\mathbf{x} + \int_{\Gamma_N} \mathbf{u}^* \cdot \mathbf{F}_d \, d\mathbf{x}, \quad \forall \mathbf{u}^* \tag{3}$$

where $\mathbf{K}$ is the generalized 6×6 Hooke tensor, $\mathbf{f}_d$ represents the volumetric body forces while $\mathbf{F}_d$ represents the traction applied on the boundary Γ_N. The separation of variables introduced in Eq. (1) yields the following expression for the derivatives of the displacement components u_i, $i = 1,2,3$:

$$\frac{\partial u_i}{\partial x_j} \approx \sum_{k=1}^{k=N} \frac{\partial P_i^k}{\partial x_j} \cdot T_i^k \tag{4}$$

for $j = 1,2$; and

$$\frac{\partial u_i}{\partial x_3} \approx \sum_{k=1}^{k=N} P_i^k \cdot \frac{\partial T_i^k}{\partial x_3} \tag{5}$$

from which we can obtain the separated vector form of the strain tensor $\boldsymbol{\epsilon}$:

$$\boldsymbol{\epsilon}(\mathbf{u}(x_1, x_2, x_3)) \approx \sum_{k=1}^{N} \begin{pmatrix} \frac{\partial P_1^k}{\partial x_1} \cdot T_1^k \\ \frac{\partial P_2^k}{\partial x_2} \cdot T_2^k \\ P_3^k \cdot \frac{\partial T_3^k}{\partial x_3} \\ \frac{\partial P_1^k}{\partial x_2} \cdot T_1^k + \frac{\partial P_2^k}{\partial x_1} \cdot T_2^k \\ \frac{\partial P_3^k}{\partial x_1} \cdot T_3^k + P_1^k \cdot \frac{\partial T_1^k}{\partial x_3} \\ \frac{\partial P_3^k}{\partial x_2} \cdot T_3^k + P_2^k \cdot \frac{\partial T_2^k}{\partial x_3} \end{pmatrix}. \tag{6}$$

Depending on the number of non-zero elements in the $\mathbf{K}$ matrix, the development of $\boldsymbol{\epsilon}(\mathbf{u}^*)^T \cdot \mathbf{K} \cdot \boldsymbol{\epsilon}(\mathbf{u})$ involves different number of terms, 21 in the case of an isotropic material and 41 in the case of general anisotropic behaviors.

The separated representation constructor proceeds by computing a term of the sum at each iteration. Assuming that the first $n-1$ modes (terms of the finite sum) of the solution were already computed, $\mathbf{u}^{n-1}(x_1, x_2, x_3)$ with $n \geq 1$, the solution enrichment reads:

$$\mathbf{u}^n(x_1, x_2, x_3) = \mathbf{u}^{n-1}(x_1, x_2, x_3) + \mathbf{P}^n(x_1, x_2) \circ \mathbf{T}^n(x_3) \tag{7}$$

where both vectors $\mathbf{P}^n$ and $\mathbf{T}^n$ containing functions P_i^n and T_i^n ($i = 1, 2, 3$) depending on (x_1, x_2) and x_3 respectively, are unknown at the present iteration. The test function $\mathbf{u}^*$ reads $\mathbf{u}^* = \mathbf{P}^* \circ \mathbf{T}^n + \mathbf{P}^n \circ \mathbf{T}^*$.

The introduction of Eq. (7) into (3) results in a non-linear problem. We proceed by considering the simplest linearization strategy, an alternated directions fixed point algorithm, that proceeds by calculating $\mathbf{P}^{n,k}$ from $\mathbf{T}^{n,k-1}$ and then by updating $\mathbf{T}^{n,k}$ from the just calculated $\mathbf{P}^{n,k}$ where k refers to the step of the non-linear solver. The iteration procedure continues until convergence, that is, until reaching the fixed point $\|\mathbf{P}^{n,k} \circ \mathbf{T}^{n,k} - \mathbf{P}^{n,k-1} \circ \mathbf{T}^{n,k-1}\| < \epsilon$, that results in the searched functions $\mathbf{P}^{n,k} \rightarrow \mathbf{P}^n$ and $\mathbf{T}^{n,k} \rightarrow \mathbf{T}^n$. Then, the enrichment step continues by looking for the next mode $\mathbf{P}^{n+1} \circ \mathbf{T}^{n+1}$. The enrichment stops when the model residual becomes small enough.

When $\mathbf{T}^n$ is assumed known, we consider the test function $\mathbf{u}^\star$ given by $\mathbf{P}^\star \circ \mathbf{T}^n$. By introducing the trial and test functions into the weak form and then integrating in $\mathcal{I}$ because all the functions depending on the thickness coordinate are known, we obtain a 2D weak formulation defined in Ω whose discretization (by using a standard discretization strategy, e.g. finite elements) allows computing $\mathbf{P}^n$.

Analogously, when $\mathbf{P}^n$ is assumed known, the test function $\mathbf{u}^\star$ is given by $\mathbf{P}^n \circ \mathbf{T}^\star$. By introducing the trial and test functions into the weak form and then integrating in Ω because all the functions depending on the in-plane coordinates (x_1, x_2) are at present known, we obtain a 1D weak formulation defined in $\mathcal{I}$ whose discretization (using any technique for solving standard ODE equations) allows computing $\mathbf{T}^n$.

The problems related to the solution of functions $\mathbf{P}^n$ and $\mathbf{T}^n$ are defined in Appendix A.

As discussed in [24] this separated representation allows computing 3D solutions while keeping a computational complexity characteristic of 2D solution procedures. If we consider a hexahedral domain discretized using a regular structured grid with N_1, N_2 and N_3 nodes in the x_1, x_2 and x_3 directions respectively, usual mesh-based discretization strategies imply a challenging issue because the number of nodes involved in the model scales with $N_1 \cdot N_2 \cdot N_3$, however, by using the separated representation and assuming that the solution involves N modes, one must solve about N 2D problems related to the functions involving the in-plane coordinates (x_1, x_2) and the same number of 1D problems related to the functions involving the thickness coordinate x_3. The computing time related to the solution of the one-dimensional problems can be neglected with respect to the one required for solving the two-dimensional ones. Thus, the resulting complexity scales as $N \cdot N_1 \cdot N_2$. By comparing both complexities we can notice that as soon as $N_3 \gg N$ the use of separated representations leads to impressive computing time savings, making possible the solution of models never until now solved, and even using light computing platforms.

In [24] we considered the simplest approximations of functions involving the in-plane coordinates $P_k^i(x_1, x_2)$ by considering bilinear quadrilateral finite elements and piecewise linear 1D elements for approximating functions involving the thickness coordinate x_3, $T_k^i(x_3)$. Richer approximations were analyzed in [26].

In the present work we are generalizing the just described separated representation for solving 3D models defined in shell domains.

Methods

3D elastic problem in a shell domain: Shell representation

The shell domain Ω^S, assumed with constant thickness, can be described from a reference surface $\mathbf{X}$, that in what follows will be identified to the shell middle surface but that in the general case could be any other one, parametrized by the coordinates ξ, η, that is $\mathbf{X}(\xi, \eta)$, where:

$$\mathbf{X}(\xi,\eta) = \begin{pmatrix} X_1(\xi,\eta) \\ X_2(\xi,\eta) \\ X_3(\xi,\eta) \end{pmatrix} \tag{8}$$

Being $\mathbf{n}$ the unit vector normal to the middle surface, the shell domain Ω^S can be parametrized from:

$$\mathbf{x}(\xi,\eta,\zeta) = \mathbf{X}(\xi,\eta) + \zeta \cdot \mathbf{n} \tag{9}$$

The geometrical transformation $(\xi,\eta,\zeta) \to (x_1,x_2,x_3)$ involves

$$\tilde{\mathbf{F}} = \left[\frac{\partial \mathbf{x}}{\partial \xi} \ \frac{\partial \mathbf{x}}{\partial \eta} \ \mathbf{n} \right] \tag{10}$$

whose expression is given in the Appendix B (Eqs. (73) and (74) that involve Eqs. (42), (43) and (49)).

The inverse transformation $(x_1,x_2,x_3) \to (\xi,\eta,\zeta)$, described by $\tilde{\mathbf{F}}^{-1}$ is also given in the appendix (Eq. (75)).

3D elastic problem in a shell domain: Weak form

The weak form of the elastic problem defined in the shell domain Ω^S writes:

$$\int_{\Omega^S} \boldsymbol{\epsilon}(\mathbf{u}^*)^T \cdot \mathbf{K} \cdot \boldsymbol{\epsilon}(\mathbf{u}) \, d\mathbf{x} = \int_{\Omega^S} \mathbf{u}^* \cdot \mathbf{f}_d \, d\mathbf{x} + \int_{\Gamma_N^S} \mathbf{u}^* \cdot \mathbf{F}_d \, d\mathbf{x} \tag{11}$$

Now we are considering the coordinates transformation introduced in the previous section and deeply developed in the appendix, mapping $\mathbf{x} \in \Omega^S$ into $(\xi,\eta,\zeta) \in \Xi = \Omega \times \mathcal{I}$, with $(\xi,\eta) \in \Omega \subset \mathcal{R}^2$ and $\zeta \in \mathcal{I} \subset \mathcal{R}$.

The geometric transformation requires to transform the differential operator as well as the different volume and surface elements. Knowing that under the small displacements and strains assumption, the strain tensor consists of the symmetric part of the gradient of displacement tensor, i.e.

$$\boldsymbol{\epsilon}(\mathbf{u}) = \frac{1}{2} \cdot \left(\nabla \mathbf{u} + (\nabla \mathbf{u})^T \right) \tag{12}$$

it can be transformed taking into account the transformation of the gradient differential operator

$$\nabla(\cdot) = \nabla_\xi(\cdot) \cdot \tilde{\mathbf{F}}^{-1} \tag{13}$$

where $\nabla_\xi(\cdot)$ denotes the gradient in the parametric space.

The volume element involved in the integral in Ω^S writes according to Eq. (70)

$$d\mathbf{x} = dx_1 \cdot dx_2 \cdot dx_3 = \sqrt{a} \cdot (1 - 2 \cdot H \cdot \zeta + K \cdot \zeta^2) \cdot d\xi \cdot d\eta \cdot d\zeta \tag{14}$$

with a is the determinant of the metric tensor related to the middle surface mapping (46) and H and K the curvatures given by Eqs. (56) and (57) respectively. Finally the integrals applying on the domain boundary (where tractions apply) are transformed according to Eq. (68).

3D elastic problem in a shell domain: In-plane-out-of-plane separated representation

With the weak form defined in $\Xi = \Omega \times \mathcal{I}$ the situation is quite similar to the one encountered in the analysis of elastic problems in plate geometries, that was addressed in [24] and we just summarized in section 'Background'.

We could perform an in-plane-out-of-plane separated representation of the displacement field, similar to (1) but now involving the coordinates (ξ, η, ζ)

$$\mathbf{u}(\xi, \eta, \zeta) = \begin{pmatrix} u_1(\xi, \eta, \zeta) \\ u_2(\xi, \eta, \zeta) \\ u_3(\xi, \eta, \zeta) \end{pmatrix} \approx \sum_{i=1}^{N} \begin{pmatrix} P_1^i(\xi, \eta) \cdot T_1^i(\zeta) \\ P_2^i(\xi, \eta) \cdot T_2^i(\zeta) \\ P_3^i(\xi, \eta) \cdot T_3^i(\zeta) \end{pmatrix} \tag{15}$$

or in a more compact form

$$\mathbf{u}(\xi, \eta, \zeta) \approx \sum_{i=1}^{N} \mathbf{P}^i(\xi, \eta) \circ \mathbf{T}^i(\zeta) \tag{16}$$

As explained in section 'Background' the construction of such a separated representation is performed sequentially, thus assuming known the solution at iteration $n-1$, the solution at iteration n is sought as

$$\mathbf{u}^n(\xi, \eta, \zeta) = \mathbf{u}^{n-1}(\xi, \eta, \zeta) + \mathbf{P}^n(\xi, \eta) \circ \mathbf{T}^n(\zeta) \tag{17}$$

By introducing (17) in the weak form and using the alternated directions fixed point algorithm we can calculate $\mathbf{P}^n(\xi, \eta)$ by assuming $\mathbf{T}^n(\zeta)$ known and then updated $\mathbf{T}^n(\zeta)$ from the just calculated $\mathbf{P}^n(\xi, \eta)$. The iteration continues until reaching the convergence (the fixed point) that determines both functions $\mathbf{P}^n(\xi, \eta)$ and $\mathbf{T}^n(\zeta)$.

However, the decomposition in a problem defined in Ω for calculating function $\mathbf{P}^n(\xi, \eta)$, obtained by integrating the weak form in $\mathcal{I}$, and in another problem defined in $\mathcal{I}$ for calculating function $\mathbf{T}^n(\zeta)$, obtained by integrating the weak form in Ω, requires the separated representation of all the operators, variables, coefficients and functions involved in the weak form.

For the displacement (the trial $\mathbf{u}$ and the test $\mathbf{u}^*$ displacements) we just indicated the separated in-plane-out-of-plane representation. This representation allows defining a separated representation of the associated strain tensors $\boldsymbol{\epsilon}(\mathbf{u})$ and $\boldsymbol{\epsilon}(\mathbf{u}^*)$ as illustrated in Eq. (6) but for this purpose we must define a separated representation of the transformation gradient involved in Eq. (13) $\tilde{\mathbf{F}}^{-1}$.

In the case of small thickness and curvatures $\tilde{\mathbf{F}}^{-1}$ can be approximated, according to (73), from:

$$\tilde{\mathbf{F}}^{-1} \approx (\mathbf{I} - \zeta \cdot \mathbf{b}_n + \zeta^2 \cdot \mathbf{b}_n^2) \cdot \mathbf{F}^{-1} \tag{18}$$

where $\mathbf{b}_n$ and $\mathbf{F}$ only depend on the middle surface parametrization.

Thus, we can define a direct separated representation of $\tilde{\mathbf{F}}^{-1}$:

$$\tilde{\mathbf{F}}^{-1} = \sum_{i=1}^{i=3} \mathbf{G}_i^P(\xi, \eta) \circ \mathbf{G}_i^T(\zeta) \tag{19}$$

with

$$\begin{cases} \mathbf{G}_1^P = \mathbf{F}^{-1}; & \mathbf{G}_1^T = \mathbf{U}_3 \\ \mathbf{G}_2^P = \mathbf{b}_n \cdot \mathbf{F}^{-1}; & \mathbf{G}_2^T = -\zeta \cdot \mathbf{U}_3 \\ \mathbf{G}_3^P = \mathbf{b}_n^2 \cdot \mathbf{F}^{-1}; & \mathbf{G}_3^T = \zeta^2 \cdot \mathbf{U}_3 \end{cases} \tag{20}$$

where $\mathbf{U}_3$ is the 3×3 matrix with unit components, that is:

$$\mathbf{U}_3 = \begin{pmatrix} 1 & 1 & 1 \\ 1 & 1 & 1 \\ 1 & 1 & 1 \end{pmatrix} \tag{21}$$

When this approximation (small thickness and curvatures) does not apply we should consider the separated representation of $\tilde{\mathbf{F}}^{-1}$ by applying a singular value decomposition –SVD– .

The elasticity tensor and the applied forces must be also expressed in a separated form. Most of time this decomposition is direct and in the more complex cases the application of a singular value decomposition allows such decomposition.

Finally the volume and surface elements must be also written in a separated form. In the case of the volume element $d\mathbf{x}$ expressed by Eq. (14)

$$d\mathbf{x} = \sqrt{a} \cdot (1 - 2 \cdot H \cdot \zeta + K \cdot \zeta^2) \cdot d\xi \cdot d\eta \cdot d\zeta \tag{22}$$

the following separated representation can be assumed

$$d\mathbf{x} = \sum_{i=1}^{i=3} A_i^P(\xi, \eta) \cdot A_i^T(\zeta) \tag{23}$$

with

$$\begin{cases} A_1^P = \sqrt{a}; & A_1^T = 1 \\ A_2^P = -2 \cdot H \cdot \sqrt{a}; & A_2^T = \zeta \\ A_3^P = K \cdot \sqrt{a}; & A_3^T = \zeta^2 \end{cases} \tag{24}$$

From Eq. (68) we can expect a similar decomposition of the surface element.

3D elastic problem in a shell domain: Composite laminates

In this section we consider composite laminates composed of P anisotropic (generally orthotropic) plies of thickness h (for the sake of simplicity and without loss of generality we assume all the plies with the same thickness).

We consider the mechanical behavior of each ply i expressed by its elasticity tensor $\mathcal{K}^i$ whose form is quite simple when it is expressed in the basis related to its mechanical principal directions.

Now, for the sake of simplicity we are also considering a local orthonormal basis defined at each point on the middle surface $(\mathbf{t}_1, \mathbf{t}_2, \mathbf{n})$, where the normal vector $\mathbf{n}$ is the one previously considered, and the tangent vectors $\mathbf{t}_1$ and $\mathbf{t}_2$ to the middle surface at the considered location can be chosen arbitrarily under the constraints: $\mathbf{t}_1 \cdot \mathbf{t}_2 = 0$, $\mathbf{t}_1 \cdot \mathbf{n} = 0$, $\mathbf{t}_2 \cdot \mathbf{n} = 0$, $\|\mathbf{t}_1\| = 1$ and $\|\mathbf{t}_2\| = 1$.

If we define $\mathbf{R}^i$ as the rotation tensor allowing the expression of the elastic tensor in the orthonormal local system $(\mathbf{t}_1, \mathbf{t}_2, \mathbf{n})$ and $\mathbf{Q}^i$ the one allowing to express finally its behavior in the cartesian basis (both expressed using the Voigt notation) it results

$$\mathbf{K}^i = \mathbf{Q}^{i^T} \cdot \mathbf{R}^{i^T} \cdot \mathcal{K}^i \cdot \mathbf{R}^i \cdot \mathcal{Q}^i \tag{25}$$

If we assume that the elastic properties of each ply are constant along the ply thickness, then $\mathbf{K}^i$ only depends on the in-plane coordinates. Thus, the laminate elastic tensor can be written in the separated form

$$\mathbf{K}(\xi, \eta, \zeta) = \sum_{i=1}^{i=P} \chi^i(\zeta) \cdot \mathbf{K}^i(\xi, \eta) \tag{26}$$

or using the Hadamard notation

$$\mathbf{K}(\xi, \eta, \zeta) = \sum_{i=1}^{i=P} \mathbf{k}^i(\zeta) \circ \mathbf{K}^i(\xi, \eta) \tag{27}$$

with $\mathbf{k}^i(\zeta) = \chi^i(\zeta) \cdot \mathbf{U}_6$ ($\mathbf{U}_6$ being the 6×6 matrix with unit components) and $\chi^i(\zeta)$ the characteristic function related to the i-ply:

$$\chi^i(\zeta) = \begin{cases} 1 \ \ if \ \zeta \in ((i-1) \cdot h, i \cdot h) \\ 0 \ \ elsewhere \end{cases} \tag{28}$$

Results and discussion

Strategy verification

First we verify the solution computed using the proposed strategy by comparing it with the exact solution of a linear and isotropic elastic problem defined in an infinite tube subjected to an internal pressure. Figure 1 depicts the tube cross-section. The problem parameters are given by: $a = 2$, $b = 3$, $P_i = 1$, $E = 150000$ and $\nu = 0.3$ (all the units in the metric system).

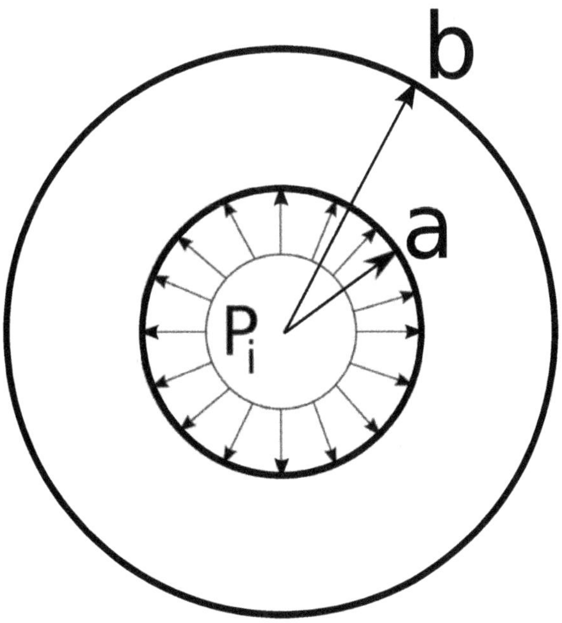

Figure 1 Cross-section of a tube of infinite length subjected to an internal pressure.

We consider the middle surface representation

$$\begin{cases} X_1 = \frac{\overline{R}}{2\cdot\pi}\sin(\frac{\xi}{\overline{R}}) \\ X_2 = \eta \\ X_3 = \frac{\overline{R}}{2\cdot\pi}\cos(\frac{\xi}{\overline{R}}) \end{cases} \tag{29}$$

with $\overline{R} = 2.5$, $\xi \in [0, 2\cdot\pi\cdot\overline{R}]$ and $\eta \in [0,1]$ (we are considering an arbitrary tube length due to the plane state of strain).

The shell geometry is defined by introducing the thickness coordinate $\zeta \in [-0.5, 0.5]$.

The exact elastic solution only involves radial displacements, u_r, depending on the radial coordinate r:

$$u_r(r) = \frac{1+\nu}{E\cdot(b^2 - a^2)}\left[(1-2\cdot\nu)(a^2\cdot P_i\cdot r) + \frac{a^2\cdot b^2\cdot P_i}{r}\right] \tag{30}$$

In the numerical solution the domain $(\xi,\eta) \in \Omega = [0, 2\cdot\pi\cdot\overline{R}]\times[0,1]$ is discretized by using a regular mesh of rectangular bilinear elements whereas the thickness interval $\zeta \in \mathcal{I} = [-0.5, 0.5]$ is discretized by using standard one-dimensional linear elements (the simplest choices, but higher order approximations can be used in Ω and/or $\mathcal{I}$). The displacement is constrained at some appropriate locations in order to avoid rigid body movements.

A convergence analysis is performed by considering the 3 different meshes reported in Table 1 where N_ξ, N_η and N_ζ are the number of elements along the angular, axial and thickness coordinates respectively.

The problem is also solved numerically by using a 3D finite element discretization operating on the domain Ξ by considering regular meshes of 3D trilinear elements, compatible with the description given by the meshes in Table 1 consisting of 32, 128 and 384 3D-elements respectively.

Radial displacements calculated with finite elements and separated representations for equivalent meshes are compared with the exact solution (30) in Figure 2. A superiority of the separated representation solution with respect to the standard finite element solution can be noticed. This superiority can be associated to a better representation of the metrics and curvature.

Because the solution of the considered problem only involves radial (thickness) displacements depending on the radial coordinate, a 3D elastic solution seems to be the most appropriate and simplest choice, more natural than using computational shell theories. This problem can be solved easily with quadratic finite elements or isogeometric ones by only one element across the thickness. Also with axisymmetric representation only one 8 nodes element gives a nearly perfect solution. However here we prefer using linear finite elements in order to compare with the linear representation of functions involved in the PGD representation here considered, even if we could use higher order representations within the PGD framework. The objective

Table 1 Meshes considered in the numerical solution

	N_ξ	N_η	N_ζ
MESH 1	8	2	2
MESH 2	16	2	4
MESH 3	32	2	8

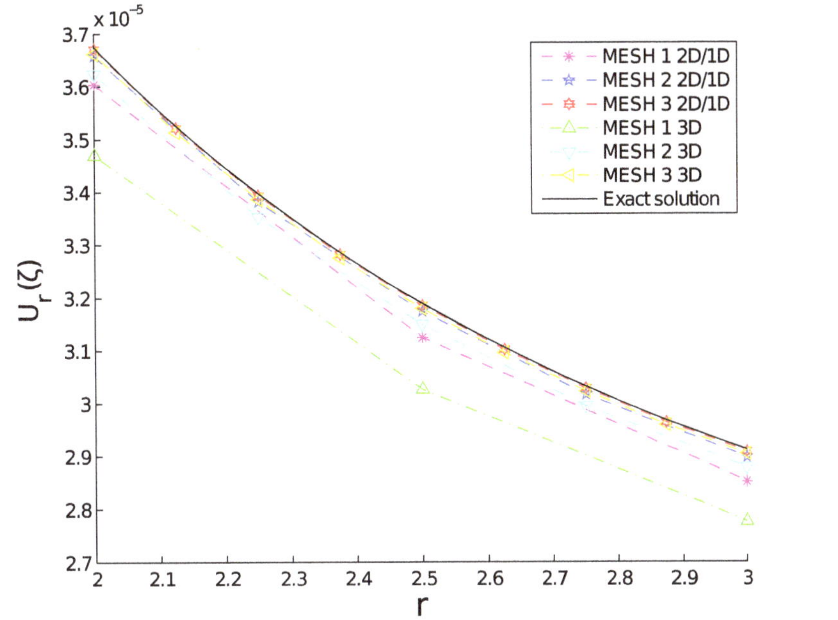

Figure 2 Comparison of separated representation, finite elements and exact radial displacements.

here is proving the convergence and comparing the simplest finite elements and PGD formulations.

As the solution only depends on the radial direction one could expect to capture the solution with a single mode of the separated representation. This expectation is confirmed by the numerical solution that converges with a single term in the decomposition. Figures 3, 4 and 5 depict the computed mode for each component of the displacement field $(P_1^1(\xi,\eta), T_1^1(\zeta))$, $(P_2^1(\xi,\eta), T_2^1(\zeta))$ and $(P_3^1(\xi,\eta), T_3^1(\zeta))$ associated with the decomposition (15) respectively, all them associated with the finest mesh (MESH 3) reported in Table 1.

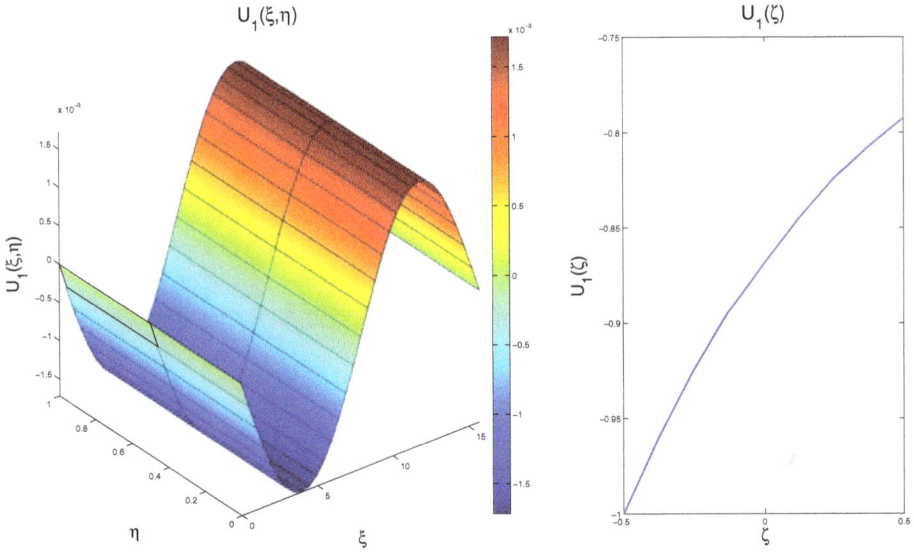

Figure 3 Functions related to $u_1(\xi,\eta,\zeta) : P_1^1(\xi,\eta)$ and $T_1^1(\zeta)$.

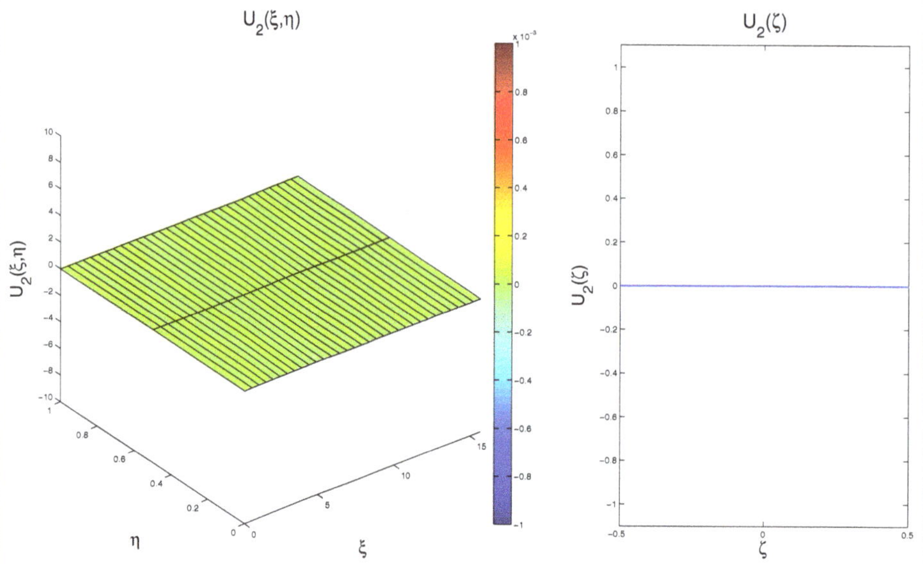

Figure 4 Functions related to $u_2(\xi,\eta,\zeta)$: $P_2^1(\xi,\eta)$ and $T_2^1(\zeta)$.

This numerical example served to validate the proposed approach based on the use of separated representations for solving 3D elastic problems in shell domains. It is important to notice that we compute the fully 3D solution with a prescribed accuracy. Thus, the proposed procedure should be viewed as a 3D solver that due to the separated representation of the 3D fields involved in the model allows a reduction of the computational complexity to the one characteristic of 2D solutions.

Elliptic tube

In order to address a problem whose solution consists of more than a single mode, we consider in this section a slightly different problem. It concerns a tube of infinite length but now with an elliptical cross section and again subjected to an internal pressure.

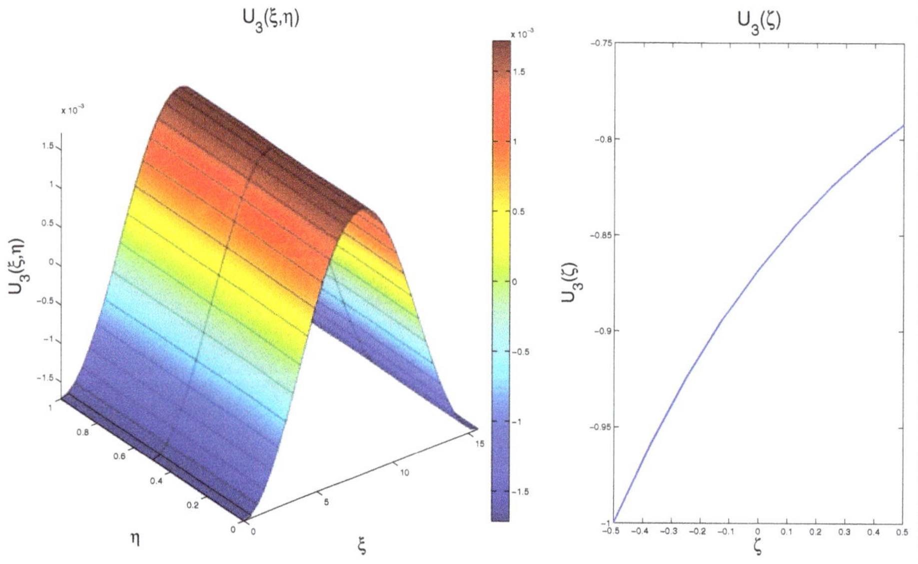

Figure 5 Functions related to $u_3(\xi,\eta,\zeta)$: $P_3^1(\xi,\eta)$ and $T_3^1(\zeta)$.

The surface representation in given by:

$$\begin{cases} X_1 = 0.9\frac{\overline{R}}{2\cdot\pi}\sin(\frac{\xi}{\overline{R}}) \\ X_2 = \eta \\ X_3 = 1.1\frac{\overline{R}}{2\cdot\pi}\cos(\frac{\xi}{\overline{R}}) \end{cases} \tag{31}$$

where $\overline{R} = 2.5$, $\xi \in [0, 2\cdot\pi\cdot\overline{R}]$ and $\eta \in [0,1]$ (we are considering an arbitrary tube length due to the plane strain assumption). The tube thickness is defined from $\zeta \in [-0.5, 0.5]$. Material parameters are the same than in the previous test.

We consider the same discretization than in the previous example, bilinear rectangles in Ω and standard one-dimensional linear elements in the thickness (again the simplest choice).

The problem solution is depicted in Figure 6 and the most significant associated modes in Figures 7, 8, 9, 10, 11, 12, 13, 14, 15, 16, 17 and 18.

Tube with finite length

Now in order to remove the plane strain hypothesis we consider a circular tube of finite length ($L = 10$) with $\overline{R} = 5$ and unit thickness, clamped at both ends and again subjected to an internal pressure. The most significant modes in the present case are depicted inFigures 19, 20, 21 and 22 and the 3D solution in Figure 23.

Twisted tourus

In this section we consider a quarter of a twisted torus of thickness $h = 0.2$ depicted in Figure 24. Again we consider the linear and isotropic elastic problem related to an internal unit pressure and we constraint the displacement of both bases with respect to their normal directions. Again the displacement is constrained at some points for avoiding rigid body displacements.

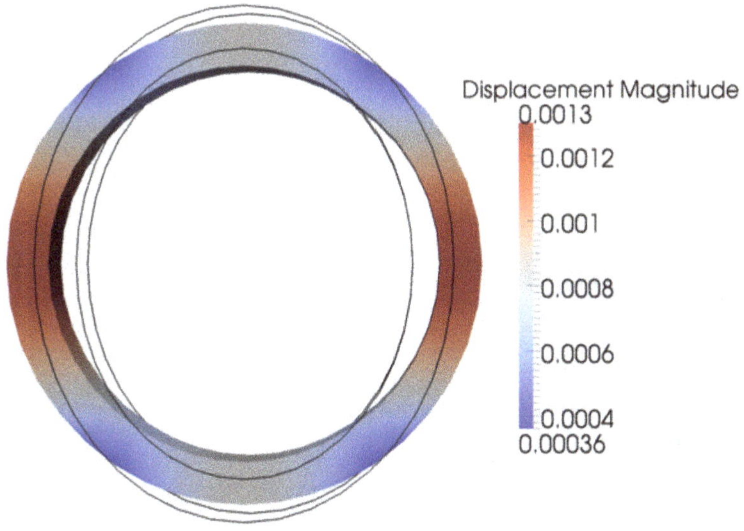

Figure 6 Undeformed (black) and deformed configuration.

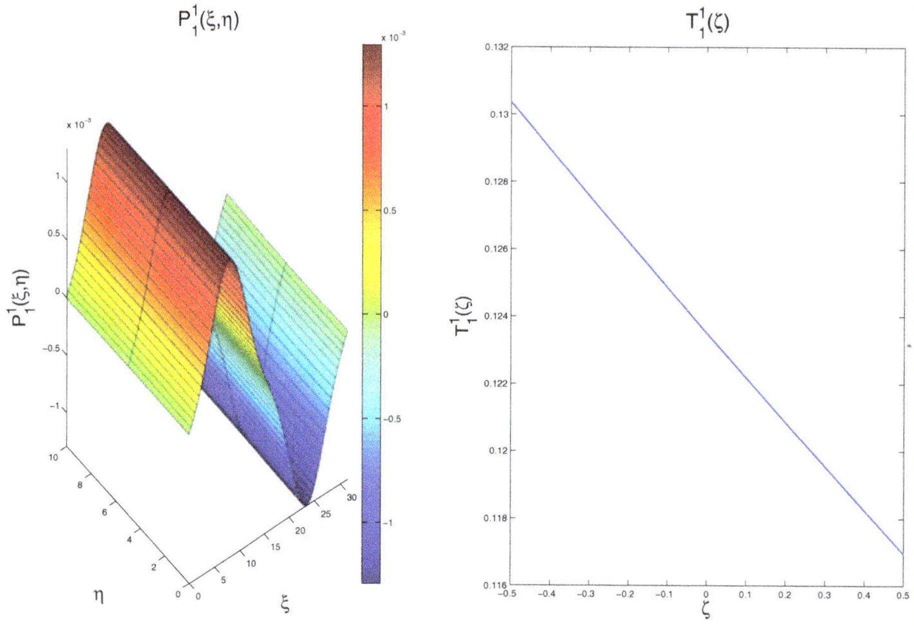

Figure 7 Functions related to $u_1(\xi, \eta, \zeta) : P_1^1(\xi, \eta)$ and $T_1^1(\zeta)$.

We consider the middle surface representation

$$\begin{cases} X_1 = (r + R \cdot \cos(\xi + 2 \cdot \eta) + e \cdot cos(6 \cdot \xi)) \cos(\eta) \\ X_2 = (r + R \cdot \cos(\xi + 2 \cdot \eta) + e \cdot cos(6 \cdot \xi)) \sin(\eta) \\ X_3 = R \cdot \sin(\xi + 2 \cdot \eta) + e \cdot sin(6 \cdot \xi) \end{cases} \tag{32}$$

where $R = 2$, $r = 4$ and $e = 0.15$.

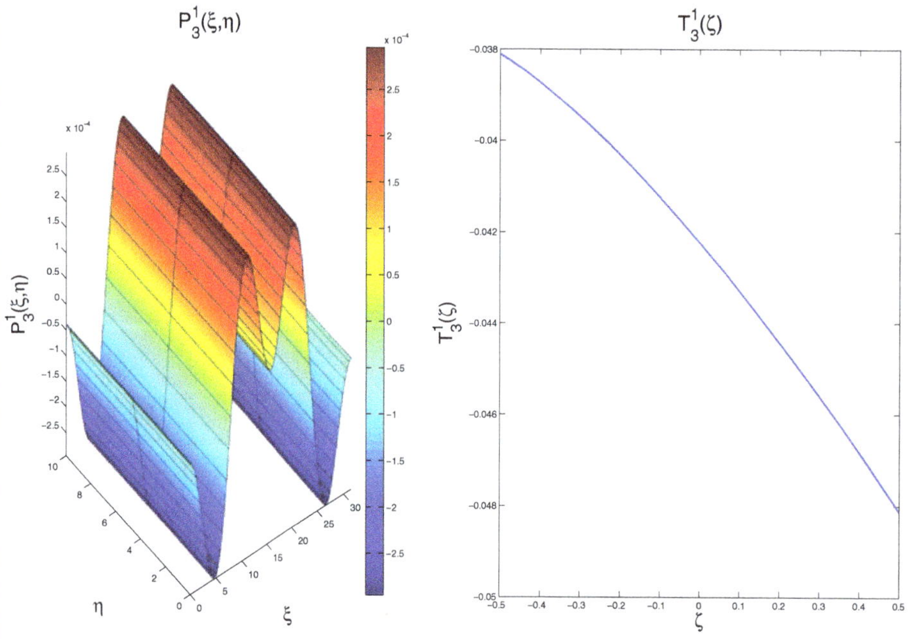

Figure 8 Functions related to $u_3(\xi, \eta, \zeta) : P_3^1(\xi, \eta)$ and $T_3^1(\zeta)$.

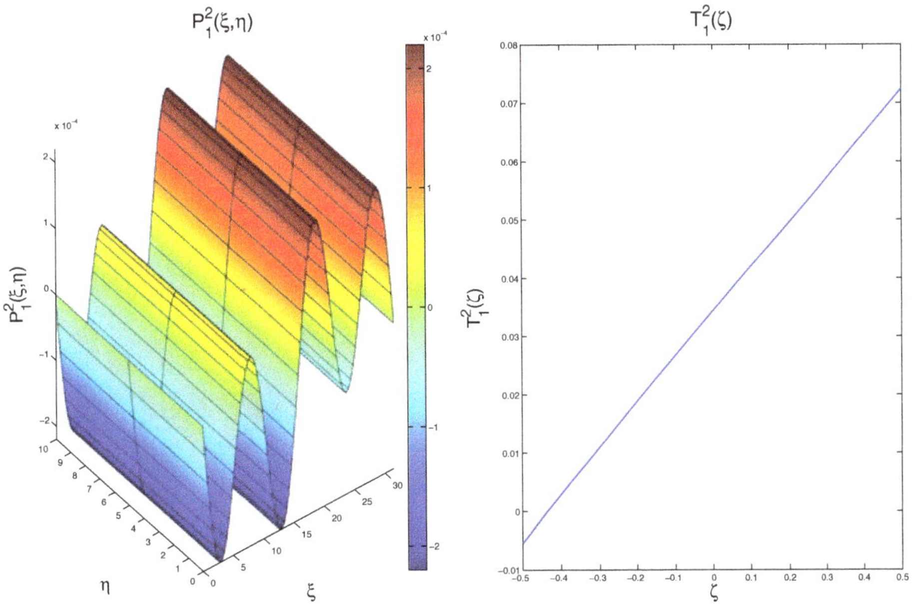

Figure 9 Functions related to $u_1(\xi,\eta,\zeta)$: $P_1^2(\xi,\eta)$ and $T_1^2(\zeta)$.

The parametric space consists of $\Xi = \Omega \times \mathcal{I}$, with $\mathcal{I} = [-h/2, h/2]$. The 2D domain Ω is discretized by using a regular mesh consisting of 64×30 2D bilinear finite elements whereas the thickness interval $\mathcal{I}$ is discretized from 50 one-dimensional linear elements (again higher order approximation are possible). Figure 25 depicts the first component of the strain ϵ_{11}.

Figure 26 depicts the components of the displacement along the normal direction at the different locations shown in Figure 25. It can be noticed that the solution evolves significantly from one point to other. The separated representation captures this rich

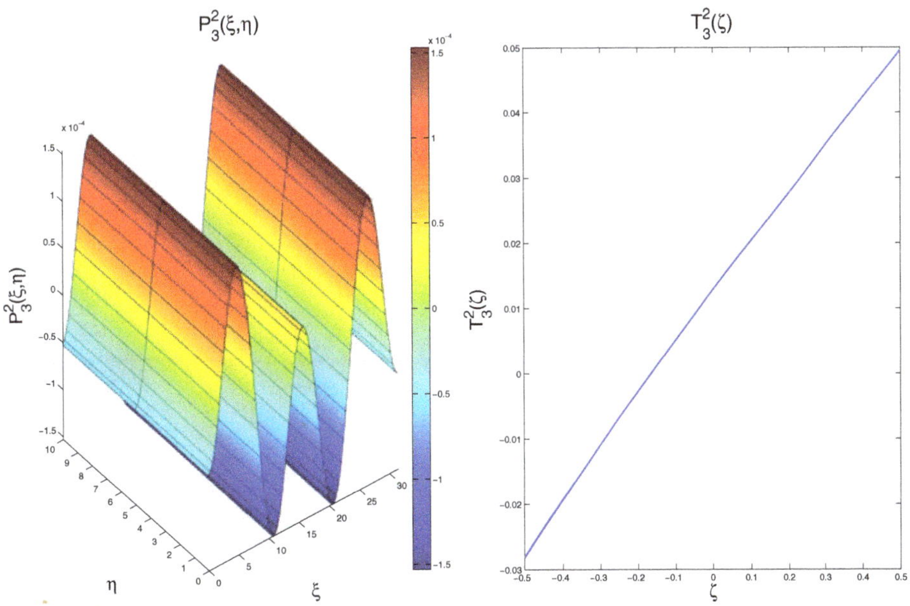

Figure 10 Functions related to $u_3(\xi,\eta,\zeta)$: $P_3^2(\xi,\eta)$ and $T_3^2(\zeta)$.

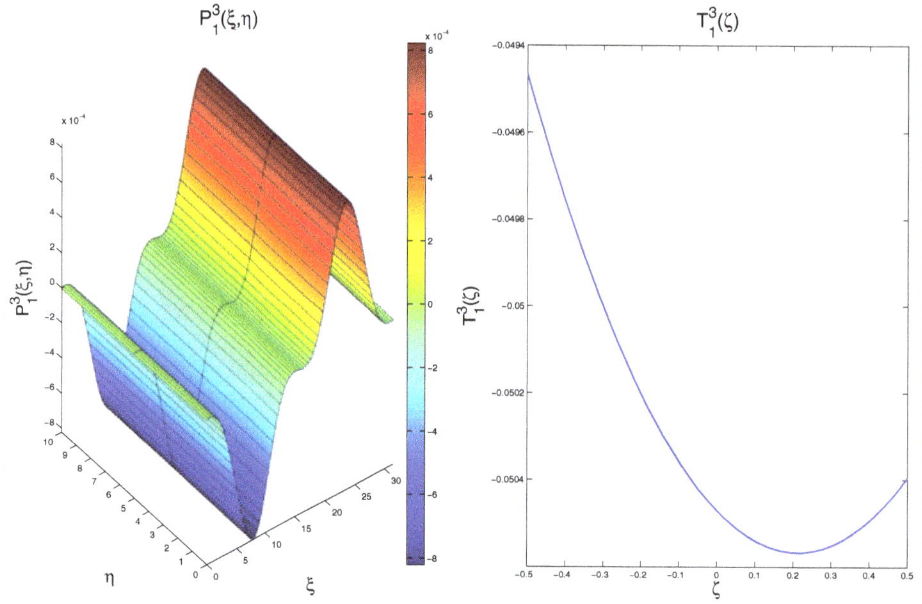

Figure 11 Functions related to $u_1(\xi,\eta,\zeta) : P_1^3(\xi,\eta)$ and $T_1^3(\zeta)$.

evolution justifying the interest of performing a fully 3D solution as soon as the computational complexity can be reduced to the one characteristic of 2D solutions.

Scordelis-Lo roof

In this section we address a critical scenario to check convergence and locking issues: the Scordelis-Lo roof depicted in Figure 27. Other than the symmetry conditions indicated in that figure we constraint the first and third components of the displacement, i.e. $u_1 = u_3 = 0$ on the end sections (in particular BC in the domain of calculation).

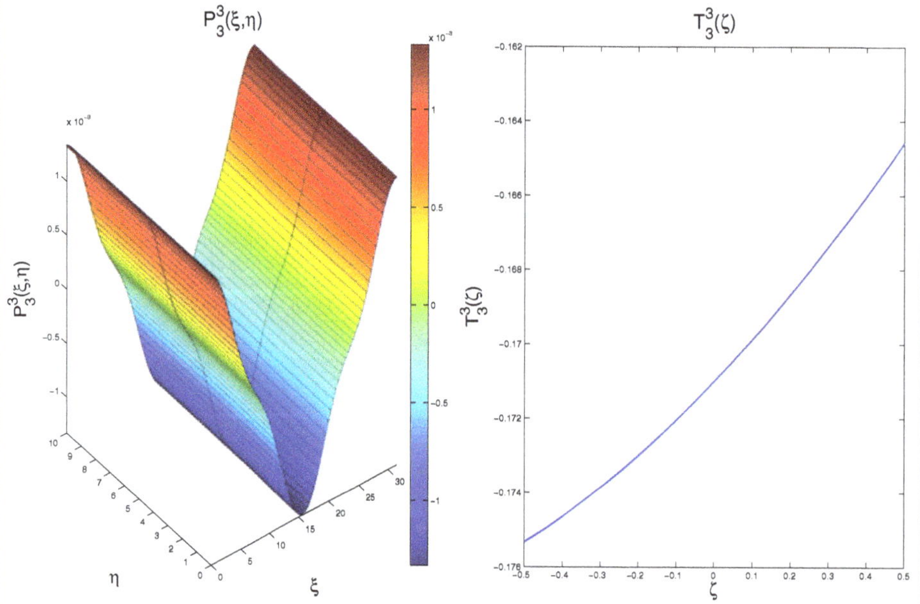

Figure 12 Functions related to $u_3(\xi,\eta,\zeta) : P_3^3(\xi,\eta)$ and $T_3^3(\zeta)$.

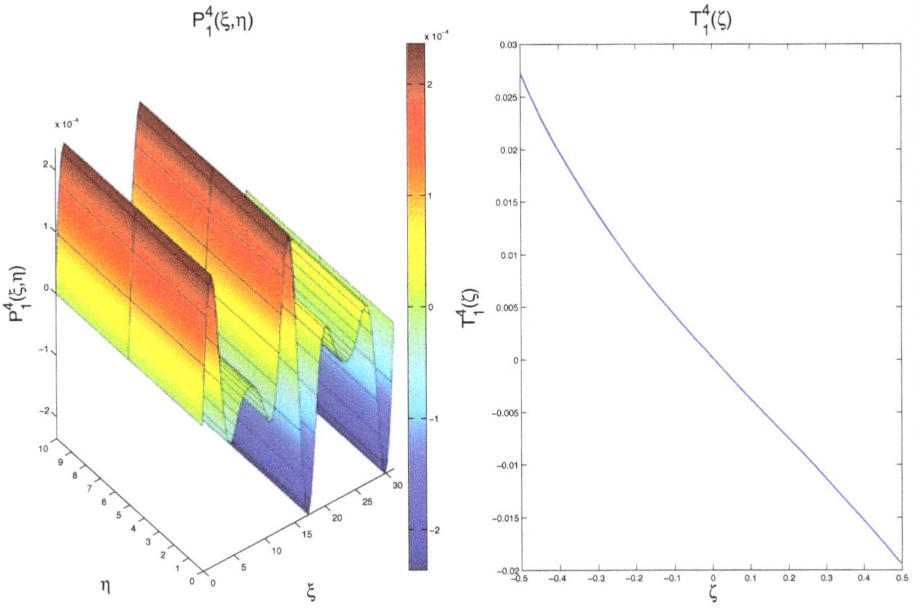

Figure 13 Functions related to $u_1(\xi, \eta, \zeta)$: $P_1^4(\xi, \eta)$ and $T_1^4(\zeta)$.

The problem material and geometrical parameters are the following $E = 3\ 10^6$, $\nu = 0$, $a = 600$, $r = 300$, $e = 3$ and $\alpha = \pi/3$. The shell is subjected to a surface load of 0.625 that induces a vertical displacement of 3.6288 at position A in Figure 27 [27].

The computed results obtained by using the in-plane-out-of-plane separated representations are again compared to the ones obtained by using standard 3D linear finite elements (better choices related to higher order approximations exist, but here again we focus in the simplest choice). The comparison is depicted in Figure 28.

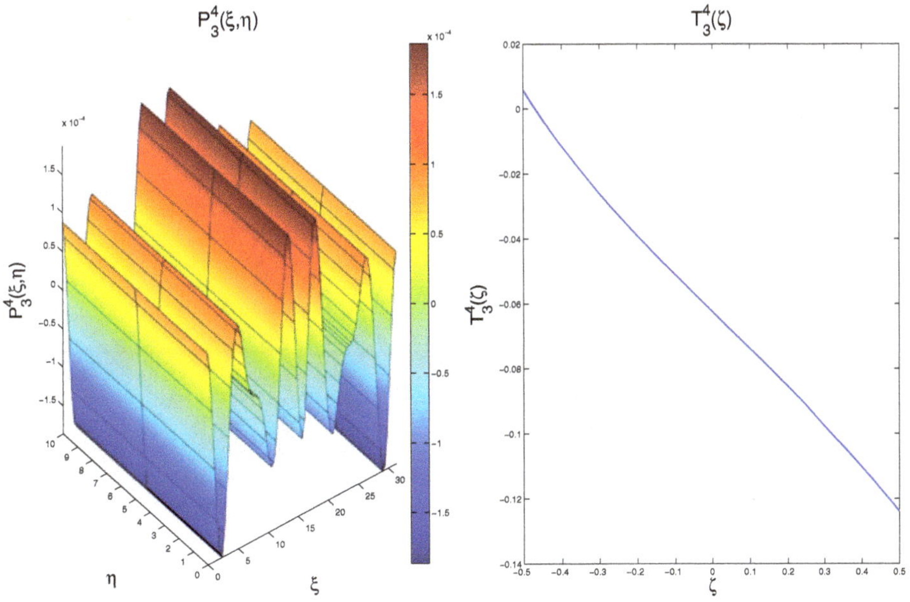

Figure 14 Functions related to $u_3(\xi, \eta, \zeta)$: $P_3^4(\xi, \eta)$ and $T_3^4(\zeta)$.

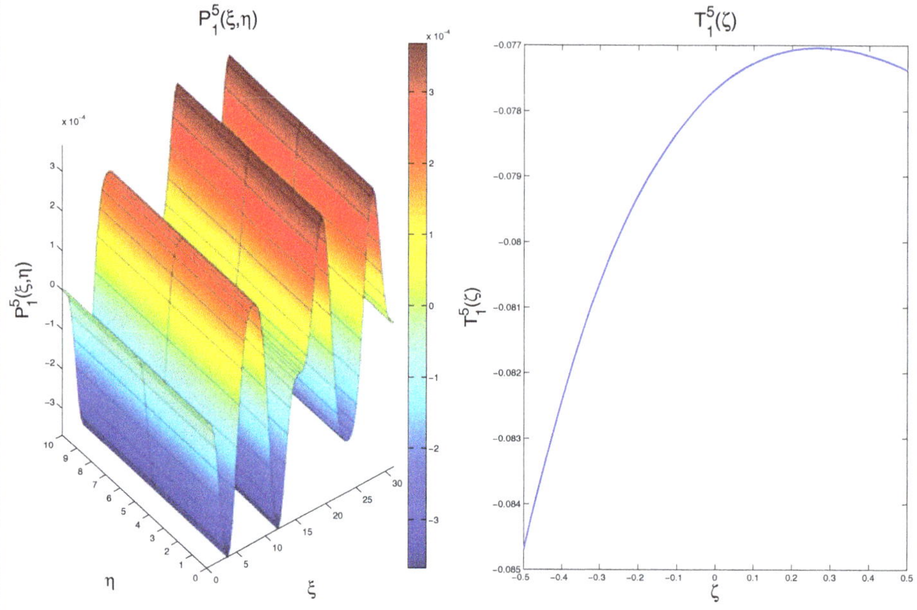

Figure 15 Functions related to $u_1(\xi, \eta, \zeta) : P_1^5(\xi, \eta)$ and $T_1^5(\zeta)$.

From this comparison we can conclude that our approach based on a separated representation converges and that it exhibits similar convergence behavior than standard linear 3D finite elements.

Multilayered tube

We consider now a composite tubular part composed of $P = 8$ plies of thickness 0.125 mm with a stacking sequence $[90, 45, 0, -45]_s$ in the local orthonormal basis $(\mathbf{t}_1, \mathbf{t}_2, \mathbf{n})$, with

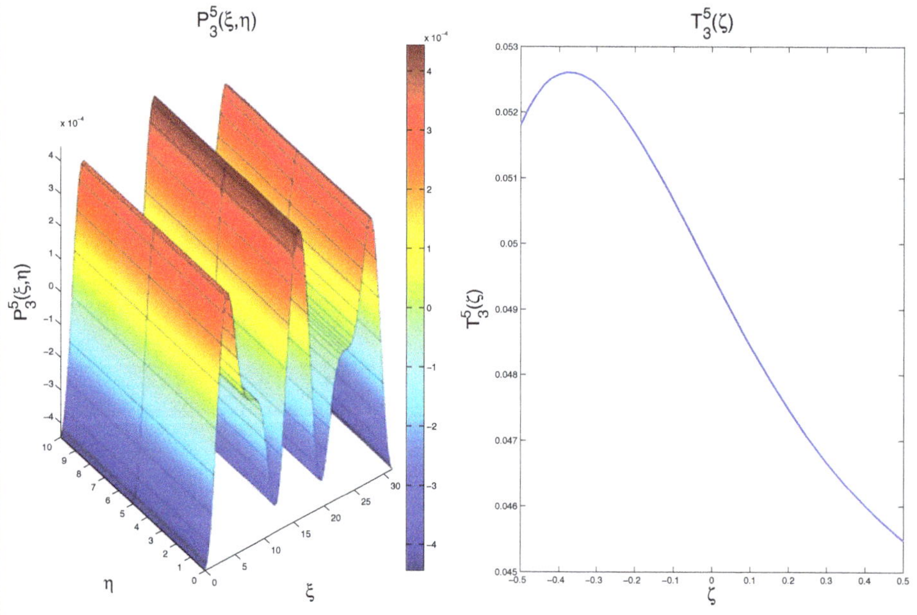

Figure 16 Functions related to $u_3(\xi, \eta, \zeta) : P_3^5(\xi, \eta)$ and $T_3^5(\zeta)$.

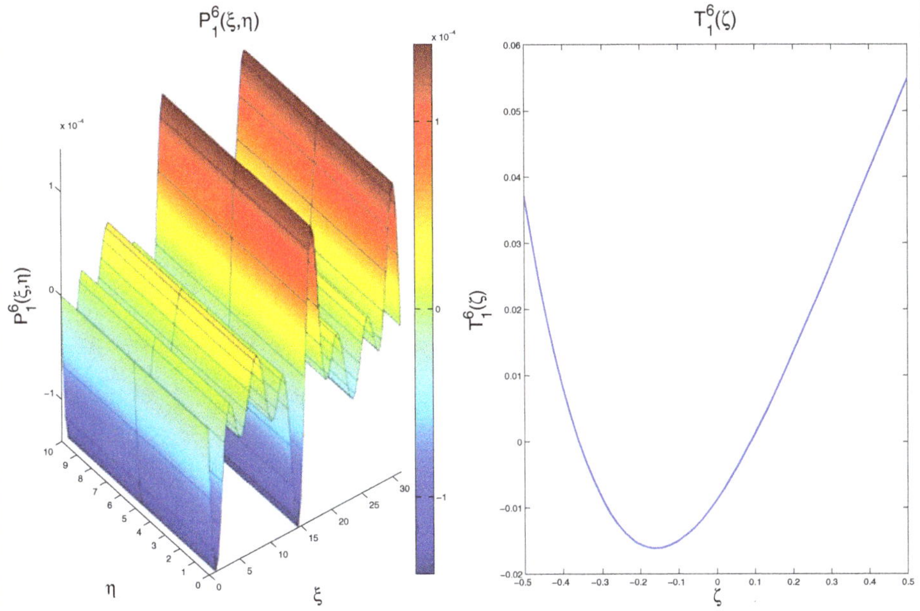

Figure 17 Functions related to $u_1(\xi,\eta,\zeta)$: $P_1^6(\xi,\eta)$ and $T_1^6(\zeta)$.

$\mathbf{t}_1$, $\mathbf{t}_2$ and $\mathbf{n}$ defining the axial, the circumferential and the thickness direction respectively as depicted in Figure 29.

We solved the same elastic problem than in section 'Strategy verification', being the only difference the one related to the material mechanical behavior. The in-plane discretization of Ω was performed by considering a regular mesh composed of 40×2 2D bilinear finite elements, whereas 80 one-dimensional linear elements were considered for discretizing the thickness interval $\mathcal{I}$. Thus each ply is represented by 10 elements ensuring a rich description (perhaps too rich in the case of elastic behaviors but that could be

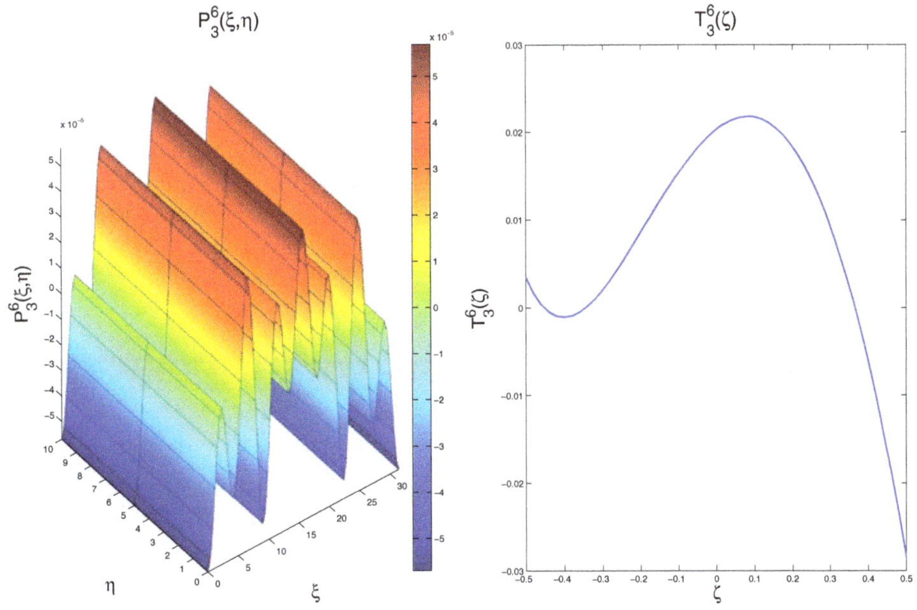

Figure 18 Functions related to $u_3(\xi,\eta,\zeta)$: $P_3^6(\xi,\eta)$ and $T_3^6(\zeta)$.

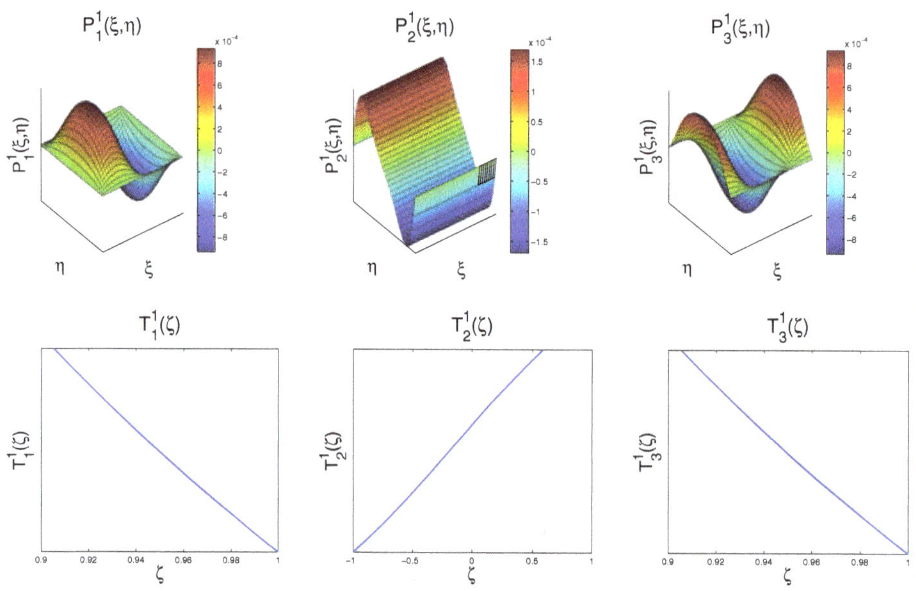

Figure 19 Tube of finite length: Mode 1.

compulsory for addressing more complex mechanical behaviors). In this work we would like emphasizing that the proposed strategy allows very rich descriptions of the thickness coordinate without a significant impact on the numerical efficiency.

The displacement field depicted in Figure 30 remains close to the one obtained in section 'Strategy verification' with some slight differences. On the other hand, the radial component of the strain represented in Figure 31 results, as expected, discontinuous across the interplies.

Analysis of large composite structural parts

In this section we address the elastic analysis of a large composite structural part similar to the ones involved in aircraft fuselages. The whole part, 5 m of diameter, and a detail of it

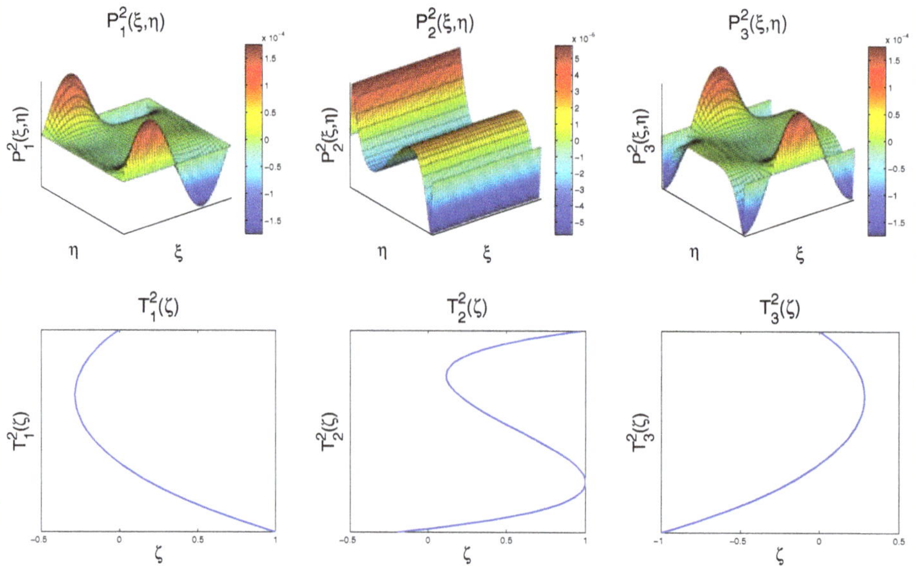

Figure 20 Tube of finite length: Mode 2.

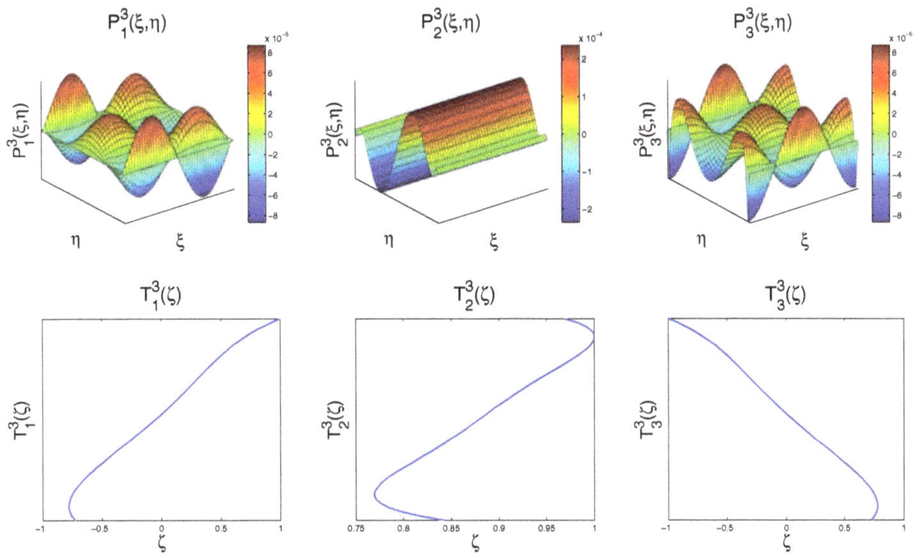

Figure 21 Tube of finite length: Mode 3.

are depicted in Figures 32 and 33 respectively. In both figures can be noticed the presence of stiffeners distributed on the laminate surface, all them reinforced along their longitudinal direction.The structure was subjected to a unit internal pressure and an elastic analysis was carried out by assuming the separated representations described in this paper.

The laminate is composed of 12 plies, each 0.125 *mm* thick and involving an unidirectional reinforcement. The stacking sequence is $[0, 45, 0, -45]_{3S}$.

The 3D solution was obtained by using an in-plane-out-of-plane separated representation of the elastic fields. The approximation of functions involving the thickness direction was performed by considering 10 elements per ply. An equivalent finite element mesh of the whole part would have implied $7 \cdot 10^7$ elements with $2 \cdot 10^8$ nodes, each one involving 3 degrees of freedom.

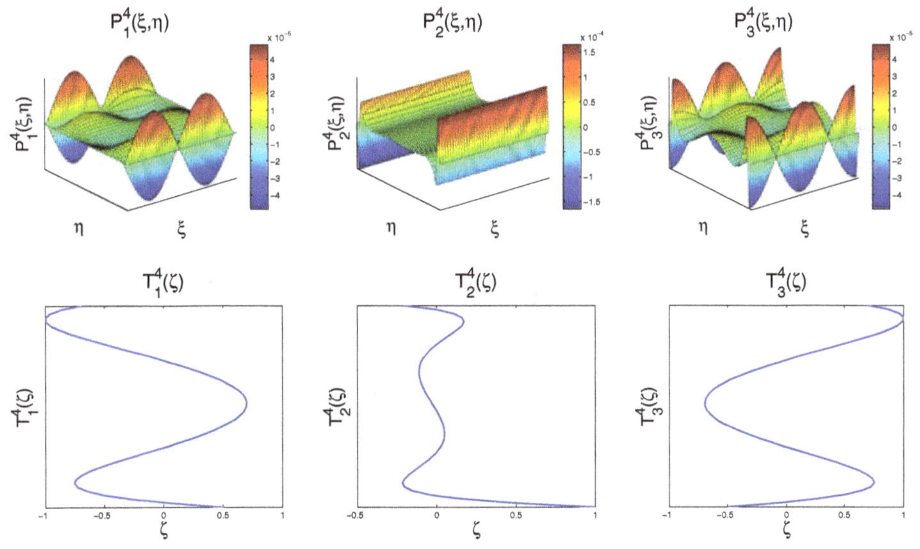

Figure 22 Tube of finite length: Mode 4.

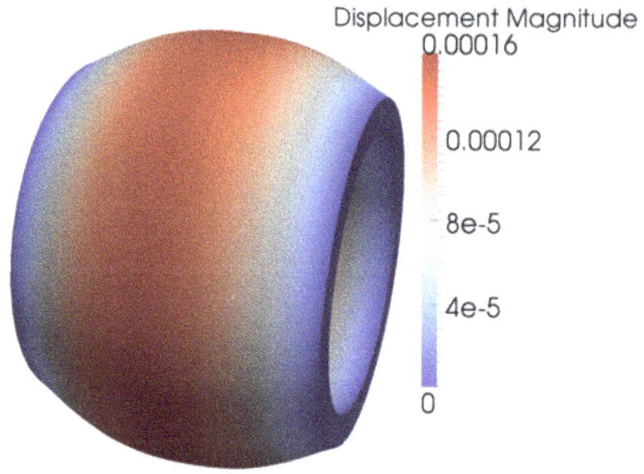

Figure 23 Tube deformation.

Figure 34 depicts the magnitude of the normal displacement (*mm*) on the deformed configuration. Even if we considered a laminate consisting of 12 plies, we could consider hundreds of plies without degrading the computing time thanks to the in-plane-out-of-plane separated representation. In fact the mesh considered in the thickness direction only affects the one-dimensional problems defined in that direction that are decoupled of the ones involving the in-plane coordinates. The ones involving the in-plane coordinates being 2D determine the computational complexity of the whole solution procedure.

The main advantage of the procedure that we propose does not concern the possibility of addressing laminates involving several plies, because when elastic behaviors are considered they exist advanced shell elements able to address efficiently such configurations. However, our technique allows considering in an unified description the laminates and the stiffeners placed on it, by using the same in-plane-out-of-plane representation, as was widely discussed in [24]. The consideration of all these stiffeners within a standard shell theory requires special treatments, however in our fully 3D approach, both the shell laminate and the stiffeners are integrated in an unified hypotheses-free description.

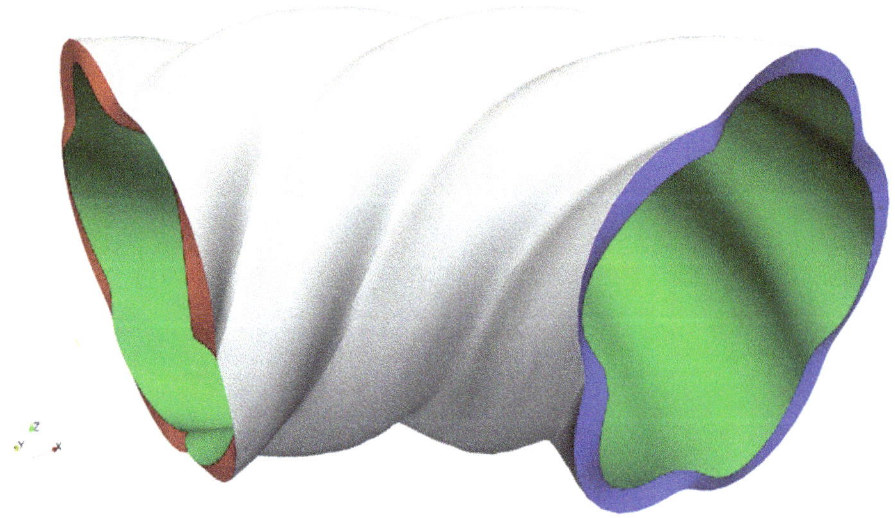

Figure 24 Twisted torus shell geometry.

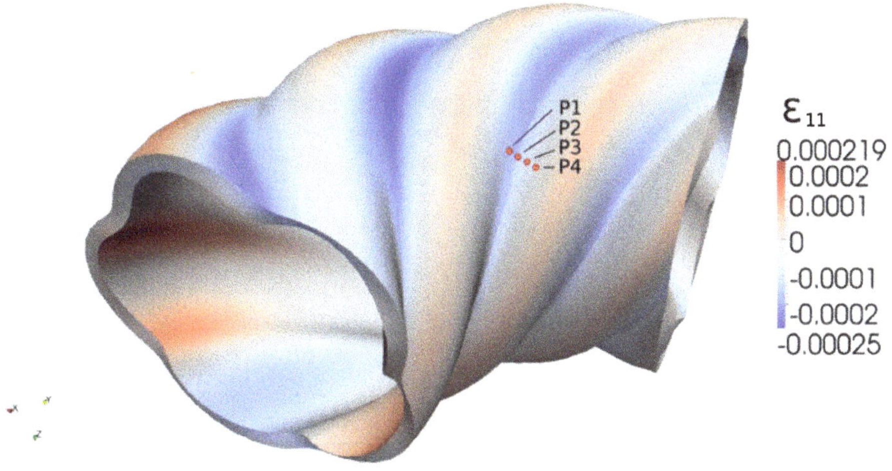

Figure 25 First strain component ϵ_{11}.

Conclusions

We proposed in our former works a procedure based on the separated representation of the displacement field involved in elastic problems defined in plate domains that allowed calculating fully 3D elastic solutions by solving a sequence of 2D and 1D problems, the former ones define in the plane and the last ones in the thickness. Thus, we calculated 3D solutions with a computational cost characteristic of 2D problems, as the ones related to standard plate theories, however here, as we compute directly the 3D solution, we do not need introducing any hypothesis or correction.

In the present paper we extended that methodology for solving elastic linear problems defined in shell geometries. For that purpose we proposed to use a standard mapping of the real geometry to a plate-type parametric domain in which the procedure proposed in our former works was applied. However, the main difficulty when using this procedure concerns the necessity of performing a separated representation of displacement and

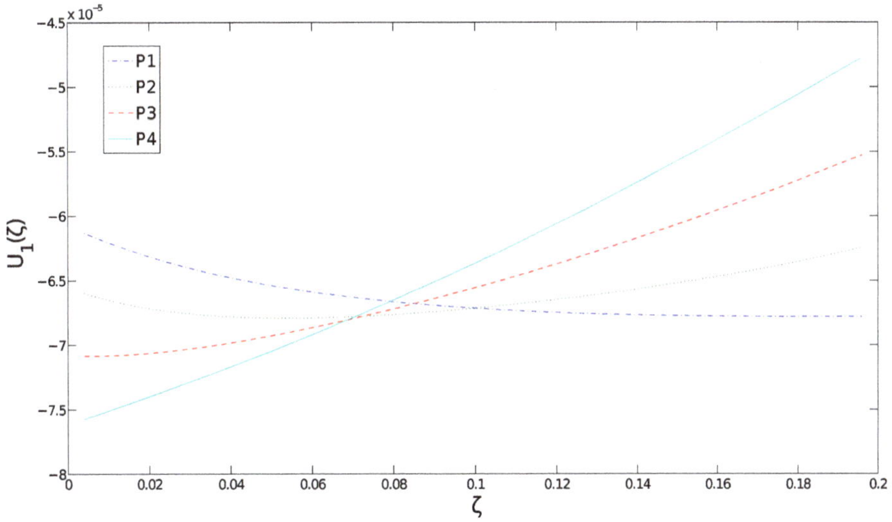

Figure 26 Evolution in the thickness direction of the first component of the displacement at the four locations indicated in Figure 25.

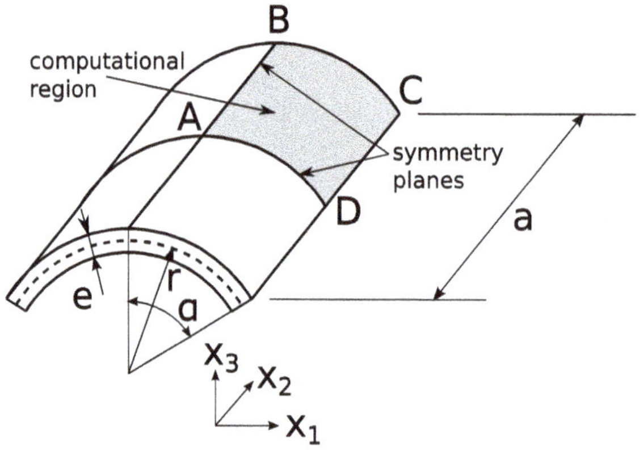

Figure 27 Scordelis-Lo roof.

strain fields, loads, material behavior (elasticity tensor), the differential operators as well as the surface and volume elements. This decomposition is for some problems quite direct and in all cases it can be performed by invoking the singular value decomposition.

The analyzed examples prove the potentiality and efficiency of the proposed strategy, where the computational complexity was found evolving as reported in [24], proving that 3D solutions can be computed at a 2D cost. On the other hand the fact of solving fully 3D models avoids the locking issue that was never found in our analysis.

The extension of this strategy for addressing more complex scenarios involving material or geometrical non-linearities or for considering more complex geometries in which the domain thickness could vary from one point to other represent some of the works in progress.

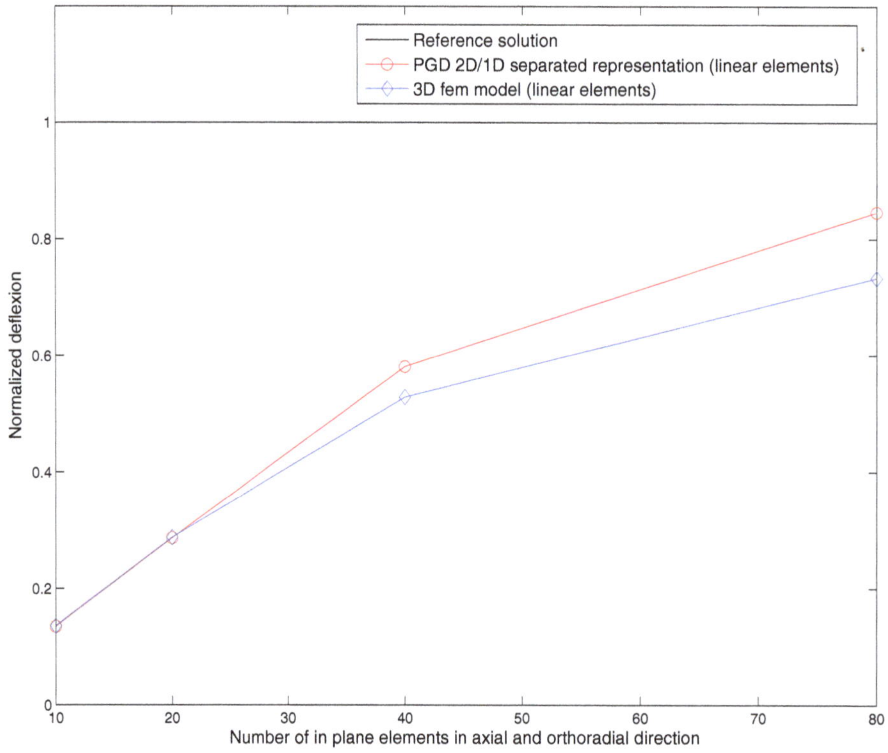

Figure 28 Normalized displacement at position A.

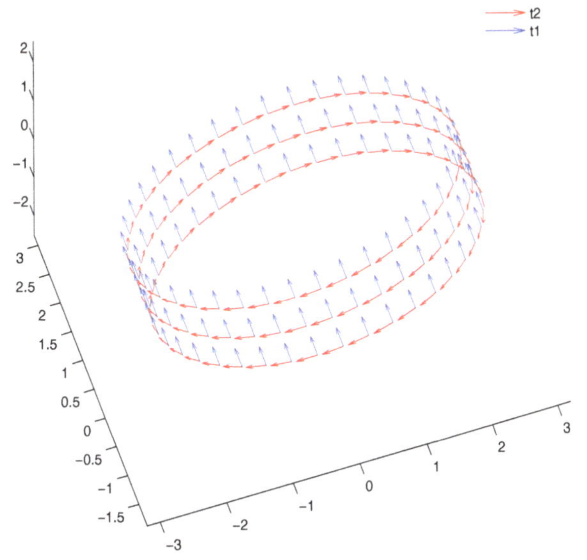

Figure 29 Local orthonormal basis at the nodes on the middle surface.

Appendix A

In-plane and out-of-plane problems

The problem related to the solution of $\mathbf{P}^n$ is given by

$$\int_{\Xi} \boldsymbol{\epsilon}^T(\mathbf{P}^* \circ \mathbf{T}^n) \cdot \mathbf{K} \cdot \boldsymbol{\epsilon}(\mathbf{P}^n \circ \mathbf{T}^n) \; d\mathbf{x} =$$
$$- \int_{\Xi} \boldsymbol{\epsilon}^T(\mathbf{P}^* \circ \mathbf{T}^n) \cdot \mathbf{K} \cdot \boldsymbol{\epsilon}(\mathbf{u}^{n-1}) \; d\mathbf{x} + \int_{\Xi} (\mathbf{P}^* \circ \mathbf{T}^n) \cdot \mathbf{f}_d \, d\mathbf{x} + \int_{\Gamma_N} (\mathbf{P}^* \circ \mathbf{T}^n) \cdot \mathbf{F}_d \, d\mathbf{x} \tag{33}$$

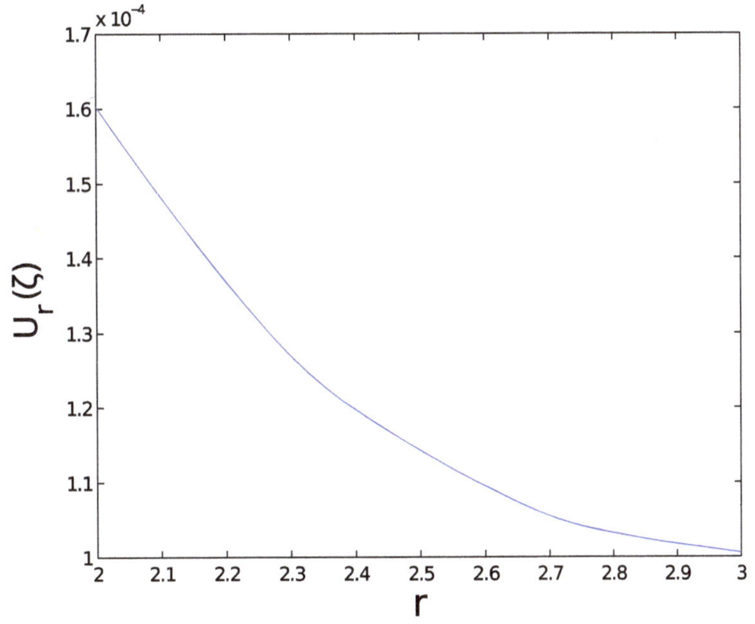

Figure 30 Radial displacement along the radius.

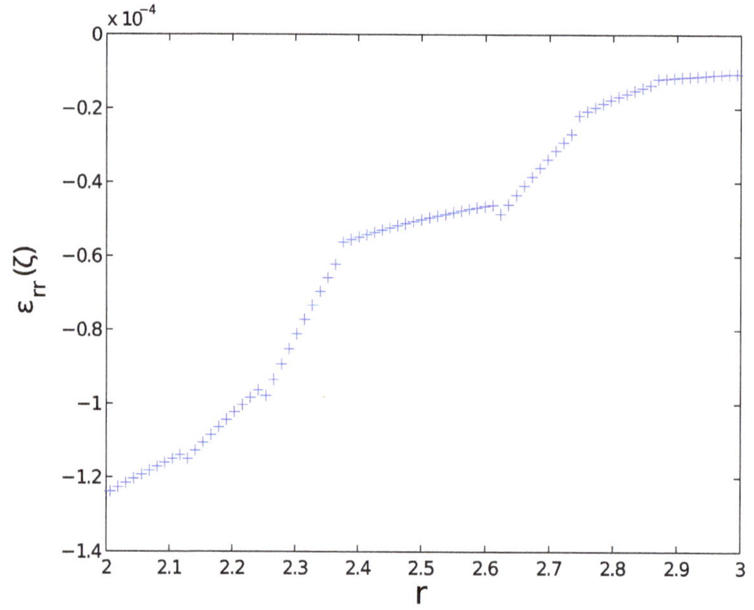

Figure 31 Evolution of the radial component of the strain along the radial direction.

Figure 32 Large composite structural part involving many stiffeners placed on its surface.

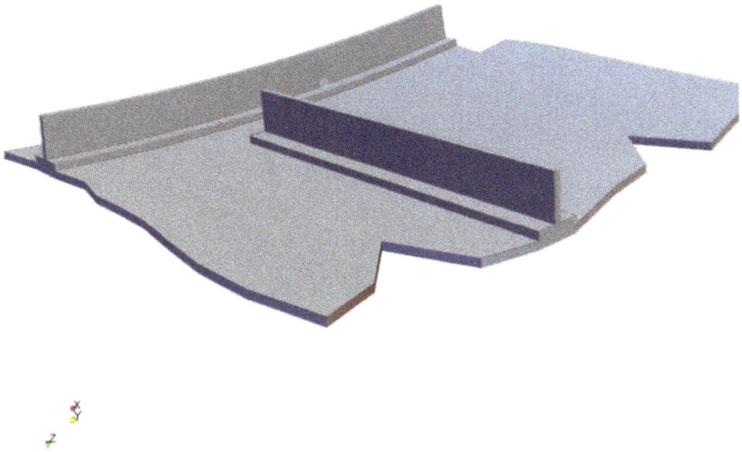
Figure 33 Detail exhibiting the stiffeners geometry.

with

$$\epsilon(\mathbf{P}^n \circ \mathbf{T}^n) = \begin{pmatrix} \frac{\partial P_1^n}{\partial x_1} \cdot T_1^n \\ \frac{\partial P_2^n}{\partial x_2} \cdot T_2^n \\ P_3^n \cdot \frac{\partial T_3^n}{\partial x_3} \\ \frac{\partial P_1^n}{\partial x_2} \cdot T_1^n + \frac{\partial P_2^n}{\partial x_1} \cdot T_2^n \\ \frac{\partial P_3^n}{\partial x_1} \cdot T_3^n + P_1^n \cdot \frac{\partial T_1^n}{\partial x_3} \\ \frac{\partial P_3^n}{\partial x_2} \cdot T_3^n + P_2^n \cdot \frac{\partial T_2^n}{\partial x_3} \end{pmatrix} \quad (34)$$

$$\epsilon(\mathbf{P}^* \circ \mathbf{T}^n) = \begin{pmatrix} \frac{\partial P_1^*}{\partial x_1} \cdot T_1^n \\ \frac{\partial P_2^*}{\partial x_2} \cdot T_2^n \\ P_3^* \cdot \frac{\partial T_3^n}{\partial x_3} \\ \frac{\partial P_1^*}{\partial x_2} \cdot T_1^n + \frac{\partial P_2^*}{\partial x_1} \cdot T_2^n \\ \frac{\partial P_3^*}{\partial x_1} \cdot T_3^n + P_1^* \cdot \frac{\partial T_1^n}{\partial x_3} \\ \frac{\partial P_3^*}{\partial x_2} \cdot T_3^n + P_2^* \cdot \frac{\partial T_2^n}{\partial x_3} \end{pmatrix} \quad (35)$$

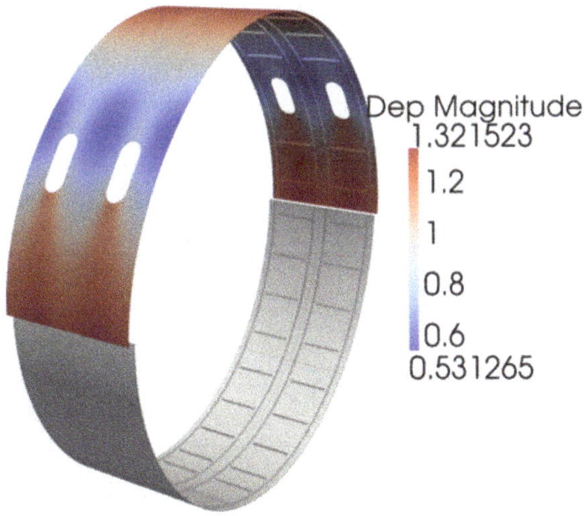

Figure 34 Normal component of the elastic displacement.

and

$$\mathbf{K}=\begin{pmatrix} K_{11} & K_{12} & K_{13} & 0 & 0 & K_{16} \\ K_{12} & K_{22} & K_{23} & 0 & 0 & K_{26} \\ K_{13} & K_{23} & K_{33} & 0 & 0 & K_{36} \\ 0 & 0 & 0 & K_{44} & K_{45} & 0 \\ 0 & 0 & 0 & K_{45} & K_{55} & 0 \\ K_{16} & K_{26} & K_{36} & 0 & 0 & K_{66} \end{pmatrix} \tag{36}$$

By integrating Eq. (37) along the coordinate x_3 we obtain a discrete system from which results $\mathbf{P}^n$.

By considering now

$$\int_{\Xi} \boldsymbol{\epsilon}^T(\mathbf{P}^n \circ \mathbf{T}^*) \cdot \mathbf{K} \cdot \boldsymbol{\epsilon}(\mathbf{P}^n \circ \mathbf{T}^n)\ d\mathbf{x} = \\ -\int_{\Xi} \boldsymbol{\epsilon}^T(\mathbf{P}^n \circ \mathbf{T}^*) \cdot \mathbf{K} \cdot \boldsymbol{\epsilon}(\mathbf{u}^{n-1})\ d\mathbf{x} + \int_{\Xi} (\mathbf{P}^n \circ \mathbf{T}^*) \cdot \mathbf{f}_d\, d\mathbf{x} + \int_{\Gamma_N} (\mathbf{P}^n \circ \mathbf{T}^*) \cdot \mathbf{F}_d\, d\mathbf{x} \tag{37}$$

with

$$\boldsymbol{\epsilon}(\mathbf{P}^n \circ \mathbf{T}^*) = \begin{pmatrix} \frac{\partial P_1^n}{\partial x_1} \cdot T_1^* \\ \frac{\partial P_2^n}{\partial x_2} \cdot T_2^* \\ P_3^n \cdot \frac{\partial T_3^*}{\partial x_3} \\ \frac{\partial P_1^n}{\partial x_2} \cdot T_1^* + \frac{\partial P_2^n}{\partial x_1} \cdot T_2^* \\ \frac{\partial P_3^n}{\partial x_1} \cdot T_3^* + P_1^n \cdot \frac{\partial T_1^*}{\partial x_3} \\ \frac{\partial P_3^n}{\partial x_2} \cdot T_3^* + P_2^n \cdot \frac{\partial T_2^*}{\partial x_3} \end{pmatrix} \tag{38}$$

and by integrating in Ω it results the linear problem from with it results $\mathbf{T}^n$.

Appendix B

Elements of differential geometry

Consider a shell domain Ξ whose middle surface is parametrized by the coordinates (ξ, η), that is $\mathbf{X}(\xi, \eta)$

$$\mathbf{X}(\xi,\eta) = \begin{pmatrix} X_1(\xi,\eta) \\ X_2(\xi,\eta) \\ X_3(\xi,\eta) \end{pmatrix} \tag{39}$$

as depicted in Figure 35.

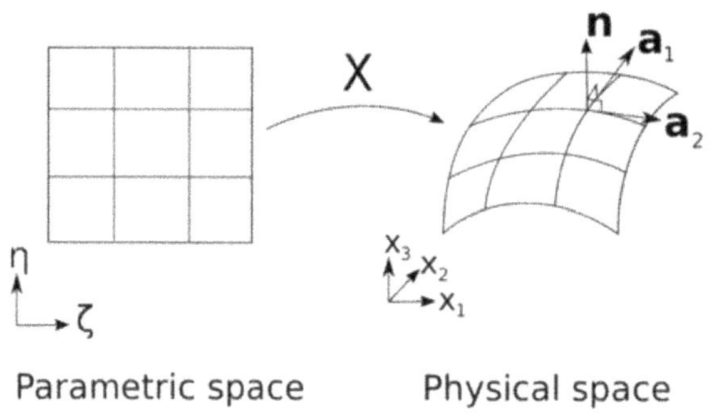

Figure 35 Definition of the mapping.

From this mapping we can define the covariant basis $\mathbf{a}_1$ and $\mathbf{a}_2$, both vectors being tangent to the middle surface:

$$\begin{cases} \mathbf{a}_1 = \frac{\partial \mathbf{X}}{\partial \xi} \\ \mathbf{a}_2 = \frac{\partial \mathbf{X}}{\partial \eta} \end{cases} \tag{40}$$

The unit normal vector to the middle surface can be defined from the cross product of $\mathbf{a}_1$ and $\mathbf{a}_2$ according to:

$$\mathbf{n} = \frac{\mathbf{a}_1 \times \mathbf{a}_2}{\| \mathbf{a}_1 \times \mathbf{a}_2 \|} \tag{41}$$

Now, we define tensor $\mathbf{F}$ containing the Cartesian components of the covariant basis:

$$\mathbf{F} = [\mathbf{a}_1\ \mathbf{a}_2\ \mathbf{n}] \tag{42}$$

We can also define the contravariant basis by considering:

$$\mathbf{F}^{-T} = \left[\mathbf{a}^1\ \mathbf{a}^2\ \mathbf{n}\right] \tag{43}$$

Figures 36 and 37 show the covariant and the contravariant basis respectively on a torus.

Surface fundamental forms

The length ds defines the first fundamental form I from

$$I = (ds)^2 = d\mathbf{X} \cdot d\mathbf{X} = (\mathbf{a}_1\, d\xi + \mathbf{a}_2\, d\eta) \cdot (\mathbf{a}_1\, d\xi + \mathbf{a}_2\, d\eta) = (d\xi\ d\eta) \cdot \mathbf{a} \cdot \begin{pmatrix} d\xi \\ d\eta \end{pmatrix} \tag{44}$$

where $\mathbf{a}$ is the metric tensor defined from

$$\mathbf{a} = \begin{pmatrix} \mathbf{a}_1 \cdot \mathbf{a}_1 & \mathbf{a}_1 \cdot \mathbf{a}_2 \\ \mathbf{a}_2 \cdot \mathbf{a}_1 & \mathbf{a}_2 \cdot \mathbf{a}_2 \end{pmatrix} \tag{45}$$

We denote by a the metric tensor determinant, i.e.:

$$a = \det(\mathbf{a}) \tag{46}$$

The surface element $d\mathbf{A}$ writes:

$$d\mathbf{A} = \mathbf{a}_1\, d\xi \times \mathbf{a}_2\, d\eta = \|\mathbf{a}_1 \times \mathbf{a}_2\| \cdot d\xi \cdot d\eta \cdot \mathbf{n} \tag{47}$$

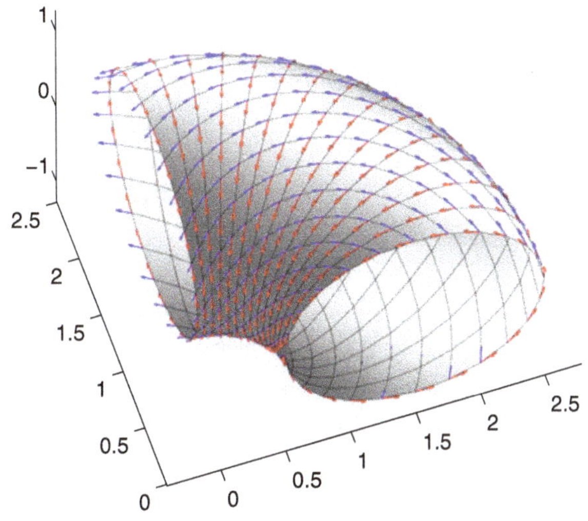

Figure 36 Representation of the covariant basis on a torus.

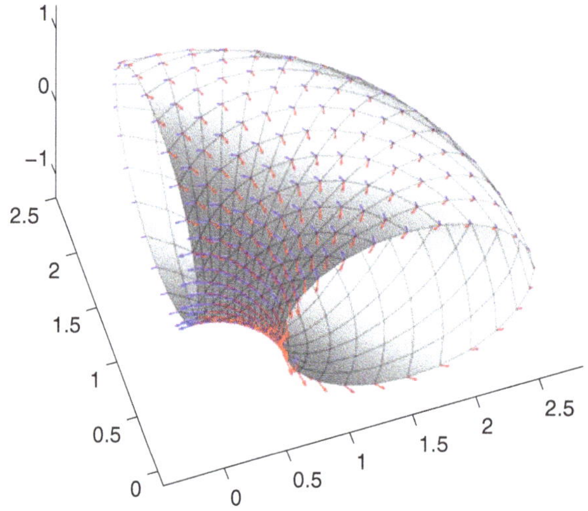

Figure 37 Representation of the contravariant basis on a torus.

and using the cross product property $\|\mathbf{a}_1 \times \mathbf{a}_2\|^2 = \|\mathbf{a}_1\|^2 \cdot \|\mathbf{a}_2\|^2 - (\mathbf{a}_1 \cdot \mathbf{a}_2)^2$ it results:

$$d\mathbf{A} = \sqrt{a} \cdot d\xi \cdot d\eta \cdot \mathbf{n} \tag{48}$$

In the case of a developable surface, the metric is constant. Figures 38 and 39 depicts a on two surfaces, a developable one in which the metric tensor is constant and another having double curvature.

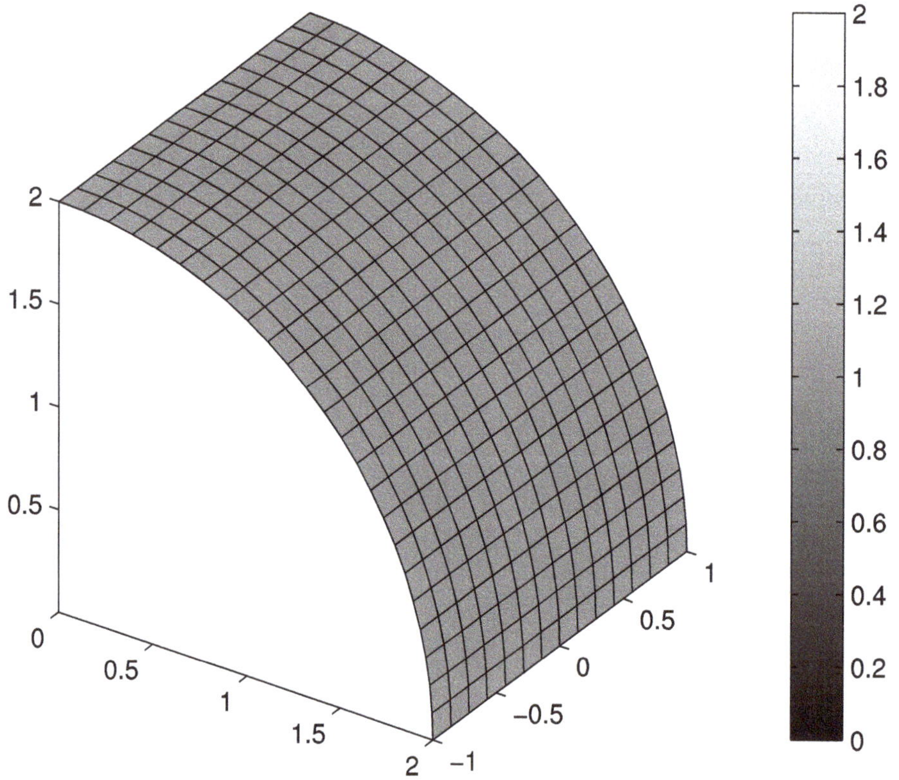

Figure 38 Value of a on a cylindrical surface.

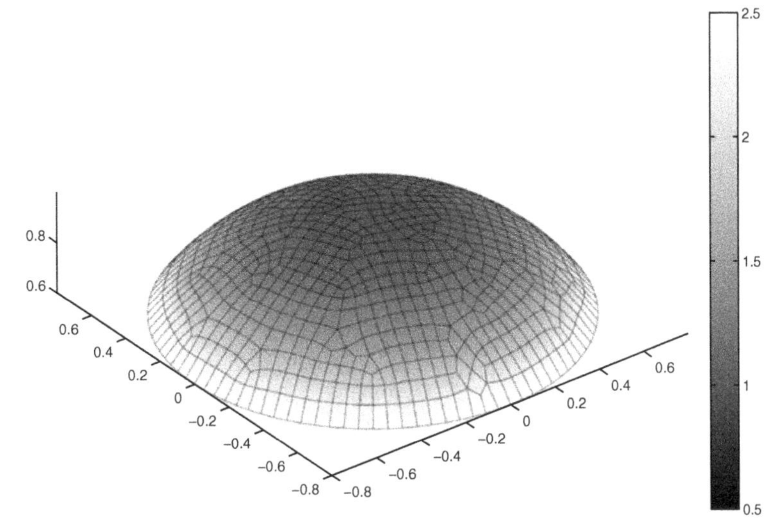

Figure 39 Value of the *a* on a spherical surface.

The second fundamental form II is defined by the projection on the normal direction of the second derivative of the mapping:

$$II = \mathbf{n} \cdot d^2\mathbf{X} = (d\xi\ d\eta) \cdot \begin{pmatrix} \mathbf{n} \cdot \frac{\partial \mathbf{a}_1}{\partial \xi} & \mathbf{n} \cdot \frac{\partial \mathbf{a}_2}{\partial \xi} \\ \mathbf{n} \cdot \frac{\partial \mathbf{a}_1}{\partial \eta} & \mathbf{n} \cdot \frac{\partial \mathbf{a}_2}{\partial \eta} \end{pmatrix} \cdot \begin{pmatrix} d\xi \\ d\eta \end{pmatrix} = (d\xi\ d\eta) \cdot \mathbf{b} \cdot \begin{pmatrix} d\xi \\ d\eta \end{pmatrix} \tag{49}$$

Being $\mathbf{n} \cdot d\mathbf{X} \equiv 0$, it results:

$$0 = d(\mathbf{n} \cdot d\mathbf{X}) = d\mathbf{n} \cdot d\mathbf{X} + \mathbf{n} \cdot d^2\mathbf{X} = d\mathbf{n} \cdot d\mathbf{X} + II \tag{50}$$

from which it results

$$II = -d\mathbf{n} \cdot d\mathbf{X} \tag{51}$$

and by developing we can identify

$$\mathbf{b} = -\begin{pmatrix} \frac{\partial \mathbf{n}}{\partial \xi} \\ \frac{\partial \mathbf{n}}{\partial \eta} \end{pmatrix} \cdot (\mathbf{a}_1\ \mathbf{a}_2) \tag{52}$$

or:

$$\mathbf{b} \cdot \begin{pmatrix} \mathbf{a}^1 \\ \mathbf{a}^2 \end{pmatrix} = -\begin{pmatrix} \frac{\partial \mathbf{n}}{\partial \xi} \\ \frac{\partial \mathbf{n}}{\partial \eta} \end{pmatrix} \cdot (\mathbf{a}_1\ \mathbf{a}_2) \cdot \begin{pmatrix} \mathbf{a}^1 \\ \mathbf{a}^2 \end{pmatrix} \tag{53}$$

that taking into account that $\mathbf{a}_i \cdot \mathbf{a}^j = \delta_{i,j}$ it results

$$\begin{pmatrix} \frac{\partial \mathbf{n}}{\partial \xi} \\ \frac{\partial \mathbf{n}}{\partial \eta} \end{pmatrix} = -\mathbf{b} \cdot \begin{pmatrix} \mathbf{a}^1 \\ \mathbf{a}^2 \end{pmatrix} \tag{54}$$

With $\overline{\mathbf{b}}$ given by

$$\overline{\mathbf{b}} = \mathbf{b} \cdot \mathbf{a}^{-1} \tag{55}$$

we can define the mean curvature H:

$$H = \frac{1}{2} \cdot \mathrm{trace}(\overline{\mathbf{b}}) \tag{56}$$

and the Gaussian curvature K:

$$K = \det(\overline{\mathbf{b}}) \tag{57}$$

Finally, the third fundamental form is defined by:

$$III = d\mathbf{n} \cdot d\mathbf{n} = (d\xi \; d\eta) \cdot \mathbf{c} \cdot \begin{pmatrix} d\xi \\ d\eta \end{pmatrix} \tag{58}$$

with

$$\mathbf{c} = \begin{pmatrix} \frac{\partial \mathbf{n}}{\partial \xi} \\ \frac{\partial \mathbf{n}}{\partial \eta} \end{pmatrix} \cdot \begin{pmatrix} \frac{\partial \mathbf{n}}{\partial \xi} & \frac{\partial \mathbf{n}}{\partial \eta} \end{pmatrix} = \mathbf{b} \cdot \mathbf{a}^{-1} \cdot \mathbf{b} \tag{59}$$

Shell description

We consider now a generic point of the shell $\mathbf{x}$, whose position can be given from:

$$\mathbf{x}(\xi, \eta, \zeta) = \mathbf{X}(\xi, \eta) + \zeta \cdot \mathbf{n} \tag{60}$$

that establishes a mapping between (ξ, η, ζ) and (x_1, x_2, x_3).

The covariant basis results now:

$$\begin{cases} \tilde{\mathbf{a}}_1 = \frac{\partial \mathbf{x}}{\partial \xi} = \mathbf{a}_1 + \zeta \cdot \frac{\partial n}{\partial \xi} \\ \tilde{\mathbf{a}}_2 = \frac{\partial \mathbf{x}}{\partial \eta} = \mathbf{a}_2 + \zeta \cdot \frac{\partial \mathbf{n}}{\partial \eta} \end{cases} \tag{61}$$

The differential element in a point of the shell writes:

$$d\mathbf{x} = d\mathbf{X} + \zeta \cdot d\mathbf{n} + \mathbf{n} \cdot d\zeta = \tilde{\mathbf{a}}_1 \cdot d\xi + \tilde{\mathbf{a}}_2 \cdot d\eta + \mathbf{n} \cdot d\zeta \tag{62}$$

from which the first fundamental form $I = d\mathbf{x} \cdot d\mathbf{x}$ involves the metric tensor $\tilde{\mathbf{a}}$:

$$\tilde{\mathbf{a}} = \begin{pmatrix} \tilde{\mathbf{a}}_1 \cdot \tilde{\mathbf{a}}_1 & \tilde{\mathbf{a}}_1 \cdot \tilde{\mathbf{a}}_2 & 0 \\ \tilde{\mathbf{a}}_2 \cdot \tilde{\mathbf{a}}_1 & \tilde{\mathbf{a}}_2 \cdot \tilde{\mathbf{a}}_2 & 0 \\ 0 & 0 & 1 \end{pmatrix} \tag{63}$$

where it is easy to verify that

$$\begin{pmatrix} \tilde{\mathbf{a}}_1 \cdot \tilde{\mathbf{a}}_1 & \tilde{\mathbf{a}}_1 \cdot \tilde{\mathbf{a}}_2 \\ \tilde{\mathbf{a}}_2 \cdot \tilde{\mathbf{a}}_1 & \tilde{\mathbf{a}}_2 \cdot \tilde{\mathbf{a}}_2 \end{pmatrix} = \mathbf{a} - 2 \cdot \zeta \cdot \mathbf{b} + \zeta^2 \cdot \mathbf{c} \tag{64}$$

Now we define

$$\tilde{a} = \det(\tilde{\mathbf{a}}) = \det(\mathbf{a} - 2 \cdot \zeta \cdot \mathbf{b} + \zeta^2 \cdot \mathbf{c}) \tag{65}$$

that taking into account (55) it results

$$\tilde{a} = a \cdot \det(\mathbf{I} - 2 \cdot \zeta \cdot \overline{\mathbf{b}} + \zeta^2 \cdot \overline{\mathbf{b}} \cdot \overline{\mathbf{b}}) = a \cdot \left(\det(\mathbf{I} - \zeta \cdot \overline{\mathbf{b}})\right)^2 \tag{66}$$

where, by developping the determinant and by using the definition of both the mean and the Gaussian curvatures previously defined, it results:

$$\det(\mathbf{I} - \zeta \cdot \overline{\mathbf{b}}) = 1 - 2 \cdot H \cdot \zeta + K \cdot \zeta^2 \tag{67}$$

The area element $d\tilde{A}$ oriented in the normal direction writes:

$$d\tilde{A} = \|\tilde{\mathbf{a}}_1 \times \tilde{\mathbf{a}}_2\| \cdot d\xi \cdot d\eta = \sqrt{\tilde{a}} \cdot d\xi \cdot d\eta \tag{68}$$

being the volume element dv:

$$dv = d\tilde{A} \cdot d\zeta \tag{69}$$

that taking into account the previous developments results in:

$$dv = \sqrt{a} \cdot (1 - 2 \cdot H \cdot \zeta + K \cdot \zeta^2) \cdot d\xi \cdot d\eta \cdot d\zeta \tag{70}$$

Finally the transformation gradient $\tilde{\mathbf{F}}$ results

$$\tilde{\mathbf{F}} = \left[\tilde{\mathbf{a}}_1 \, \tilde{\mathbf{a}}_2 \, \mathbf{n}\right] = \mathbf{F} + \zeta \cdot \mathbf{F}_n \tag{71}$$

where:

$$\mathbf{F}_n = \left[\frac{\partial \mathbf{n}}{\partial \xi} \; \frac{\partial \mathbf{n}}{\partial \eta} \; \mathbf{0}\right] \tag{72}$$

that can be rewritten as:

$$\tilde{\mathbf{F}} = \mathbf{F} \cdot (\mathbf{I} + \zeta \cdot \mathbf{b}_n) \tag{73}$$

with

$$\mathbf{b}_n = \mathbf{F}^{-1} \cdot \mathbf{F}_n = \begin{pmatrix} \mathbf{a}^1 \cdot \frac{\partial \mathbf{n}}{\partial \xi} & \mathbf{a}^1 \cdot \frac{\partial \mathbf{n}}{\partial \eta} & 0 \\ \mathbf{a}^2 \cdot \frac{\partial \mathbf{n}}{\partial \xi} & \mathbf{a}^2 \cdot \frac{\partial \mathbf{n}}{\partial \eta} & 0 \\ 0 & 0 & 0 \end{pmatrix} \tag{74}$$

The gradient of the inverse transformation is given by $\tilde{\mathbf{F}}^{-1}$. In the case of small thickness and curvatures $\tilde{\mathbf{F}}^{-1}$ can be approximated from (73) according to

$$\tilde{\mathbf{F}}^{-1} \approx (\mathbf{I} - \zeta \cdot \mathbf{b}_n + \zeta^2 \cdot \mathbf{b}_n^2) \cdot \mathbf{F}^{-1} \tag{75}$$

When this approximation fails we should consider the full inverse of $\tilde{\mathbf{F}}$.

Competing interests

The authors declare that they have no competing interests.

Authors' contributions

All authors contributed to each research task. All authors read and approved the final manuscript.

References

1. Timoshenko SP, Woinowsky-Krieger S (1959) Theory of plates and shells. McGraw-Hill classic textbook reissue, New York, second edition
2. Qatu MS (2012) Review of recent literature on static analyses of composite shells: 2000–2010. Open J Composite Mater 2(3): 61–86
3. Zhang YX, Yang CH (2009) Recent developments in finite element analysis for laminated composite plates. Composite Struct 88(1): 147–157
4. Reddy JN, Arciniega RA (2004) Shear deformation plate and shell theories: from Stavsky to present. Mech Adv Mater Struct 11(6): 535–582
5. Viola E, Tornabene F, Fantuzzi N (2013) Static analysis of completely doubly-curved laminated shells and panels using general higher-order shear deformation theories. Composite Struct 101: 59–93
6. Carrera E (2003) Historical review of Zig-Zag theories for multilayered plates and shells. Appl Mech Rev 56(3): 287
7. Sedira L, Ayad R, Sabhi H, Hecini M, Sakami S (2012) An enhanced discrete Mindlin finite element model using a zigzag function. Eur J Comput Mech 21(1–2): 122–140
8. Carrera E (2002) Theories and finite elements for multilayered, anisotropic, composite plates and shells. Arch Comput Methods Eng 9(2): 87–140
9. Carrera E (2003) Theories and finite elements for multilayered plates and shells: a unified compact formulation with numerical assessment and benchmarking. Arch Comput Methods Eng 10(3): 215–296
10. Kratzig WB, Jun D (2002) Multi-layer multi-director concepts for D-adaptivity in shell theory. Comput Struct 80(9): 719–734
11. Naceur H, Shiri S, Coutellier D, Batoz JL (2013) On the modeling and design of composite multilayered structures using solid-shell finite element model. Finite Elem Anal Design 70: 1–14
12. Trinh V-D, Abed-Meraim F, Combescure A (2011) A new assumed strain solid-shell formulation "SHB6" for the six-node prismatic finite element. J Mech Sci Technol 25(9): 2345–2364
13. Quatmann M, Aswini N, Reimerdes HG, Gupta NK (2013) Superelements for a computationally efficient structural analysis of elliptical fuselage sections. Aerosp Sci Technol 27(1): 76–83
14. Wisnom MR, Gigliotti M, Ersoy N, Campbell M, Potter KD (2006) Mechanisms generating residual stresses and distortion during manufacture of polymer-matrix composite structures. Composites Part A: Appl Sci Manuf 37(4): 522–529
15. Hochard Ch, Ladeveze P, Proslier L (1993) A simplified analysis of elastic structures. Eur J Mech A/Solids 12(4): 509–535
16. Ladeveze P, Arnaud L, Rouch P, Blanz C (2001) The variational theory of complex rays for the calculation of medium-frequency vibrations. Eng Comput 18(1–2): 193–221
17. Ladeveze P (1999) Nonlinear computational structural mechanics. Springer, New York

18. Ammar A, Mokdad B, Chinesta F, Keunings R (2006) A new family of solvers for some classes of multidimensional partial differential equations encountered in kinetic theory modeling of complex fluids. J Non-Newtonian Fluid Mech 139: 153–176
19. Chinesta F, Ammar A, Cueto E (2010) Recent advances and new challenges in the use of the Proper Generalized Decomposition for solving multidimensional models. Arch Comput Methods Eng 17: 327–350
20. Chinesta F, Ammar A, Leygue A, Keunings R (2011) An overview of the Proper Generalized Decomposition with applications in computational rheology. J Non-Newtonian Fluid Mech 166: 578–592
21. Chinesta F, Ladeveze P, Cueto E (2011) A short review in model order reduction based on Proper Generalized Decomposition. Arch Comput Methods Eng 18: 395–404
22. Chinesta F, Leygue A, Bordeu F, Aguado JV, Cueto E, Gonzalez D, Alfaro I, Ammar A, Huerta A (2013) Parametric PGD based computational vademecum for efficient design, optimization and control. Arch Comput Methods Eng 20: 31–59
23. Leygue A, Chinesta F, Beringhier M, Nguyen TL, Grandidier JC, Pasavento F, Schrefler B (2013) Towards a framework for non-linear thermal models in shell domains. Int J Numerical Methods Heat Fluid Flow 23(1): 55–73
24. Bognet B, Leygue A, Chinesta F, Poitou A, Bordeu F (2012) Advanced simulation of models defined in plate geometries: 3D solutions with 2D computational complexity. Comput Methods Appl Mechanics Eng 201: 1–12
25. Chinesta F, Leygue A, Bognet B, Ghnatios Ch, Poulhaon F, Bordeu F, Barasinski A, Poitou A, Chatel S, Maison-Le-Poec S (2014) First steps towards an advanced simulation of composites manufacturing by automated tape placement. Int J Mater Forming. http://www.springerlink.com/index/10.1007/s12289-012-1112-9
26. Vidal P, Gallimard L, Polit O (2013) Proper Generalized Decomposition and layer-wise approach for the modeling of composite plate structures. Int J Solids Struct 50(14–15): 2239–2250
27. Macneal RH, Harder RL (1985) A proposed standard set of problems to test finite element accuracy. Finite Elem Anal Design 1: 3–20

4

Estimation of error in observables of coarse-grained models of atomic systems

John Tinsley Oden*, Kathryn Farrell and Danial Faghihi

*Correspondence:
oden@ices.utexas.edu
Institute for Computational Engineering and Sciences The University of Texas at Austin, 201 East 24th St, Stop C0200 POB 4.102, 78712 Austin, TX, USA

Abstract

Background: The use of coarse-grained approximations of atomic systems is the most common methods of constructing reduced-order models in computational science. However, the issue of central importance in developing these models is the accuracy with which they approximate key features of the atomistic system. Many methods have been proposed to calibrate coarse-grained models so that they qualitatively mimic the atomic systems, but these are often based on heuristic arguments.

Methods: A general framework for deriving a *posteriori* estimates of modeling error in coarse–grained models of key observables in atomistic systems is presented. Such estimates provide a new tool for model validation analysis. The connection of error estimates with relative information entropy of observables and model predictions is explained for so-called misspecified models. The relationship between model plausibilities and Kullback-Leibler divergence between the true parameters and model predictions is summed up in several theorems.

Results: Numerical examples are presented in this paper involving a family of coarse-grained models of a polyethylene chain of united atom monomers. Numerical results suggest that the proposed methods of error estimation can be very good indications of the error inherent in coarse-grained models of observables in the atomistic systems. Also, new theorems relating the Kullback-Leibler divergence between model predictions and observations to measures of model plausibility are presented.

Conclusions: A formal structure for estimating errors produced by coarse-graining atomistic models is presented. Numerical examples confirm that the estimates are in agreement with exact errors for a simple class of materials. Errors measured in the D_{KL}-divergence can be related to computable model plausibilities. The results should provide a powerful framework for assessing the validity and accuracy of coarse-grained models.

Keywords: Molecular dynamics, Coarse–grained models, Adjoint systems, Information entropy

Background

Coarse-grained-reduced order models

The most common method of constructing reduced-order models in all of computational science involves the use of coarse-grained models of atomic systems, whereby systems of atoms are aggregated into "beads", or "super atoms", or molecules to reduce the number of degrees of freedom and to lengthen the time scales in which the evolution of events are simulated.

The use of coarse-grained (CG) approximations has been prevalent in molecular dynamics (MD) simulations for many decades Comprehensive reviews of a large segment of the literature on CG models was recently given by Noid [1] and Li *et al.* [2], and an application to semiconductor nano-manufacturing is discussed in Farrell *et al.* [3]. The issue of central importance in developing CG models is the accuracy with which they approximate key features of the atomistic system. Many methods have been proposed to calibrate CG models so that they qualitatively mimic the all-atom (AA) systems, but these are often based on heuristic arguments.

In this paper, we develop *a posteriori* estimates of error in CG approximations of observables in the AA system. We focus on standard molecular dynamics models of micro-canonical ensemble (NVE) thermodynamics, and we call upon the theory of model adaptivity and error estimation laid down in [4] and [5]. In this particular setting, new estimates are also obtained when the information entropy of Shannon [6] is used as a quantity of interest. This leads to methods for estimating CG-model parameters that involve the Kullback-Leibler divergence between probability densities of observables in the AA and CG systems.

In the final *Results and discussion* section of this presentation, we review several statistical properties of parametric models, including asymptotic properties of misspecified models and generalizations of the Bernstein-von Mises theorem advanced by Kleijn and van der Vaart [7]. There, the fundamental role of the Kullback-Leibler distance (the D_{KL}) between the true probability distribution and the observations accessible by the model is reviewed. We present results in the form of theorems that relate the D_{KL} to measures of model plausibility that arise from Bayesian approaches to model selection. The relationships of the *a posteriori* estimates to the statistical interpretations are summarized in concluding remarks.

Preliminaries, conventions and notations

We generally approach the problem of developing computer models of large atomic systems through the use of any of several hardened molecular dynamics (MD) codes or through equivalent Monte Carlo approximations invoking the ergodic hypothesis. For a system of n atoms, the Hamiltonian is

$$H\left(\mathbf{r}^n, \mathbf{p}^n\right) = \sum_{\alpha=1}^{n} \frac{m_\alpha}{2} \mathbf{p}_\alpha \cdot \mathbf{p}_\alpha + u(\mathbf{r}^n), \tag{1}$$

where $\mathbf{r}^n = \{\mathbf{r}_1, \mathbf{r}_2, \ldots, \mathbf{r}_n\}$ is the set of atomic coordinate vectors, $\mathbf{p}^n = \{\mathbf{p}_1, \mathbf{p}_2, \ldots, \mathbf{p}_n\}$ is the set of particle momentum vectors, m_α the atomic mass of the $\alpha-$th atom and $u(\mathbf{r}^n)$ is the potential energy or interaction potential. Then $(\mathbf{r}^n, \mathbf{p}^n)$ defines a point or microstate in the phase space Γ_{AA} of the all atom (AA) model. In typical MD simulations, the potential is, for example, of the form

$$u(\mathbf{r}^n) = V_{\text{bond}}(\mathbf{r}^n) + V_{\text{angle}}(\mathbf{r}^n) + V_{\text{dihedral}}(\mathbf{r}^n) + V_{\text{non-bonded}}(\mathbf{r}^n) + V_{\text{coulomb}}(\mathbf{r}^n), \tag{2}$$

where

$$V_{\text{bond}}(\mathbf{r}^n) = \sum_{i=1}^{N_b} \frac{1}{2} k_{ri} |\mathbf{r}_i - \mathbf{r}_{0i}|^2, \tag{3a}$$

$$V_{\text{angle}}(\mathbf{r}^n) = \sum_{i=1}^{N_a} \frac{1}{2} k_{\theta i} (\theta_i - \theta_{0i})^2, \tag{3b}$$

$$V_{\text{dihedral}}(\mathbf{r}^n) = \sum_{i=1}^{N_{nb}-1} \sum_{j=1}^{N_{nb}} \frac{V_{ji}}{2} \left[1 + (-1)^{j-1} \cos(j\phi_i)\right], \tag{3c}$$

$$V_{\text{non-bonded}}(\mathbf{r}^n) = \sum_{i=1}^{N_{nb}} \sum_{j>i} 4\epsilon_{ij} \left(\left(\frac{\sigma_{ij}}{r_{ij}}\right)^{\alpha} - \left(\frac{\sigma_{ij}}{r_{ij}}\right)^{\beta} \right), \qquad r_{ij} \le r_2, \tag{3d}$$

$$V_{\text{columb}}(\mathbf{r}^n) = \sum_{i=1}^{N_q-1} \sum_{j>i}^{N_q} 4\epsilon_0 \frac{q_i q_j}{r_{ij}}. \tag{3e}$$

Here covalent bonds are represented by the harmonic potential (1a), changes in bond angles by (1b), torsional potentials by changes in dihedral angles (1c), Lennard-Jones non-bonded potentials by (1d), with $r_{ij} = |\mathbf{r}_i - \mathbf{r}_j|$ and r_c the cut-off radius, (α, β) typically $= (12, 6)$, and Coulomb potentials between charges q_i at $\mathbf{r}_i$ and q_j at $\mathbf{r}_j$ (1e). These forms are typical of those implemented in popular MD codes, although several other common potentials could be added. The parameters of the potential model are given by the vector of physical coefficients: $\{k_i, k_{\theta i}, V_{ji}, \phi_i, \epsilon_{ij}, \sigma_{ij}, r_c, \dots\}$. In general, atomic properties and values of parameters for the full all-atom system are supplied by systems calibrated using experimental data or quantum mechanics predictions (see, e.g. the OPLS data in [8,9]).

Given the Hamiltonian (1), Hamilton's equations of motion are:

$$\frac{\partial H}{\partial \mathbf{p}_\alpha} = \dot{\mathbf{r}}_\alpha, \qquad \frac{\partial H}{\partial \mathbf{r}_\alpha} = -\dot{\mathbf{p}}_\alpha, \qquad 1 \le \alpha \le n. \tag{4}$$

In MD, it is assumed that the atomic system evolves according to the laws of Newtonian mechanics, so we set $\mathbf{p}_\alpha = m_{(\alpha)} \dot{\mathbf{r}}_\alpha$, and the second Hamiltonian equation in (4) reduces to the system of equations

$$m_{\alpha\beta} \ddot{r}_{\beta i}(t) + \partial_{\alpha i} u(\mathbf{r}^n(t)) - f_{\alpha i}(t) = 0, \qquad 1 \le \alpha, \beta \le n\,,\ 1 \le i \le 3, \tag{5}$$

where repeated indices are summed throughout their range, $m_{\alpha\beta} = m_{(\alpha)} \delta_{\alpha\beta}$ is the mass of atom α, superimposed dots indicate time derivations, $r_{\beta i}$ is the component of $\mathbf{r}_\beta$ in direction i, $\partial_{\alpha i} = \partial / \partial r_{\alpha i} u(\mathbf{r}^n(t))$ is the total interatomic potential of the system given, e.g., by (2), and $f_{\alpha i}(t)$ is the ith component of applied force on atom α at time t. We will add initial conditions, $\dot{r}_{\beta i}(0) = v_{\beta i}$, and $r_{\beta i}(0) = r^0_{\beta i}$, where $v_{\beta i}$ and $r^0_{\beta i}$, for now, are assumed to be given.

Molecular dynamical equations of the form (5) are typical of those in standard MD codes that are numerically integrated with randomly-sampled initial conditions over time intervals to approximate systems with constant energy and fixed volume and fixed number of particles corresponding to so-called micro-canonical ensembles. Without loss in generality, we confine this development to such thermodynamic scenarios noting that straightforward extensions to, say, constant temperature settings, are covered by replacing (5) with appropriate "thermostat" models, such as the Langevin or Nose–Hoover formulations (see e.g. [10]). The general approach is then applicable to canonical ensembles and more general statistical thermodynamics settings.

A fundamental concept in the molecular theory of matter is that macroscopic properties actually observed in experiments, the so-called *observables*, must be the result of averages over a time interval τ of some phase function $q(\mathbf{r}^n, \mathbf{p}^n)$ that depends on the phase-point positions $(\mathbf{r}^n, \mathbf{p}^n)$ in phase space Γ_{AA}, as all measurements require a finite duration. Moreover, for thermodynamic systems in equilibrium, this average, denoted $\langle q \rangle$, must be independent of the starting time t_0 and it must attain a value from an essentially infinite time duration. Thus, our goal in constructing the all-atom model is to compute observables of the form (cf [10,11]),

$$\langle q \rangle = \lim_{\tau \to \infty} \tau^{-1} \int_{t_0}^{t_0+\tau} q(\mathbf{r}^n(t), \mathbf{p}^n(t)) \mathrm{d}t. \tag{6}$$

Here we shall confine our attention to phase functions that depend only on the configurations of systems in thermodynamic equilibrium. Our *quantities of interest* are then written,

$$Q_{\mathbf{r}} = \lim_{\tau \to \infty} \tau^{-1} \int_{t_0}^{t_0+\tau} q(\mathbf{r}^n(t)) \mathrm{d}t. \tag{7}$$

In all but the simplest applications, it is impossible to solve the dynamic system (5) owing to its enormous size. Therefore, reduced-order models must be developed. The process involves aggregating groups of atoms into beads or molecules or "super atoms" so as to create a coarse-grained (CG) molecular model. The CG model has N coordinate vectors $\mathbf{R}^N(t) = \{\mathbf{R}_1(t), \mathbf{R}_2(t), \ldots, \mathbf{R}_N(t)\}$, $N < n$; and the corresponding equations of motion are

$$M_{AB}\ddot{R}_{Bi}(t) + \partial_{Ai} U(\mathbf{R}^N(t), \boldsymbol{\theta}) - F_{Ai}(t) = 0, \qquad 1 \leq A, B \leq N\,,\ 1 \leq i \leq 3. \tag{8}$$

M_{AB} defining the CG mass matrix, $\partial_{Ai} = \partial / \partial R_{Ai}$, $U(\cdot, \cdot)$ the interaction potential energy of the CG system, $\boldsymbol{\theta}$ a vector of parameters defining the CG model, and $F_{Ai}(t)$ the ith component of applied force at bead A at time t. Initial conditions are $\dot{R}_{Ai}(0) = V_{Ai}$, and $R_{Ai}(0) = R^0_{Ai}$. The unknown parameters with a potential of the form (2) are denoted, for example,

$$\boldsymbol{\theta} = (K_{Ri}, R_{0i}, K_{\theta i}, \theta_{0i}, V_{0i}, \varepsilon_{ii}, \sigma_{ii}, \cdots), \tag{9}$$

the notation, in analogy with (1), being chosen to indicate parameters of the CG model.

It is important to establish a kinematic and algebraic relation between coordinates of particles in the AA system and those in the CG system. A very large literature exists on various coarse-graining mapping schemes, and choices of the appropriate map from the AA to the CG system or vice versa are often based on heuristic methods (see e.g. [2]). Our general approach can be adapted to any such well-defined AA-to-CG or CG-to-AA map, but for definiteness, we describe one such family of mappings.

Let $\mathcal{J}_A$ be the index set of AA-coordinate labels of atoms aggregated into a single bead A with CG-coordinate vector $\mathbf{R}_A$ emanating from the origin to a reference point labeled A within the bead (e.g. $\mathbf{R}_A$ could be chosen to define the center of mass, $\mathbf{R}_A = \sum_{\alpha \in \mathcal{J}_A} m_\alpha \mathbf{r}_\alpha / M_A, M_A = \sum_{\alpha \in \mathcal{J}_A} m_\alpha$). Let $\mathbf{a}_{A\alpha}(t)$ be a vector from the reference center of bead A to the end point of AA-coordinate vector $\mathbf{r}_\alpha(t), \alpha \in \mathcal{J}_A$. Let $G^{\cdot A}_{\alpha}$ be component of the $n \times N$ array,

$$G^{\cdot A}_{\alpha} = \begin{cases} 1 \text{ if } & \alpha \in \mathcal{J}_A \\ 0 \text{ if otherwise.} \end{cases} \tag{10}$$

Then we have,

$$\mathbf{r}_\alpha = \sum_{A=1}^{N} G_\alpha^{\cdot A} \left(\mathbf{R}_A + \mathbf{a}_{A\alpha}\right), \qquad 1 \leq \alpha \leq n. \tag{11}$$

Here we assume that each AA coordinate vector $\mathbf{r}_\alpha$ belongs to only one bead identified with CG vector $\mathbf{R}_A$, but this is not a necessary assumption. In the case of bonded systems in which $\mathbf{r}_\alpha$ is associated with, say, two index sets $\mathcal{J}_A$ and $\mathcal{J}_B$, we simply choose either $\mathcal{J}_A$ or $\mathcal{J}_B$ as the representative of $\mathbf{r}_\alpha$ and associate $\mathbf{r}_\alpha$ with only one bead to avoid double counting.

These kinematical conventions are illustrated in Figure 1. The AA coordinates $\mathbf{r}_\alpha$ of atoms assigned to molecular bead A remain with that bead throughout all possible motions of the CG system. The Boolean array $G_\alpha^{\cdot A}$ merely adjusts labels of the AA system to agree with labels assigned beads in the CG system.

Returning to (7), it is clear that the CG approximation of the quantities of interest (QoI) are of the form,

$$Q_{\mathbf{R}}(\boldsymbol{\theta}) = \lim_{\tau\to\infty} \tau^{-1} \int_{t_0}^{t_0+\tau} q\left(G\left(\mathbf{R}^N(t), \boldsymbol{\theta}\right)\right) \mathrm{d}t \tag{12}$$

where G is the AA-to-CG map defined in (11), where we denote $\mathbf{r}^n = G(\mathbf{R}^N(t), \boldsymbol{\theta})$, and where we specifically present the dependence of the QoI on the CG potential parameters $\boldsymbol{\theta}$.

Now it is obvious that the evolution of the CG system defined by the coordinate vectors $\mathbf{R}_A(t)$ satisfying (8) do not satisfy the "true" equations of motion (5). Indeed, if $\mathbf{R}^N(t) = \{\mathbf{R}_1(t), \mathbf{R}_2(t), \ldots, \mathbf{R}_N(t)\}$ is the set of N-vectors satisfying (8), they constrain the motion

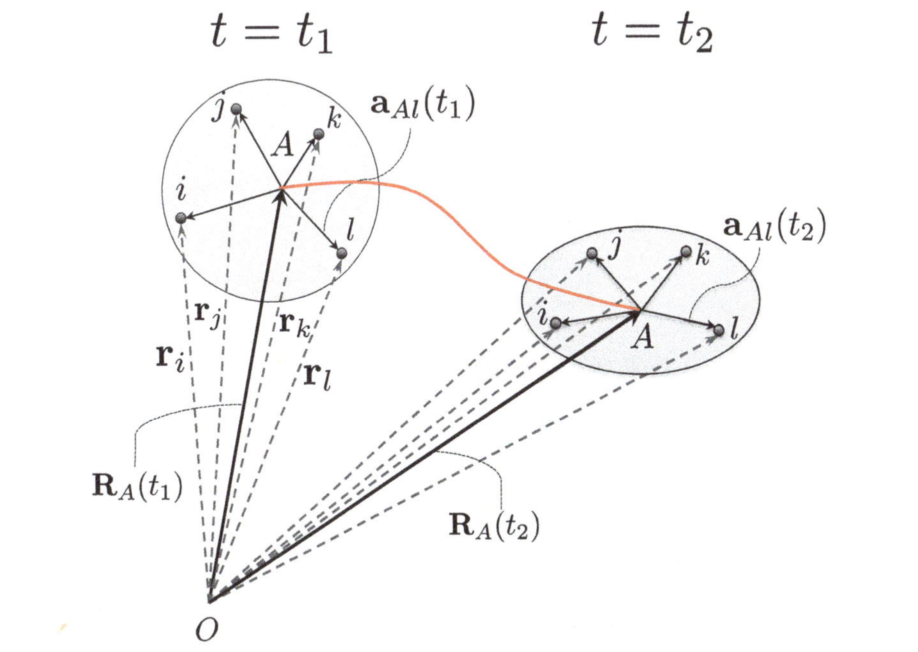

Figure 1 A Lagrangian description of the motion of a molecular bead A and its associated AA atoms $(i, j, k, l) \in \mathcal{J}_A$.

of the AA system via (11), so upon introducing the constrained motions into (5) we obtain the residual,

$$\rho_{\alpha i}(\boldsymbol{\theta},t) = m_{\alpha\beta}\sum_{A=1}^{N} G_{\beta}^{\cdot A}\left(\ddot{R}_{Ai}+\ddot{a}_{A\beta i}\right)(t)$$
$$+\partial_{\alpha i}u\left(G(\mathbf{R}^{N}(t),\mathbf{a}^{N}(t))\right)-\sum_{A=1}^{N} G_{\beta}^{\cdot A}\omega_{\alpha}^{\beta}F_{Ai}(t),$$
$$1\le\alpha,\beta\le n\,,\;1\le i\le 3, \tag{13}$$

where $\sum_{\beta}\omega_{\alpha}^{\beta}=1, \sum_{\alpha\in\mathcal{J}_A}\mathbf{f}_{\alpha}=\mathbf{F}_A$, and

$$G(\mathbf{R}^{N}(t),\mathbf{a}^{N}(t)) = \left\{\sum_{A} G_{1}^{\cdot A}(\mathbf{R}_A+\mathbf{a}_{A1}), G_{2}^{\cdot A}(\mathbf{R}_A+\mathbf{a}_{A2})\right.$$
$$\left.\ldots, G_{n}^{\cdot A}(\mathbf{R}_A+\mathbf{a}_{An})\right\},$$
$$=\mathbf{r}^{n}(t). \tag{14}$$

On the left-hand side of (13), we have taken note that the residual depends on the CG-model parameters $\boldsymbol{\theta}$, not explicitly presented on the right for simplicity. The AA force $\mathbf{f}_{\alpha}$ at coordinate α is a fraction ω_{α}^{β} of the CG force F_{Ai}. In general, as a first-order approximation, one can take the vectors $\mathbf{a}_{A\alpha}$ as time-independent constant vectors equal to their value in a given reference configuration ($\mathbf{a}_{A\alpha}(t)=\mathbf{a}_{A\alpha}(t_0); \dot{\mathbf{a}}_{A\alpha}=\ddot{\mathbf{a}}_{A\alpha}=\mathbf{0}$). Then the approximate residual results of the form

$$\hat{\rho}_{\alpha i}(\boldsymbol{\theta},t) = G_{\alpha}^{\cdot A}m_{\alpha\beta}\ddot{\mathbf{R}}_{Ai}+\partial_{\alpha i}u\left(G(\mathbf{R}^{N}(t)),\mathbf{a}^{N}(t_0)\right)-f_{\alpha i}, \tag{15}$$

with $f_{\alpha i}=\omega_{\alpha}^{\beta}G_{\beta}^{\cdot A}F_{Ai}$, and with repeated indices summed, $1\le\alpha,\beta\le n, 1\le A\le N$.

It is possible to define a reverse or "push back" relationship that assigns to every AA coordinate vector $\mathbf{r}_{\alpha}$ the CG coordinate vector $\mathbf{R}_A$ of the bead to which $\mathbf{r}_{\alpha}$ belongs. Given $\mathbf{r}_{\alpha}$, set

$$\mathbf{R}_A = G_{A}^{\cdot\alpha}(\mathbf{r}_{\alpha}-\mathbf{a}_{A\alpha}), \qquad 1\le A\le N, \tag{16}$$

where $G_{A}^{\cdot\alpha}$ is the transpose of $G_{\alpha}^{\cdot A}$ (no sum on α).

Thus, for each time $t\in[t_0,\tau]$, one can select a sample $\omega(t)$ of AA coordinates $\{\mathbf{r}_1^*(t),\mathbf{r}_2^*(t),\ldots,\mathbf{r}_n^*(t)\}$ and employ (16) to generate the corresponding CG coordinates $\{\mathbf{R}_1^*(t),\mathbf{R}_2^*(t),\ldots,\mathbf{R}_N^*(t)\}$. We use the simplified notation,

$$\mathbf{R}^{N}(t)=G(\omega(t)) \tag{17}$$

to define the image of this sample in the CG system.

Methods

Weak forms of the dynamical problem

It is convenient to recast the molecular dynamics problem into a weak or variational form. Toward this end, we introduce the Banach spaces

$$\mathcal{V}=\left\{\mathbf{r}^{n}(t);\mathbf{r}_{\alpha}\in\Omega\subset\mathbb{R}^3\times H^2(0,\tau);1\le\alpha\le n;\right.$$
$$\left.\|\mathbf{r}^{n}(\cdot)\|^2=\int_0^{\tau}[\ddot{\mathbf{r}}_{\alpha}\cdot\ddot{\mathbf{r}}_{\alpha}+\dot{\mathbf{r}}_{\alpha}\cdot\dot{\mathbf{r}}_{\alpha}+\mathbf{r}_{\alpha}\cdot\mathbf{r}_{\alpha}]\,d\tau<\infty\right\}, \tag{18}$$

and the semilinear and linear forms $\mathcal{B} : \mathcal{V} \times \mathcal{V} \to \mathbb{R}, \mathcal{F} : \mathcal{V} \to \mathbb{R}$, given by

$$\mathcal{B}\left(\mathbf{r}^n;\mathbf{v}^n\right) = \int_0^\tau \left[m_{\alpha\beta}\ddot{\mathbf{r}}_\alpha(t) \cdot \mathbf{v}_\beta(t) + \partial_{\alpha i} u\left(\mathbf{r}^n(t)\right) v_{\alpha i}(t)\right] \mathrm{d}t + I_0(\mathbf{v}^n), \tag{19}$$

$$\mathcal{F}\left(\mathbf{v}^n\right) = \int_0^\tau f_{\alpha i}(t) v_{\alpha i}(t)\mathrm{d}t + v_{\alpha i}(0) m_{\alpha\beta} V_\beta - \dot{v}_{\alpha i}(0) m_{\alpha\beta} r^0_\beta, \tag{20}$$

where

$$I_0(\mathbf{v}^n) = v_{\alpha i}(0) m_{\alpha\beta}\dot{r}_\beta(0) - \dot{v}_{\alpha i}(0) m_{\alpha\beta} r_\beta(0). \tag{21}$$

The notation $\mathcal{B}(\cdot;\cdot)$ is intended to mean that $\mathcal{B}(\cdot;\cdot)$ is possibly nonlinear in entries to the left of the semi-colon and linear in the entries to the right of it.

The problem of finding $\mathbf{r}^n \in \mathcal{V}$ such that

$$\mathcal{B}\left(\mathbf{r}^n;\mathbf{v}^n\right) = \mathcal{F}\left(\mathbf{v}^n\right) \qquad \forall \mathbf{v}^n \in \mathcal{V}, \tag{22}$$

is equivalent to (8) in the sense that every solution of (8) with appropriate initial conditions, satisfies (22), and any sufficiantly smooth solution of (22) satisfies (8).

The adjoint problem

Let

$$\mathcal{B}'(\mathbf{r}^n;\mathbf{z}^n,\mathbf{v}^n) = \lim_{\theta\to 0} \theta^{-1}\left[\mathcal{B}(\mathbf{r}^n + \theta\mathbf{z}^n;\mathbf{v}^n) - \mathcal{B}(\mathbf{r}^n;\mathbf{v}^n)\right] \tag{23}$$

and

$$Q'(\mathbf{r}^n;\mathbf{v}^n) = \lim_{\theta\to 0} \theta^{-1}\left[Q(\mathbf{r}^n + \theta\mathbf{v}^n) - Q(\mathbf{r}^n)\right], \tag{24}$$

where Q is a functional on $\mathcal{V}$, and both $\mathcal{B}'(\cdot;\cdot)$ and $Q'(\cdot;\cdot)$ are assumed to exist and be finite (i.e. $\mathcal{B}(\cdot;\cdot)$ and $Q(\cdot)$ are Gateaux differentiable). Then the adjoint or dual problem associated with (22) is

Find $\mathbf{z}^n = \{\mathbf{z}_1, \mathbf{z}_2, \ldots, \mathbf{z}_n\} \in \mathcal{V}$ *such that*

$$\mathcal{B}'(\mathbf{r}^n;\mathbf{z}^n,\mathbf{v}^n) = Q'(\mathbf{r}^n;\mathbf{v}^n) \qquad \forall \mathbf{v}^n \in \mathcal{V}. \tag{25}$$

Introducing (19) into (23) gives, after some manipulations,

$$\begin{aligned}\mathcal{B}'(\mathbf{r}^n;\mathbf{z}^n,\mathbf{v}^n) = & \int_0^\tau \left(m_{\alpha\beta}\ddot{z}_{\beta i} - H_{\alpha i\beta j}(\mathbf{r}^n(t)) z_{\beta j}\right) v_{\alpha i}\mathrm{d}t \\ & + \dot{m}_{\alpha\beta} z_{\beta i}(\tau) v_{\alpha i}(\tau) \\ & - m_{\alpha\beta}\dot{z}_{\beta i}(\tau) v_{\alpha i}(\tau),\end{aligned} \tag{26}$$

where $H_{\alpha i\beta j}$ is the Hessian,

$$H_{\alpha i\beta j}(\mathbf{r}^n(t)) = \frac{\partial^2 u(\mathbf{r}^n(t))}{\partial r_{\alpha i}\partial r_{\beta j}}. \qquad 1 \le \alpha, \beta \le n\,,\ 1 \le i, j \le 3. \tag{27}$$

Likewise, if $Q(\mathbf{r}^n) = \int_0^\tau q(\mathbf{r}^n(t))\mathrm{d}t$, then

$$Q'(\mathbf{r}^n(t);\mathbf{v}^n) = \int_0^\tau \partial_{\alpha i} q(\mathbf{r}^n(t)) v_{\alpha i}\mathrm{d}t. \tag{28}$$

Note that (25) is solved "backward in time;" the forward problem (22) is solved for $\mathbf{r}^n(t)$, which determines the coefficients in $\mathcal{B}'(\mathbf{r}^n;\mathbf{z}^n,\mathbf{v}^n)$ which marches the adjoint solution from $t = \tau$ to $t = 0$. The dynamical system corresponding to (25) is:

$$m_{\alpha\beta}\ddot{z}_{\beta i} + H_{\alpha i\beta j}\left(\mathbf{r}^n(t)\right) z_{\beta j}(t) = \partial_{\alpha i} q\left(\mathbf{r}^n(t)\right). \tag{29}$$

Theory of a posteriori estimation of modeling error

Let us now review the theory of *a posteriori* estimation of modeling error advanced in [4] and expanded in [5]. We consider an abstract variational problem of finding an element u in a topological vector space $\mathcal{V}$ such that,

$$\mathcal{B}(u; v) = \mathcal{F}(v), \qquad \forall v \in \mathcal{V}, \tag{30}$$

where $\mathcal{B}(\cdot;\cdot)$ is a semilinear form from $\mathcal{V} \times \mathcal{V}$ into $\mathbb{R}$ and $\mathcal{F}$ is a linear functional on $\mathcal{V}$. Problem (30) is equivalent to the problem of finding a solution u of the problem $A(u) = F$ in the dual space $\mathcal{V}'$, where A is the map induced by $\mathcal{B}(\cdot;\cdot) : \langle A(u), v\rangle = \mathcal{B}(u; v) = \mathcal{F}(v) = \langle \mathcal{F}, v\rangle$, $\langle\cdot;\cdot\rangle$ denoting duality pairing in $\mathcal{V}' \times \mathcal{V}$. Assuming (30) is solvable for u, we wish to compute the value $Q(u)$ of a functional $Q : \mathcal{V} \to \mathbb{R}$ representing a quantity of interest, or an observable of interest.

We assume that the semilinear form $\mathcal{B}(\cdot;\cdot)$ and the functional $Q(\cdot;\cdot)$ are three times Gateaux differentiable on $\mathcal{V}$ with respect to u. In particular, the following limits exist (recall (23) and (24)),

$$\left.\begin{aligned} \mathcal{B}'(u; w, v) &= \lim_{\theta\to 0} \theta^{-1} [\mathcal{B}(u + \theta w, v) - \mathcal{B}(u, v)] \\ Q'(u; v) &= \lim_{\theta\to 0} \theta^{-1} [Q(u + \theta v) - Q(u)] \end{aligned}\right\}. \tag{31}$$

with similar definitions of higher-order derivatives, e.g. $\mathcal{B}''(u; w_1, w_2, v)$, $\mathcal{B}'''(u; w_1, w_2, w_3, v)$, $Q''(u; w, v)$, $Q''(u; v_1, v_2, v_3)$, etc. See [5] for details.

The adjoint problem associated with (30) and the quantity of interest Q consists of finding $z \in \mathcal{V}$ such that

$$\mathcal{B}'(u; z, v) = Q'(u; v), \qquad \forall v \in \mathcal{V}. \tag{32}$$

Now let u_0 be an arbitrary element selected in $\mathcal{V}$. The residual functional (or "residuum") associated with u_0 is defined as the semilinear functional $\mathcal{R} : \mathcal{V} \times \mathcal{V} \to \mathbb{R}$,

$$\mathcal{R}(u_0; v) = \mathcal{F}(v) - \mathcal{B}(u_0; v), \tag{33}$$

which, for each $u_0 \in \mathcal{V}$, is a linear functional on $\mathcal{V}$.

Obviously, if $u_0 = u$, the solution of (30), $\mathcal{R}(u; v) = 0 \ \ \forall v \in \mathcal{V}$. Thus, $\mathcal{R}(u_0; v)$ describes the degree to which the vector u_0 fails to satisfy the central problem (30).

We now recall the basic theorem in [5]:

Theorem 1. *Let the semilinear form $\mathcal{B}(\cdot;\cdot)$ in (30) and the quantity of interest Q be three-times continuously Gateaux differentiable on $\mathcal{V}$. Let u_0 be an arbitrary element of $\mathcal{V}$. Then the error in $Q(u)$ produced by replacing u by u_0 is given by:*

$$Q(u) - Q(u_0) = \mathcal{R}(u_0; z) + \Delta \tag{34}$$

where Δ is a remainder involving higher-order terms in $e_0 = u - u_0$ and $\varepsilon_0 = z - z_0$, z_0 being an approximation of z.

An explicit form of Δ is given in the appendix.

If u_0 is not an arbitrary vector taken from $\mathcal{V}$ but is a solution of a surrogate problem approximating (30) (such as a coarse-grained model approximating an AA model), then it often happens that Δ is negligible compared to the residual. Then (34) reduces to the approximation,

$$Q(u) - Q(u_0) \approx \mathcal{R}(u_0; z). \tag{35}$$

This relation is the basis for many successful methods of *a posteriori* error estimation of both modeling error and numerical error. Whenever $\mathcal{B}(\cdot;\cdot)$ is a bilinear form and $Q(\cdot)$ is linear, $\Delta \equiv 0$.

A-posteriori estimation of error in CG approximations

The CG approximations of the "ground truth" AA system are characterized by a parametric class $\mathcal{P}(\boldsymbol{\theta})$ of molecular dynamics models, one model corresponding to each choice of the vector $\boldsymbol{\theta}$ in a space Θ of parameters defining the CG intermolecular potential $\mathcal{U}(\mathbf{R}^N(t), \boldsymbol{\theta})$. For a given value of $\boldsymbol{\theta}$, observables of interest in states of thermodynamic equilibrium of the CG system are typically generated as averages of samples of the observables taken over subintervals $[t_k, t_{k+1}] \subset [0, \tau]$, for a distribution of initial conditions (see, e.g. [10]).

If we employ the approximation (35) to the AA and CG models, then an estimate of the error in CG approximations of the observable is immediate. Let $q(\mathbf{r}^n)$ be a phase function whose ensemble average $\langle q \rangle$ is an observable of interest, denoted $Q_{\mathbf{r}}$ as in (7). The CG approximation is $Q_{\mathbf{R}}(\boldsymbol{\theta})$ and the error, given by (35), is

$$\varepsilon(\boldsymbol{\theta}) = Q_{\mathbf{r}} - Q_{\mathbf{R}}(\boldsymbol{\theta}) \approx \mathcal{R}(\mathbf{R}^N(\boldsymbol{\theta}); \mathbf{z}^n), \tag{36}$$

where $\mathcal{R}(\mathbf{R}^N(\boldsymbol{\theta}); \mathbf{z}^n) = \int_0^\tau \rho_{\alpha i}(\boldsymbol{\theta}, t) z_{\alpha i}(t) \mathrm{d}t$, $\rho_{\alpha i}(\boldsymbol{\theta}, t)$ is the residual in (13) (or (15)), and $\mathbf{z}^n$ is the solution to the corresponding adjoint problem, and is generally unknown. If $\mathbf{Z}^n$ is an approximation of $\mathbf{z}^n$, then

$$|\varepsilon(\boldsymbol{\theta})| \leq |\mathcal{R}(\mathbf{R}^N(\boldsymbol{\theta}); \mathbf{Z}^n)| + C(\boldsymbol{\theta}) \|\mathbf{z}^n - \mathbf{Z}^n\| \tag{37}$$

with

$$C(\boldsymbol{\theta}) = \sup_{\mathbf{v}^n \in \mathcal{V}} \frac{\|\mathcal{R}(\mathbf{R}^N(\boldsymbol{\theta}, t))\|_{\mathcal{V}'}}{\|\mathbf{v}^n\|_{\mathcal{V}}} \tag{38}$$

$\| \cdot \|_{\mathcal{V}'}$ being the norm on the dual space $\mathcal{V}'$. The problem of error estimation thus reduces to one of developing efficient procedures to compute the residual (ρ) and to compute reasonable approximations of $\mathbf{z}^n$.

It is clear that a quantitative estimate such as (36) (or an approximation with $\mathbf{z}^n$ replaced by $\mathbf{Z}^n$) could be a powerful tool for determining validity of the CG model or in designing validation experiments for CG models. In theory, it also provides a basis for selecting optimal parameters for a given model so as to manage $\varepsilon(\boldsymbol{\theta})$. We elaborate on this notion in the final part of the *Results and discussion* section.

Results and discussion

Numerical example: estimation of error in CG-approximation of a polyethelyne chain

We describe in this section an application of a poasterior error estimation described in the previous section, involving CG approximations of a well-known model of polyethylene. For the base "AA" model, we consider a united atom model of a polyethylene chain containing 200 CH2 (methyl) monomers, meaning we have aggregated hydrogen and carbon atoms into an "AA" bead for simplicity. The united-atom coordinates r_i define the locations of each particle on an r-axis, and the displacement is denoted $u_i(t)$. As an additional simplification, we assume that the interatomic potential is characterized by harmonic bonds of the form (1a), with parameter $k_l = k = 350\ kCal/mol$, bond length of $l = 1.5$ Å, and atomic mass $m = 14.026\ gr/mol$. Initially, each united atom is separated by bond length l, the initial velocities are zero, and the system is assigned an initial displacement field $u_i(0) = f(r_i)$, where $f(r_i) = 1.2e^{-0.1 r_i(0)}$. Under these conditions, the AA system (5) reduces to

$$\boldsymbol{m}\ddot{\boldsymbol{u}} - \boldsymbol{k}\boldsymbol{u} = 0, \qquad \boldsymbol{u}(0) = \mathbf{u}_0,\ \dot{\boldsymbol{u}}(0) = \mathbf{v}_0, \tag{39}$$

where $\boldsymbol{m}$ and $\boldsymbol{k}$ are the mass matrix and the stiffness matrix of the united atom system and are of the form

$$\boldsymbol{m} = m\begin{bmatrix} 1 & & \\ & \ddots & \\ & & 1 \end{bmatrix}, \quad \boldsymbol{k} = k\begin{bmatrix} 2 & -1 & & \\ -1 & 2 & \ddots & \\ & \ddots & \ddots & -1 \\ & & \ddots\ -1 & 2 \end{bmatrix}. \tag{40}$$

A family $\mathcal{M}$ of CG approximations of this model is obtained by aggregating the atoms into beads, with CG models in $\mathcal{M}$ distinguished by the number P of atoms per CG bead. The resulting CG system (8) is of the form

$$\boldsymbol{M}\ddot{\boldsymbol{U}} - \boldsymbol{K}\boldsymbol{U} = \mathbf{0}, \qquad \boldsymbol{U}(0) = \mathrm{U}_0,\ \dot{\boldsymbol{U}}(0) = \mathrm{V}_0, \tag{41}$$

with the mass of each bead set to $M = Pm$ and the bond stiffness $K = \alpha k/P$; $\alpha \in \mathbb{R}^+$.

Upon solving (41) for the CG displacement trajectory $\boldsymbol{U}(t)$, we compute the residual trajectory

$$\rho = \boldsymbol{m}\ddot{\boldsymbol{U}}_{\mathrm{CG}} - \boldsymbol{k}\boldsymbol{U}_{\mathrm{CG}}, \tag{42}$$

where $\boldsymbol{U}_{\mathrm{CG}}(t)$ is the projection $\Pi\boldsymbol{U}(t)$ onto AA atom locations.

The bilinear and linear forms described in (19)-(21) reduce, in this case, to

$$\mathcal{B}(\boldsymbol{u};\boldsymbol{v}) = \int_0^\tau \boldsymbol{v}^T(\boldsymbol{m}\ddot{\boldsymbol{u}} - \boldsymbol{k}\boldsymbol{u})\,\mathrm{d}t - \boldsymbol{v}^T(0)\boldsymbol{m}\dot{\boldsymbol{u}}(0) - \dot{\boldsymbol{v}}^T(0)\boldsymbol{m}\boldsymbol{u}(0), \tag{43}$$

$$\mathcal{F}(\boldsymbol{v}) = -\boldsymbol{v}^T(0)\boldsymbol{m}\mathbf{v}_0 - \dot{\boldsymbol{v}}^T(0)\boldsymbol{m}\mathbf{u}_0. \tag{44}$$

and (25) yielding

$$\begin{aligned}\mathcal{B}'(\boldsymbol{u};\boldsymbol{v},\boldsymbol{z}) &= \int_0^\tau (\boldsymbol{m}\ddot{\boldsymbol{z}} - \boldsymbol{k}\boldsymbol{z})^T\boldsymbol{v}\mathrm{d}t \\ &\quad + (\boldsymbol{m}\boldsymbol{z}(\tau))^T\dot{\boldsymbol{v}}(\tau) - (\boldsymbol{m}\dot{\boldsymbol{z}}(\tau))^T\boldsymbol{v}(\tau) \\ &= \mathcal{Q}'(\boldsymbol{v}).\end{aligned} \tag{45}$$

As an example of a QoI, we take Q_{r} to be the locally-averaged displacement,

$$\mathcal{Q}_{\mathrm{r}} = \int_0^\tau \zeta(t)\mathrm{d}t;\ \zeta(t) = \frac{1}{N_0}\sum_{i\in\mathcal{N}} u_i(t);\ \mathcal{N} = \{i : x_i \le \beta l; \beta \in \mathbb{R}^+\},\ N_0 = card\,\mathcal{N}, \tag{46}$$

for which the strong form of the dual problem is

$$\boldsymbol{m}\ddot{\boldsymbol{z}} - \boldsymbol{k}\boldsymbol{z} = \boldsymbol{q}, \qquad m\dot{\boldsymbol{z}}(\tau) = \mathbf{0}, \quad \boldsymbol{z}(\tau) = \mathbf{0}, \tag{47}$$

where N_0 is the number of united atoms considered in set $\mathcal{N}$ and $\boldsymbol{q}$ is the vector defined such that $\mathcal{Q}(\boldsymbol{u}) = \boldsymbol{q}^T\boldsymbol{u}$. Given the QoI (46), $\boldsymbol{q}$ will be as a $n \times 1$ vector,

$$q_i = \begin{cases} 1 & \text{if } x_i \le \beta l \\ 0 & \text{otherwise} \end{cases} \qquad i = 1, \cdots, n. \tag{48}$$

The residual function $\mathcal{R}(\cdot,\cdot)$ of (36) in this example is of the form

$$\mathcal{R}(\boldsymbol{U}_{\text{CG}}, \boldsymbol{z}^n) = \int_0^\tau \eta_t(t)\mathrm{d}t; \qquad \eta_t(t) = \sum_{i=1}^{n} z_i(t) \cdot \rho_i(t)\ \mathrm{d}t. \tag{49}$$

The estimated error in the QoI is then

$$\mathcal{E}_{\text{est.}} = \mathcal{R}(\boldsymbol{U}_{\text{CG}}, \boldsymbol{z}^n) \approx \mathcal{Q}_{\mathbf{r}} - \mathcal{Q}_{\mathbf{R}}, \tag{50}$$

where $Q_{\mathbf{R}} = \int_0^\tau \eta_t(t)\mathrm{d}t$ and then the exact error is

$$\mathcal{E}_{\text{exact}} = \mathcal{E}_{\text{est.}} + \Delta, \tag{51}$$

Δ being the remainder in (34). Since the forms in (43)-(46) are linear in their respective arguments, the exact remainder Δ should be zero, but the error introduced by the numerical integration schemes employed generally leads to an additional numerical error $\Delta_{\Delta t} \neq 0$. We employ a converted Runga-Kutta algorithm here to integrate (39), (41), and (47).

The results of the coarse-grained model for the case of $P = 4$ are presented in Figure 2. Figure 2a shows the coarse-scale displacement $\boldsymbol{U} = \boldsymbol{U}(t)$ at different times, obtained from solution of (41) over the time domain $t \in [0, \tau]$. The local residual of Figure 2b is then computed from (42). Figure 2c shows the solution of $\boldsymbol{z}(t)$ at different times. It observed that the adjoint solution propagates in time in the opposite direction to the primal solution, (47) being integrated backward in time.

It is known that in general, the solution of the base model is not available. However, in order to show the effectiveness of the method presented here, the equations of motion for the united atom system is also solved in this example. Having the solution of the united atom model, $\boldsymbol{u}(t)$, the evolution of the exact ζ and estimated η_t over time is shown in Figure 2d.

Numerical approximations to the exact error are compared with the estimated error for various CG approximation of the united atom model in Figure 3a. The estimated error $\mathcal{R} = \mathcal{R}(\boldsymbol{U}_{CG}(\boldsymbol{\theta}); \boldsymbol{z}^n)$ for various values of P, with $\theta = \alpha k/P$, are indicated in Figure 3b for $\alpha = 1$. The computed estimated error ($\mathcal{E}_{\text{est}} = \mathcal{R}$) versus the parameter α are indicated in Figure 3c.

In general, the solution of the base model is not available, but the effectiveness of the method presented here is determined by comparing the CG solutions with the exact united atom model. The exact ζ and estimated η_t over time are shown in Figure (2d).

Maximum entropy principle for atomic systems

Among features of the AA system that could qualify as quantities of interest, we consider a special measure of uncertainty content embodied in the so-called information entropy. In 1948, Shannon [6] introduced the concept of information entropy as a real-valued function $H(p)$ of probability distributions (densities) p as a logical measure of uncertainty content in p that satisfied four rather straight forward "common-sense" desiderata (see also [12] for full details). For a discrete pdf $p = \{p_1, p_2, \ldots, p_n\}$, the entropy is defined by

$$H(p) = -\sum_{i=1}^{n} p_i \log p_i, \tag{52}$$

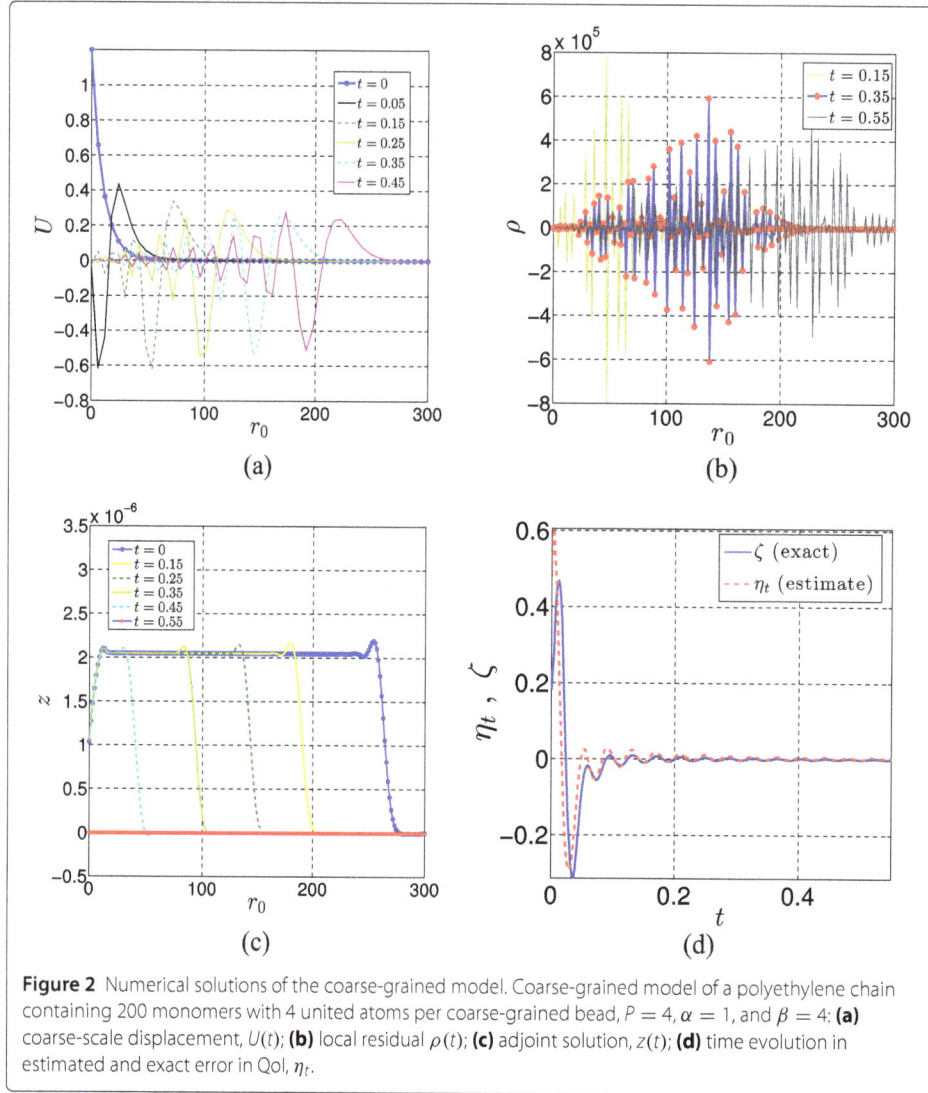

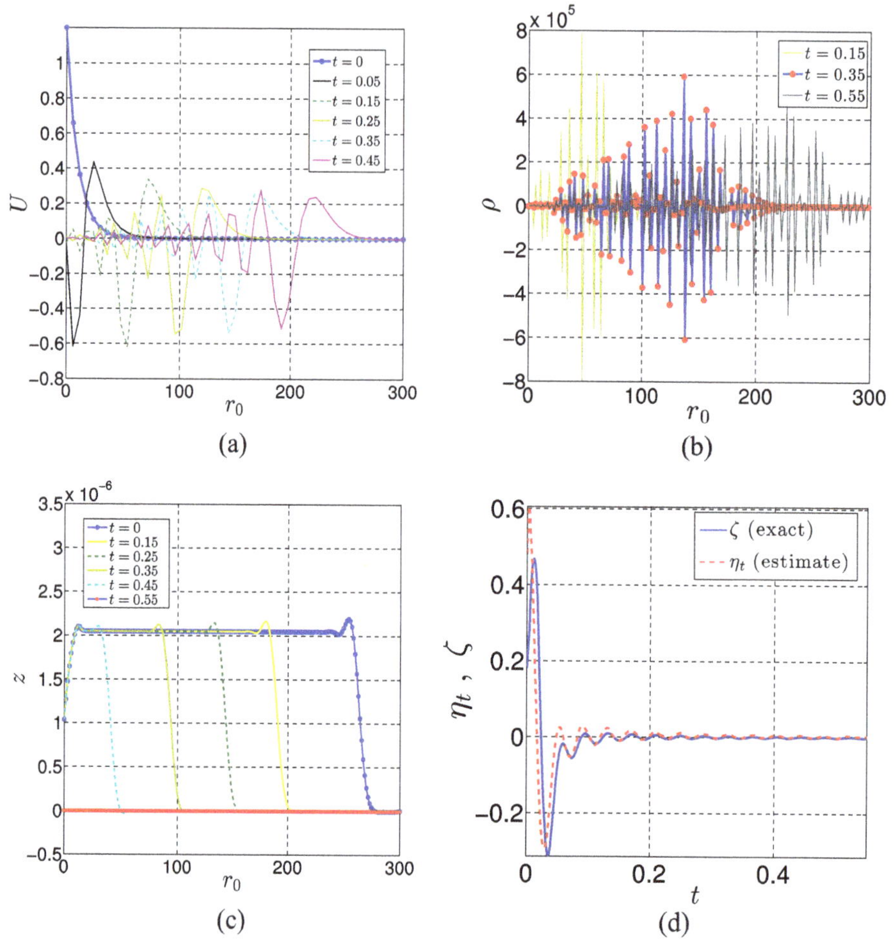

Figure 2 Numerical solutions of the coarse-grained model. Coarse-grained model of a polyethylene chain containing 200 monomers with 4 united atoms per coarse-grained bead, $P = 4$, $\alpha = 1$, and $\beta = 4$: **(a)** coarse-scale displacement, $U(t)$; **(b)** local residual $\rho(t)$; **(c)** adjoint solution, $z(t)$; **(d)** time evolution in estimated and exact error in QoI, η_t.

and for a continuous density, $p \in L^2(\mathbb{R})$, we write

$$H(p) = -\int_{\mathbb{R}} p(y) \log p(y) dy. \tag{53}$$

Given two probability densities p and q, with non-empty support of domains, the relative entropy between p and q is given by the Kullback-Leibler divergence,

$$\begin{aligned} D_{KL}(p\|q) &= \int_{\mathbb{R}} p(y) \log \frac{p(y)}{q(y)} dy \\ &= H(p,q) - H(p), \end{aligned} \tag{54}$$

where $H(p,q)$ $(= -\int_{\mathbb{R}} p \log q dy)$ is the cross entropy and it is understood that $0 \log \frac{0}{0} = 0$ and $0 \log \frac{0}{q} = 0$.

Shannon's principle of maximum entropy asserts that in the set $\mathcal{P}$ of all possible probability distributions relevant to a random field, the correct probability p corresponds to the maximum entropy:

$$H(p) = \max_{q \in \mathcal{P}} H(q). \tag{55}$$

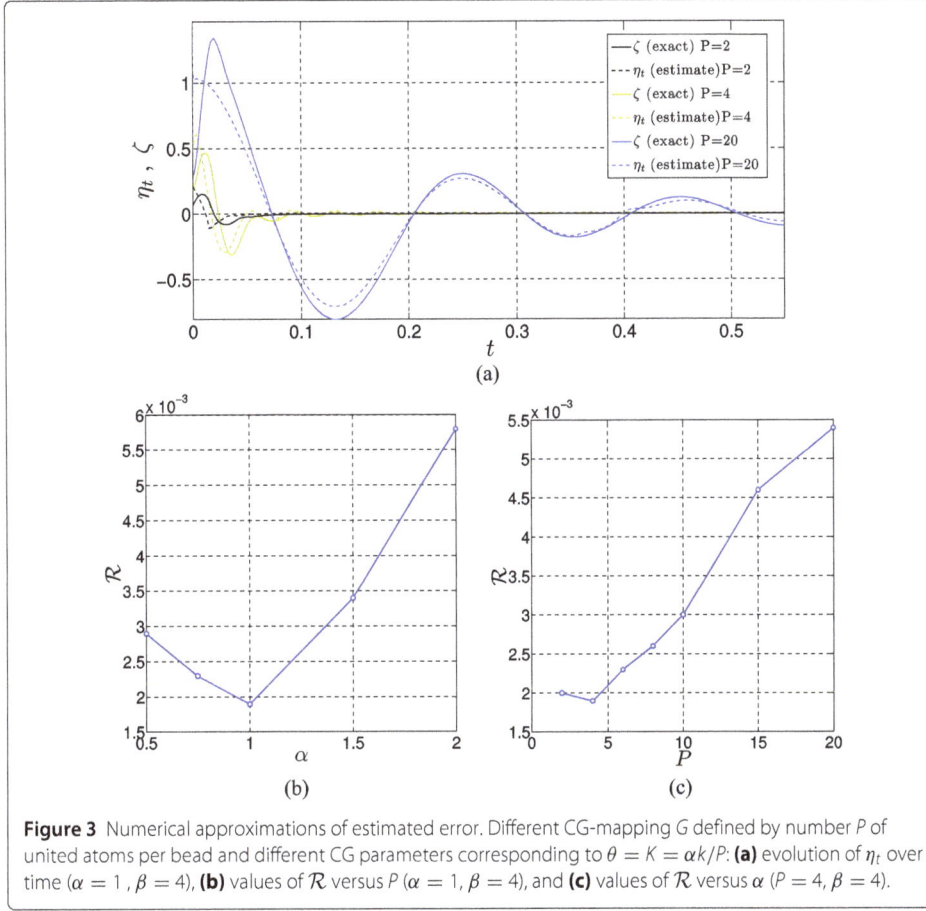

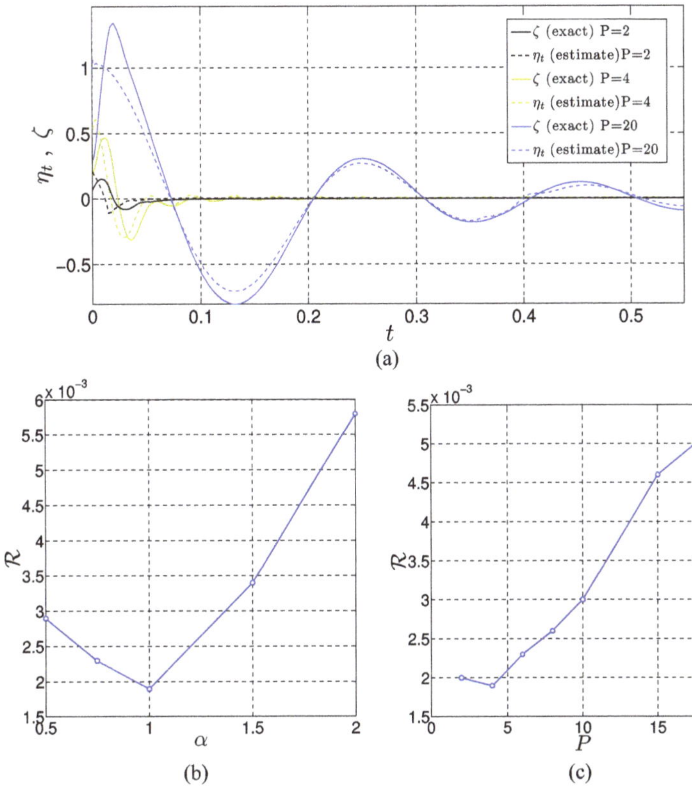

Figure 3 Numerical approximations of estimated error. Different CG-mapping G defined by number P of united atoms per bead and different CG parameters corresponding to $\theta = K = \alpha k/P$: **(a)** evolution of η_t over time ($\alpha = 1$, $\beta = 4$), **(b)** values of $\mathcal{R}$ versus P ($\alpha = 1$, $\beta = 4$), and **(c)** values of $\mathcal{R}$ versus α ($P = 4$, $\beta = 4$).

Errors in information entropy

The connection with the statistical mechanics characterization of the AA and CG models can be established by choosing as a quantity of interest the infinite–time average of the phase function $q(\mathbf{r}^n(t))$ over $[0, \infty]$. For this we invoke the ergodic hypothesis,

$$Q_{\mathbf{r}} = \lim_{\tau \to \infty} \tau^{-1} \int_{t_0}^{t_0+\tau} q(\mathbf{r}(t))\mathrm{d}t \tag{56}$$

$$= \int_{\Gamma} \rho(\mathbf{r}^n) q(\mathbf{r}^n)\mathrm{d}\mathbf{r}^n \tag{57}$$

$$= \langle q \rangle \tag{58}$$

$\rho(\mathbf{r}^n)$ being the distribution function for the ensemble under study and Γ the corresponding phase space subdomain. The corresponding CG approximation is

$$Q_{\mathbf{R}}(\boldsymbol{\theta}) = \lim_{\tau \to \infty} \tau^{-1} \int_{t_0}^{t_0+\tau} q(G(\mathbf{R}^N(t); \boldsymbol{\theta}))\mathrm{d}t \tag{59}$$

$$= \int_{\Gamma} \rho(\mathbf{r}^n) q(G(\mathbf{R}^N; \boldsymbol{\theta}))\mathrm{d}\mathbf{r}^n \tag{60}$$

where the notation $G(\mathbf{R}^N; \boldsymbol{\theta})$ represents the relation (11). Setting

$$q(\mathbf{r}^n) = \log \rho(\mathbf{r}^n), \tag{61}$$

gives immediately

$$Q_{\mathbf{r}} - Q_{\mathbf{R}}(\boldsymbol{\theta}) = D_{KL}\left(\rho(\mathbf{r}^n) \| \rho(G(\mathbf{R}^N; \boldsymbol{\theta})\right), \tag{62}$$

where $D_{KL}(\cdot\|\cdot)$ is the Kullback-Leibler divergence defined in (54). Thus, if $\mathbf{z}^n$ is the equilibrium solution of (25) with $Q'(\mathbf{r}^n; \mathbf{v}^n) = \int_0^\tau \partial_{\alpha_i} q(\mathbf{r}^n(t)) \nu_{\alpha i} dt$, then

$$D_{KL}(\rho(\mathbf{r}^n) \| \rho(G(\mathbf{R}^N; \boldsymbol{\theta}))) \cong \mathcal{R}(\mathbf{R}^N(\boldsymbol{\theta}); \mathbf{z}^n). \tag{63}$$

The specification (60) of the CG approximation (with $\rho(\mathbf{r}^n)$ as opposed to $\rho(\mathbf{R}^N(\boldsymbol{\theta}))$) requires some explanation. In interpreting (60), one assumes the role of an observer who resides in the AA system and, instead of the true phase function $q(\mathbf{r}^n)$, observes a corrupted version for each choice of $\boldsymbol{\theta}$ constrained to reside only in microstates accessible by the CG-model. This is also the interpretation of the residual described in (13) and (15). It is also noted that the estimate (63) is reminiscent of the minimum relative entropy method suggested by Shell [13].

A fundamental question arises at this point: given estimates (36) or (63), is it possible to find a special parameter vector $\boldsymbol{\theta}^*$ that makes the error $\epsilon(\boldsymbol{\theta}^*) = 0$? This question is related to the so-called well-specification or missspecification of the CG model. We believe the answer to this question is generally "no."

Model misspecification and statistical analysis

A fundamental concept in the mathematical statistics literature on parametric models is the notion of a well-specified model, one that has the property that a special parameter vector $\boldsymbol{\theta}^*$ exists that the model $\mathcal{P}$ maps into the *truth*; i.e. the true observational data. If no such parameter exists, the model is said to be *misspecified.*

More generally, we consider a space $\mathcal{Y}$ of physical observables (in our case, the values of appropriate observables sampled from the AA model) and a set $\mathbb{M}(\mathcal{Y})$ of probability measures μ on $\mathcal{Y}$. As always, a target quantity of interest $Q : \mathbb{M} \to \mathbb{R}$ is identified (e.g. $Q(\mu) = \mu[X \geq a]$, X being a random variable and a a threshold value). We seek a particular measure μ^* which yields the "true" value of the quantity of interest $Q(\mu^*)$. We wish to predict $Q(\mu^*)$ using a parametric model $\mathcal{P} : \Theta \to \mathbb{M}(\mathcal{Y})$, Θ being the space of parameters. Again, if a $\boldsymbol{\theta}^* \in \Theta$ exists such that $\mathcal{P}(\boldsymbol{\theta}^*) = \mu^*$, the model is said to be well-specified; otherwise, if $\mu^* \notin \mathcal{P}(\Theta)$, the model is misspecified. See, e.g., Geyer [14], Kleijn and van der Vaart [7], Freedman [15] , Nickl [16]. In our model discussed in Section 'Preliminaries, conventions and notations', we seek a parameter $\boldsymbol{\theta}^*$ of the CG model such that $\varepsilon(\boldsymbol{\theta}^*)$ of (36) is zero, an unlikely possibility for most choices of Q.

To recast the issue of error estimation into a statistical setting, we presume that our goal is to determine (predict) a probability distribution of a random variable, an observable q in the AA system, using a CG model $\mathcal{P}$, given a set $y_1, y_2, \cdots, y_n$ of *iid* (independent, identically-distributed) random variables representing samples $y_i = q(\omega_i)$ ($\omega_i = \mathbf{r}_i^n$ is meant to denote a particular point in phase space). We denote by $\pi(y_i|\boldsymbol{\theta})$ the conditional probability density p of the distance between the random data y_i and the parameter-to-observation map $d_i(\boldsymbol{\theta})$,

$$p(y_i - d_i(\boldsymbol{\theta})) = \pi(y_i|\boldsymbol{\theta}); \quad i = 1, 2, \cdots, n, \tag{64}$$

where $\pi(y_i|\boldsymbol{\theta})$ is the ith component of the likelihood function. The joint density of the data vector $\mathbf{y}^n = y_1, y_2, \cdots, y_n$ is then,

$$\pi_n(y_1, y_2, \cdots, y_n|\boldsymbol{\theta}) = \pi(\mathbf{y}|\boldsymbol{\theta}) = \prod_{i=1}^{n} \pi(y_i, \boldsymbol{\theta}). \tag{65}$$

The log-likelihood function is

$$L_n(\boldsymbol{\theta}) = \log \pi(\mathbf{y}|\boldsymbol{\theta}) = \sum_{i=1}^{n} \log \pi(y_i|\boldsymbol{\theta}). \tag{66}$$

Let $\pi(\boldsymbol{\theta})$ be any prior probability density on the parameters $\boldsymbol{\theta}$ (computed, for instance, using the maximum entropy method of Jaynes [12], as described for CG models in [3]); then the posterior density satisfies,

$$\pi_n(\boldsymbol{\theta}|\mathbf{y}) = \pi(y_1, y_2, \cdots, y_n|\boldsymbol{\theta})\pi(\boldsymbol{\theta})/Z(\boldsymbol{\theta}), \tag{67}$$

where $Z(\boldsymbol{\theta}) = \int_\Theta \pi(\mathbf{y}|\boldsymbol{\theta})\pi(\boldsymbol{\theta})d\boldsymbol{\theta}$ is the model evidence.

The following definitions and theorems follow from these relations:

- The Maximum Likelihood Estimate (MLE) is the parameter $\hat{\boldsymbol{\theta}}^n$ that maximizes $L_n(\boldsymbol{\theta})$:

$$\hat{\boldsymbol{\theta}}^n = \underset{\boldsymbol{\theta}\in\Theta}{\text{argmax}}\ L_n(\boldsymbol{\theta}). \tag{68}$$

- The Maximum A Posterior Estimate (MAP) is the parameter $\tilde{\boldsymbol{\theta}}^n$ that maximizes the posterior pdf:

$$\tilde{\boldsymbol{\theta}}^n = \underset{\boldsymbol{\theta}\in\Theta}{\text{argmax}}\ \pi_n(\boldsymbol{\theta}|\mathbf{y}). \tag{69}$$

- The Bayesian Central Limit Theorem for well-specified models under commonly satisfied smoothness assumptions (also called the Bernstein-von Mises Theorem [7,16,17]) asserts that

$$\pi_n(\boldsymbol{\theta}|\mathbf{y}) \xrightarrow{\mathcal{P}} \mathcal{N}(\boldsymbol{\theta}^*; \mathbf{I}^{-1}(\boldsymbol{\theta}^*)), \tag{70}$$

  where convergence is convergence in probability, $\mathcal{N}(\boldsymbol{\mu}, \boldsymbol{\Sigma})$ denotes a normal distribution with mean $\boldsymbol{\mu}$ and covariance matrix $\boldsymbol{\Sigma}$, $\hat{\boldsymbol{\theta}}$ is the generalized MLE, and $\mathbf{I}(\boldsymbol{\theta})$ is the Fisher information matrix,

$$I_{ij}(\boldsymbol{\theta}) = -\sum_{k=1}^{n} \left[\frac{\partial^2}{\partial\theta_i \partial\theta_j} \log \pi(y_k|\boldsymbol{\theta}) \right]_{\boldsymbol{\theta}=\boldsymbol{\theta}^*} \tag{71}$$

- Given a set of parametric models, $\mathcal{M} = \{\mathcal{P}_1(\boldsymbol{\theta}_1), \mathcal{P}_2(\boldsymbol{\theta}_2), \cdots, \mathcal{P}_m(\boldsymbol{\theta}_m)\}$, the posterior plausibility of model j is defined through the applications of Bayesian arguments by (see [3,18])

$$\rho_j = \pi(\mathcal{P}_j|\mathbf{y}, \mathcal{M}) = \frac{\int_{\Theta_i} \pi(\mathbf{y}|\boldsymbol{\theta}_j, \mathcal{P}_j, \mathcal{M})\pi(\boldsymbol{\theta}_j|\mathcal{P}_j, \mathcal{M})d\boldsymbol{\theta}_j \pi(\mathcal{P}_j|\mathcal{M})}{\pi(\mathbf{y}|\mathcal{M})} \tag{72}$$

  with $\sum_{j=1}^{m} \rho_j = 1$, and the largest $\rho_j \in [0, 1]$ corresponds to the most plausible model for data $\mathbf{y} \in \mathcal{Y}$.

Finally, we come to the case of misspecified parametric models in which $\mu^* \notin \mathcal{P}(\Theta)$; i.e. no parameter $\boldsymbol{\theta}^*$ exists such that the truth $\mu^* = \mathcal{P}(\boldsymbol{\theta}^*)$. This situation, we believe, is by far the most common encountered in the use of CG models.

We remark that in the (rare?) case of a well-specified CG model, for any continuous functional $Q : \Theta \to \mathbb{R}$ and if Θ is compact, if $\boldsymbol{\theta}^*$ is the unique minimizer of Q and if

$$\sup_{\boldsymbol{\theta}\in\Theta} |Q(\boldsymbol{\theta}; y_1, y_2, \cdots, y_n) - Q(\boldsymbol{\theta})| \xrightarrow{\mathcal{P}} 0, \tag{73}$$

as $n \to \infty$, then the sequence

$$\hat{\boldsymbol{\theta}}^n = \operatorname*{argmin}_{\boldsymbol{\theta}\in\Theta} Q_n(\boldsymbol{\theta}; y_1, y_2, \cdots, y_n) \tag{74}$$

converges to $\boldsymbol{\theta}^*$ in probability as $n \to \infty$. This is proved in Nickl [16]. In particular, under mild assumptions on the smoothness of the log-likelihood $L_n(\boldsymbol{\theta})$,

$$Q(\boldsymbol{\theta}^*) - Q(\boldsymbol{\theta}) = -D_{KL}(\pi(\cdot|\boldsymbol{\theta}) \parallel \pi(\cdot|\boldsymbol{\theta}^*)) \tag{75}$$

$D_{KL}(\cdot\|\cdot)$ being the Kullback-Leibler distance defined in (54). By Jensen's inequality (see, e.g. [16]), $Q(\boldsymbol{\theta}^*) \leq Q(\boldsymbol{\theta})\ \forall \boldsymbol{\theta} \in \Theta$; i.e. $\boldsymbol{\theta}^*$ is the minimizer of Q.

The asymptotic results for the finite misspecified case is summed up in the powerful result of Kleijn and van der Vaart [7,19]: let $g(\mathbf{y})$ denote the probability density associated with the true distribution μ^*. Then the posterior density $\pi_n(\boldsymbol{\theta}|\mathbf{y})$ converges in probability to the normal distribution,

$$\pi_n(\boldsymbol{\theta}|\mathbf{y}) \xrightarrow{\mathcal{P}} \mathcal{N}(\boldsymbol{\theta}^\dagger, \mathbf{V}^{-1}(\boldsymbol{\theta}^\dagger)), \tag{76}$$

where

$$\mathbf{V}_{ij}(\boldsymbol{\theta}) = -\mathbb{E}_g\left[\frac{\partial^2}{\partial\theta_i\partial\theta_j} D_{KL}\left(\cdot \mid \pi(\cdot|\boldsymbol{\theta})\right)\right]_{\boldsymbol{\theta}=\boldsymbol{\theta}^\dagger}. \tag{77}$$

Thus, the best approximation to g in $\mathcal{P}(\Theta)$ is the model with the parameter

$$\boldsymbol{\theta}^\dagger = \operatorname*{argmin}_{\boldsymbol{\theta}\in\Theta} D_{KL}\left(g\|\pi(\cdot|\boldsymbol{\theta}, \mathcal{P}, \mathcal{M})\right) \tag{78}$$

$\mathcal{M}$ being a class of parametric models to which $\mathcal{P}$ belongs.

It is easily shown that $\boldsymbol{\theta}^\dagger$ is a maximum likelihood estimate, i.e. it maximizes the expected value of the log-likelihood relative to the true density g:

$$\begin{aligned}
\boldsymbol{\theta}^\dagger &= \operatorname*{argmin}_{\Theta}\left[\int_{\mathcal{Y}^n} g(\mathbf{y}) \log g(\mathbf{y})\, d\mathbf{y} - \int_{\mathcal{Y}^n} g(\mathbf{y}) \log \pi(\mathbf{y}|\boldsymbol{\theta})\, d\mathbf{y}\right] \\
&= \operatorname*{argmin}_{\Theta}\left[-\int_{\mathcal{Y}^n} g(\mathbf{y}) \log \pi(\mathbf{y}|\boldsymbol{\theta})\, d\mathbf{y}\right] \\
&= \operatorname*{argmax}_{\Theta} \int_{\mathcal{Y}^n} g(\mathbf{y}) \log \pi(\mathbf{y}|\boldsymbol{\theta})\, d\mathbf{y} \\
&= \operatorname*{argmax}_{\Theta} \mathbb{E}_g\left[\log \pi(\mathbf{y}|\boldsymbol{\theta})\right],
\end{aligned} \tag{79}$$

where the negative self-entropy $\int g \log g\, d\mathbf{y}$ was eliminated since it does not depend on $\boldsymbol{\theta}$ and therefore does not affect the optimization.

Plausibility-D_{KL} theory

Let us now suppose that we have two misspecified models, $\mathcal{P}_1$ and $\mathcal{P}_2$. We may compare these models in the Bayesian setting through the concept of model plausibility: if $\mathcal{P}_1$ is more plausible than $\mathcal{P}_2$, $\rho_1 > \rho_2$. In the maximum likelihood setting, the model that yields a probability measure closer to μ^* is considered the "better" model. That is, if

$$D_{KL}(g\|\pi(\mathbf{y}|\boldsymbol{\theta}_1^\dagger, \mathcal{P}_1, \mathcal{M})) < D_{KL}(g\|\pi(\mathbf{y}|\boldsymbol{\theta}_2^\dagger, \mathcal{P}_2, \mathcal{M})), \tag{80}$$

it can be said that model $\mathcal{P}_1$ is better than model $\mathcal{P}_2$. The theorems presented here define the relationship between these two notions of model comparison.

However, Bayesian and frequentist methods fundamentally differ in the way they view the model parameters. Bayesian methods consider parameters to be stochastic, characterized by probability density functions, while frequentist approaches seek a single, deterministic parameter value. To bridge this gap in methodology, we note that considering parameters as deterministic vectors, for example $\boldsymbol{\theta}_0$, is akin to assigning them delta functions as their posterior probability distributions, which result from delta prior distributions. In this case, the model evidence is given by

$$\pi(\mathbf{y}|\mathcal{P}_i,\mathcal{M}) = \int_\Theta \pi(\mathbf{y}|\boldsymbol{\theta},\mathcal{P}_i,\mathcal{M})\delta(\boldsymbol{\theta}-\boldsymbol{\theta}_0)\,d\boldsymbol{\theta} = \pi(\mathbf{y}|\boldsymbol{\theta}_0,\mathcal{P}_i,\mathcal{M}). \tag{81}$$

In particular, if we consider the optimal parameter $\boldsymbol{\theta}_i^\dagger$ for model $\mathcal{P}_i$, $\pi(\mathbf{y}|\mathcal{P}_i,\mathcal{M}) = \pi(\mathbf{y}|\boldsymbol{\theta}_i^\dagger,\mathcal{P}_i,\mathcal{M})$. We can take the ratio of posterior model plausibilities,

$$\frac{\rho_1}{\rho_2} = \frac{\pi(\mathbf{y}|\mathcal{P}_1,\mathcal{M})\pi(\mathcal{P}_1|\mathcal{M})}{\pi(\mathbf{y}|\mathcal{P}_2,\mathcal{M})\pi(\mathcal{P}_2|\mathcal{M})} = \frac{\pi(\mathbf{y}|\boldsymbol{\theta}_1^\dagger,\mathcal{P}_1,\mathcal{M})\pi(\mathcal{P}_1|\mathcal{M})}{\pi(\mathbf{y}|\boldsymbol{\theta}_2^\dagger,\mathcal{P}_2,\mathcal{M})\pi(\mathcal{P}_2|\mathcal{M})} = \frac{\pi(\mathbf{y}|\boldsymbol{\theta}_1^\dagger,\mathcal{P}_1,\mathcal{M})}{\pi(\mathbf{y}|\boldsymbol{\theta}_2^\dagger,\mathcal{P}_2,\mathcal{M})}O_{12}, \tag{82}$$

where $O_{12} = \pi(\mathcal{P}_1|\mathcal{M})/\pi(\mathcal{P}_2|\mathcal{M})$ is the prior odds and is often assumed to be one. With these assumptions in force, we present the following theorems.

Theorem 2. *Let (82) hold. If $\mathcal{P}_1$ is more plausible than $\mathcal{P}_2$ and $O_{12} \leq 1$, then (80) holds.*

Proof. If $\mathcal{P}_1$ is more plausible than $\mathcal{P}_2$,

$$1 < \frac{\rho_1}{\rho_2} = \frac{\pi(\mathbf{y}|\boldsymbol{\theta}_1^\dagger,\mathcal{P}_1,\mathcal{M})}{\pi(\mathbf{y}|\boldsymbol{\theta}_2^\dagger,\mathcal{P}_2,\mathcal{M})}O_{12} \leq \frac{\pi(\mathbf{y}|\boldsymbol{\theta}_1^\dagger,\mathcal{P}_1,\mathcal{M})}{\pi(\mathbf{y}|\boldsymbol{\theta}_2^\dagger,\mathcal{P}_2,\mathcal{M})} \tag{83}$$

Equivalently, the reciprocal of the far right-hand side is less than one, so

$$\log\frac{\pi(\mathbf{y}|\boldsymbol{\theta}_2^\dagger,\mathcal{P}_2,\mathcal{M})}{\pi(\mathbf{y}|\boldsymbol{\theta}_1^\dagger,\mathcal{P}_1,\mathcal{M})} < 0. \tag{84}$$

Since $g(\mathbf{y})$ is a probability measure, it is always non-negative. Thus

$$g(\mathbf{y})\log\frac{\pi(\mathbf{y}|\boldsymbol{\theta}_2^\dagger,\mathcal{P}_2,\mathcal{M})}{\pi(\mathbf{y}|\boldsymbol{\theta}_1^\dagger,\mathcal{P}_1,\mathcal{M})} < 0 \Rightarrow \int_{\mathcal{Y}^n} g(\mathbf{y})\log\frac{\pi(\mathbf{y}|\boldsymbol{\theta}_2^\dagger,\mathcal{P}_2,\mathcal{M})}{\pi(\mathbf{y}|\boldsymbol{\theta}_1^\dagger,\mathcal{P}_1,\mathcal{M})}\,d\mathbf{y} < 0. \tag{85}$$

This can be expanded into

$$\int_{\mathcal{Y}^n} g(\mathbf{y})\log\pi(\mathbf{y}|\boldsymbol{\theta}_2^\dagger,\mathcal{P}_2,\mathcal{M})\,d\mathbf{y} - \int_{\mathcal{Y}^n} g(\mathbf{y})\log\pi(\mathbf{y}|\boldsymbol{\theta}_1^\dagger,\mathcal{P}_1,\mathcal{M})\,d\mathbf{y} < 0, \tag{86}$$

which means

$$-\int_{\mathcal{Y}^n} g(\mathbf{y})\log\pi(\mathbf{y}|\boldsymbol{\theta}_1^\dagger,\mathcal{P}_1,\mathcal{M})\,d\mathbf{y} < -\int_{\mathcal{Y}^n} g(\mathbf{y})\log\pi(\mathbf{y}|\boldsymbol{\theta}_2^\dagger,\mathcal{P}_2,\mathcal{M})\,d\mathbf{y}. \tag{87}$$

By adding the quantity $\int_{\mathcal{Y}^n} g\log g\,d\mathbf{y}$ to both sides, the desired result (80) immediately follows. □

This theorem demonstrates that if model $\mathcal{P}_1$ is "better" than model $\mathcal{P}_2$ in the Bayesian sense, it is also a "better" deterministic model in the sense of (80). However, the reverse implication requires much stronger conditions. The assertion (80) can be equivalently written as

$$\int_{\mathcal{Y}^n} g(\mathbf{y}) \log \frac{\pi(\mathbf{y}|\boldsymbol{\theta}_2^\dagger, \mathcal{P}_2, \mathcal{M})}{\pi(\mathbf{y}|\boldsymbol{\theta}_1^\dagger, \mathcal{P}_1, \mathcal{M})} d\mathbf{y} < 0. \tag{88}$$

For this inequality to hold, the relationship

$$\frac{\pi(\mathbf{y}|\boldsymbol{\theta}_2^\dagger, \mathcal{P}_2, \mathcal{M})}{\pi(\mathbf{y}|\boldsymbol{\theta}_1^\dagger, \mathcal{P}_1, \mathcal{M})} < 1 \tag{89}$$

does not necessarily need to be true for *every* point $\mathbf{y} \in \mathcal{Y}^n$.

One perhaps naive way to proceed is to invoke the Mean Value Theorem: if $|\mathcal{Y}^n| < \infty$ and under suitable smoothness conditions, there exists some $\bar{\mathbf{y}} \in \mathcal{Y}^n$ such that

$$\int_{\mathcal{Y}^n} g(\mathbf{y}) \log \frac{\pi(\mathbf{y}|\boldsymbol{\theta}_2^\dagger, \mathcal{P}_2, \mathcal{M})}{\pi(\mathbf{y}|\boldsymbol{\theta}_1^\dagger, \mathcal{P}_1, \mathcal{M})} d\mathbf{y} = |\mathcal{Y}^n| g(\bar{\mathbf{y}}) \log \frac{\pi(\bar{\mathbf{y}}|\boldsymbol{\theta}_2^\dagger, \mathcal{P}_2, \mathcal{M})}{\pi(\bar{\mathbf{y}}|\boldsymbol{\theta}_1^\dagger, \mathcal{P}_1, \mathcal{M})}. \tag{90}$$

Then, combining (88) and (90) yields,

$$|\mathcal{Y}^n| g(\bar{\mathbf{y}}) \log \frac{\pi(\bar{\mathbf{y}}|\boldsymbol{\theta}_2^\dagger, \mathcal{P}_2, \mathcal{M})}{\pi(\bar{\mathbf{y}}|\boldsymbol{\theta}_1^\dagger, \mathcal{P}_1, \mathcal{M})} < 0. \tag{91}$$

Since $|\mathcal{Y}^n| > 0$ and $g(\mathbf{y}) > 0$,

$$\log \frac{\pi(\bar{\mathbf{y}}|\boldsymbol{\theta}_2^\dagger, \mathcal{P}_2, \mathcal{M})}{\pi(\bar{\mathbf{y}}|\boldsymbol{\theta}_1^\dagger, \mathcal{P}_1, \mathcal{M})} < 0 \Rightarrow \frac{\pi(\bar{\mathbf{y}}|\boldsymbol{\theta}_2^\dagger, \mathcal{P}_2, \mathcal{M})}{\pi(\bar{\mathbf{y}}|\boldsymbol{\theta}_1^\dagger, \mathcal{P}_1, \mathcal{M})} < 1 \Rightarrow \frac{\pi(\bar{\mathbf{y}}|\boldsymbol{\theta}_1^\dagger, \mathcal{P}_1, \mathcal{M})}{\pi(\bar{\mathbf{y}}|\boldsymbol{\theta}_2^\dagger, \mathcal{P}_2, \mathcal{M})} > 1. \tag{92}$$

If $O_{12} \geq 1$,

$$\frac{\pi(\bar{\mathbf{y}}|\boldsymbol{\theta}_1^\dagger, \mathcal{P}_1, \mathcal{M})}{\pi(\bar{\mathbf{y}}|\boldsymbol{\theta}_2^\dagger, \mathcal{P}_2, \mathcal{M})} O_{12} > 1 \Rightarrow \frac{\rho_1}{\rho_2} > 1. \tag{93}$$

Thus $\mathcal{P}_1$ is more plausible than $\mathcal{P}_2$ for given data $\bar{\mathbf{y}}$.

In summary, we have:

Theorem 3. *If* $D_{KL}(g\|\pi(\mathbf{y}|\boldsymbol{\theta}_1^\dagger, \mathcal{P}_1, \mathcal{M})) < D_{KL}(g\|\pi(\mathbf{y}|\boldsymbol{\theta}_2^\dagger, \mathcal{P}_2, \mathcal{M}))$ *and if* $|\mathcal{Y}^n| < \infty$ *and if (90) holds, then there exists a* $\bar{\mathbf{y}} \in \mathcal{Y}^n$ *such that* $\mathcal{P}_1$ *is more plausible than* $\mathcal{P}_2$*, given that* $O_{12} \geq 1$.

Conclusions

The formal structure of *a posterior* estimates for errors in quantities of interest in CG approximations of atomistic systems is given by (36) if the CG model is sufficiently close to the AA model in some sense, and this error depends upon the CG model parameter $\boldsymbol{\theta}$. Numerical experiments presented in Section 'Numerical example: estimation of error in CG-approximation of a polyethelyne chain' involving a family of CG models of a polyethylene chain of united atom monomers suggest that these estimates can be very good indications of the error inherent in CG models of observables in the AA system.

In section 'Errors in information entropy' an example of special interest arises in the comparison of the information entropy of AA and CG models. This leads to estimates (62) and (63) involving the Kullback-Leibler divergence, D_{KL}.

When the CG model is misspecified in a statistical sense, which is generally the case, the "D_{KL}-distance" between the AA truth and the best possible approximation of any CG model is defined by choosing $\boldsymbol{\theta} = \boldsymbol{\theta}^\dagger$, the minimizer of the D_{KL} as indicated in (78). Under special assumptions, one can relate the D_{KL} distance to Bayesian posterior model plausibility, as stated in our Theorem 2, which provides sufficient conditions for the most plausible model among a class of models to be in fact closest to the AA model in the D_{KL} sense. The possible role of estimates such as (36), (62), and (63) in model validation should be noted.

For each map G : AA $\rightarrow$ CG of the type defined by (10), a set $\mathcal{M}$ of parametric model classes $\{\mathcal{P}_1(\boldsymbol{\theta}_1), \mathcal{P}_2(\boldsymbol{\theta}_2), \cdots, \mathcal{P}_m(\boldsymbol{\theta}_m)\}$ is defined, each with undetermined and possibly random parameter vectors $\boldsymbol{\theta}_i$. For a calibration scenario, AA calibration data $\mathbf{y}_c = \{y_1, y_2, \cdots, y_n\}$ are sampled, and a series of Bayesian updates is performed using an expanded form of Bayes's rule that recognizes prior choices of the set $\mathcal{M}$ and the class $\mathcal{P}_j$ within $\mathcal{M}$:

$$\pi(\boldsymbol{\theta}_j|\mathbf{y}_c, \mathcal{P}_j, \mathcal{M}) \propto \pi(\mathbf{y}_c|\boldsymbol{\theta}_j, \mathcal{P}_j, \mathcal{M})\pi(\boldsymbol{\theta}_j|\mathcal{P}_j, \mathcal{M}), \qquad 1 \leq j \leq m \tag{94}$$

The marginalization of the right-hand side of this relation is the model evidence, which serves as a likelihood function for a higher level of Bayes's rule. The corresponding posteriors are the model plausibilities of (72). We remark that the notion of model plausibilities is an extension of the idea of Bayes factors prevalent in statistic literature (see e.g. [12] for discussion of the ideas) and was introduced to the best of our knowledge in [18]. The development of algorithms involving Bayesian plausibilities to study model selection in CG models of complex atomic system is discussed in [3,20].

It has been demonstrated, the most plausible model in a set will, under stated assumptions, involve parameters that minimize the $D_{KL}-$distance between the model and the so-called truth parameters. Whether that "best" model is valid for the intended purpose depends on tolerances set of error in key observables, the QoIs of the validation scenario (see [3]).

Appendix

A surrogate pair of equations approximating (30) and (32) may be embodied in the problem of finding the pair $(u_0, z_0) \in \mathcal{V} \times \mathcal{V}$ such that

$$\left.\begin{aligned} \mathcal{B}_0(u_0; v) \quad &= \mathcal{F}_0(v) \qquad \forall v \in \mathcal{V} \\ \mathcal{B}_0'(u_0; z_0, v) &= \mathcal{Q}_0'(u_0; v) \;\; \forall v \in \mathcal{V} \end{aligned}\right\}. \tag{A-1}$$

The remainder Δ in (34) can, in this case, be shown (see [4]) to be given by:

$$\begin{aligned} \Delta = &\frac{1}{2}\int_0^1 \{\mathcal{B}''(u_0 + \theta e_0; e_0, z_0 + \theta\varepsilon_0) \\ &- \mathcal{Q}''(u_0 + \theta e_0; e_0, e_0, \varepsilon_0)\} \, d\theta \\ &+\frac{1}{2}\int_0^1 \{\mathcal{Q}'''(u_0 + \theta e_0; e_0, e_0, \varepsilon_0) - 3\mathcal{B}''(u_0 + \theta e_0; e_0, \varepsilon_0) \\ &-\mathcal{B}'''(u_0 + \theta e_0; e_0, e_0, e_0, z_0 + \theta) \\ &- \mathcal{B}'''(u_0 + \theta e_0; e_0, e_0, e_0, z_0 + \theta\epsilon_0)\theta(1-\theta)d\theta\}, \end{aligned} \tag{A-2}$$

where

$$e_0 = u - u_0 \qquad \text{and} \qquad \varepsilon_0 = z - z_0. \tag{A-3}$$

The theory and estimates reduce to finite element *a posterior* error estimates in the special case in which $u_0 = u_h$ and $z_0 = z_h$ are finite element approximation of solutions (u, z) to partial differential equations (see e.g. [21]).

Competing interests
The authors declare that they have no competing interests.

Authors' contributions
The theory and numerical simulation represent joint work by all authors. All authors read and approved the final manuscript.

Acknowledgements
This material is based upon work supported by the U.S. Department of Energy Office of Science, Office of Advanced Scientific Computing Research, Applied Mathematics program under Award Number DE-5C0009286. The authors benefited from suggestions of Eric Wright, who read an early draft of this work.

References
1. Noid WG (2013) Perspective: coarse-grained models for biomolecular systems. J Chem Phys 139:090901
2. Li Y, Abberton BC, Kroger M, Liu WK (2013) Challenges in multiscale modeling of polymer dynamics. Polymers 5(2):751–832. doi:10.3390/polym5020751
3. Farrell K, Oden JT (2014) Calibration and validation of coarse-grained models of atomic systems: Application to semiconductor manufacturing. Comput Mech 54(1):3–19. doi:10.1007/s00466-014-1028-y
4. Oden JT, Prudhomme S (2002) Estimation of modeling error in computational mechanics. J Comput Phys 182(2):496–515
5. Oden JT, Prudhomme S, Romkes A, Bauman PT (2006) Multiscale modeling of physical phenomena: adaptive control of models. SIAM J Sci Comput 28(6):2359–2389
6. Shannon CE (1948) A mathematical theory of communication. Bell Syst Tech J 27:379–423623656
7. Kleijn BJK, van der Vaart A (2002) The asymptotics of misspecified bayesian statistics. In: Mikosch T, Janzura M (eds). Proceedings of the 24th European Meeting of Statisticians. Prague, Czech Republic
8. Jorgensen WL, Tirado-Rives J (1988) The OPLS potential functions for proteins. Energy minimizations for crystals of cyclic peptides and crambin. J Am Chem Soc 110(6):1657–1666
9. Jorgensen WL, Maxwell DS, Tirado-Rives J (1996) Development and testing of the OPLS all-atom force field on conformational energetics and properties of organic liquids. J Am Chem Soc 118(45):11225–11236
10. Frenkel D, Smit B (2001) Understanding molecular simulation: from Algorithms to applications, Computational science. 2nd edn, Vol. 1. Academic Press, San Diego
11. Haile JM (1997) Molecular dynamics simulation. John Wiley and Sons, NY
12. Jaynes ET (2003) Probability theory: the logic of science. Cambridge University Press, Cambridge
13. Shell MS (2008) The relative entropy is fundamental to multiscale and inverse thermodynamic problems. J Chem Phys 129(14):144108
14. Geyer CJ (2003). 5601 Notes: the sandwich estimator. School of Statistics, University of Minnesota
15. Freedman DA (2006) On the so-called "Huber sandwich estimator" and "robust standard errors". Am Stat 34:299–302
16. Nickl R (2012). sTATISTICAL THEORY. Statistical Laboratory, Department of Pure Mathematics and Mathematical Statistics, University of Cambridge
17. Kleijn BJK (2004). Bayesian asymptotics under misspecification. PhD thesis, Free University Amsterdam
18. Beck JL, Yuan K-V (2004) Model selection using response measurements: Bayesian probabilistic approach. J Eng Mech 130(2):192–203
19. Kleijn BJK (2012) van der Vaart AW (2012) The Bernstein-von-Mises theorem under misspecification. Electronic J Stat 6:354–381. doi:10.1214/12-EJS675
20. Farrell K, Oden JT, Faghihi D (2015) A Bayesian framework for adaptive selection, calibration, and validation of coarse-grained models of atomistic systems. J Comput Phys 295:189–208. ISSN 0021-9991
21. Becker R, Rannacher R (2001) An optimal control approach to a posteriori error estimation in finite element methods. Acta Numerica 10:1–102. doi:10.1017/S0962492901000010

A new mortar formulation for modeling elastomer bedded structures with modal-analysis in 3D

Thomas Horger[1*], Stefan Kollmannsberger[2], Felix Frischmann[2], Ernst Rank[2] and Barbara Wohlmuth[1]

*Correspondence:
horger@ma.tum.de
[1]Institute for Numerical Mathematics, Technische Universität München, Boltzmannstraße 3, 85748 Garching b, München, Germany
Full list of author information is available at the end of the article

Abstract

Background: It is a well-known fact that cross-laminated timber structures are sensitive to rumbling noises. These transmissions are best captured by a fully three-dimensional mathematical model. Since the discretization of such models with hexahedral elements in a conforming manner is highly complex, we chose the mortar method to reduce the algorithmic complexity for the mesh generation. Moreover we consider high-order finite elements in order to deal with the high aspect ratios in three-dimensionally resolved, cross-laminated walls and slabs. The geometric models and material specification was derived from a building information model.

Methods: This paper derives a new mortar formulation designed to replace an explicitely discretized elastomer with a new coupling condition. To this end, tailored Robin conditions are applied at the interface as coupling conditions instead of the more standard continuity constraints. Having demonstrated the suitability of the mortar method for high order finite elements, we proceed with the derivation of the dimensional reduced model with the new coupling condition and to show its stability by numerical experiments. We then test the performance of the new formulation on benchmark examples and demonstrate the engineering relevance for practical applications.

Results: The newly derived mortar formulation performs well. We tested the new formulation on fully three-dimensional examples of engineering relevance discretized by high-order finite elements up to degrees of $p = 10$ and found the reproduction of both eigenvalues and eigenmodes to be accurate. Moreover, the mortar method allows for a significant reduction in the algorithmic complexity of mesh generation while simultaneously reducing the overall computational effort.

Conclusion: The newly derived modified mortar method for replacing an elastomer layer is not only an academically interesting variant but is capable of solving problems of practical importance in modal-analysis of cross-laminated timber structures.

Keywords: Mortar method; Weak coupling; High-order finite elements; Eigenvalue problem; Cross laminated timber structures; Modal-analysis

Background

The main contribution of this paper is a new dimensionally reduced model which captures eigenvalues and eigenmodes of elastomeric coupled domains in timber structures. Dimensionally reduced models are very attractive from the computational point of view. There is no need to mesh the three dimensional but thin subdomain of the elastomer within our approach. However, new challenges arise such as the formulation of a suitable coupling condition and their numerical realization. Here we use a variant of the popular mortar finite element method [1-3]. Mortar methods can be analyzed within the abstract framework of saddlepoint problems and can be regarded as a domain decomposition technique. Firstly, coupled problems are teared, meshed and discretized separately resulting, in general, in non-matching meshes at the interfaces. Secondly, these independent subproblems are interconnected in a weak form by balance equations involving, e.g., the surface traction. Thus, these techniques provide a very flexible and computationally attractive setting to handle numerically coupled multi-physics problems. Mortar methods have been applied successfully in many engineering applications, such as, e.g., contact problems [4-7], dynamic and static structural analysis [8-10], flow problems [11-13] and coupled problems in acoustics [14,15]. Further, the mortar method is used to simulate eigenvalue problems in [16,17]. Most contributions deal only with first or second order approaches. Although the theory of high order mortar methods is well understood [18,19], the implementation of higher order quadrature formulas on cut elements in 3D simulations is technical challenging. Here we apply high order, up to 20 in the polynomial degree, techniques to approximate eigenvalues and eigenmodes in cross laminated timber structures interconnected by thin elastomer structures.

Our motivation to derive such a formulation stems from the need to compute the modal-analysis which is a main part of vibro-acoustical-analysis. In order to control sound transmissions between slabs and walls, these components are often connected by elastomers which we firstly model by using the linear elasticity equation because of the very thin character. Due to the composition of timber constructions consisting of thin, layered and orthotropic material, we aim for a fully three-dimensional resolution of the slabs and walls. For this purpose, we use the p-version of the finite element method, as presented for example in [20]. Moreover, it is well suited for the computation of solid, but thin-walled structures because it is robust in terms of the large aspect ratios which arise naturally in fully three-dimensional models of plates and shells [21]. It also provides better accuracy and convergence properties than low-order finite elements. In addition, the p-version of the FEM has already been shown to lead to excellent results for the analysis of sound transition through timber floors [22].

However, the construction of conforming, three dimensional meshes, that are analysis-suitable, is non-trivial. In this paper, we utilize the mesh generation techniques presented in [23]. A conforming mesh of connected walls and slabs, increases the number of elements significantly, as a local mesh refinement, in only one of the components automatically spreads to the others.

These restrictions motivate the use of mortar methods allowing for an *independent* meshing of the individual building components, as the physically imperative coupling is carried out numerically at a later stage in a weak sense.

The mortar method was first introduced as a method to couple spectral elements with finite elements in [1] where the ansatz space was weakly constrained. The present contribution, however, views the mortar method in the more popular context of enforcing the coupling conditions by means of Lagrange multipliers, as introduced in [2], and thus resulting in a saddle point formulation.

The modeling of elastic interface boundary conditions has been the subject for low orders in [24-26]. Also the modeling of interface elements has been investigated in [27,28], with a spring boundary condition in [29] and with a Robin-type condition in [30]. We built on the work of [31], which demonstrated the excellent applicability of the mortar method for problems in structural mechanics for discretizations of high orders. We extend this concept to elastomeric coupled domains. To this end, we enforce a non-standard Robin type condition at the interface by means of Lagrange multipliers instead of the continuity requirements. Robin type interface conditions have been used to glue nonconforming grids, see, e.g., [32]. The main difference to the current paper is that our coupling condition not only aims to glue two nonconforming grids together, but is also able to replace the whole explicit discretization of an elastomer. Therefore, it goes beyond a simple domain decompositon method, it provides also a dimensionally reduced model.

The contribution at hand is organized as follows: We start by presenting the problem setting in Section 'Problem setting and conforming discretization' and introduce the classical mortar method in Section 'Mortar method'. In Section 'Modeling of the elastomer', we derive our new mortar coupling condition which is able to replace an explicitly discretized elastomer. In Section 'Results and discussion', we present our simulation results. Section 'Results and discussion' compares numerically the standard mortar method with the conforming high order method in the context of eigenvalue problems for a rigidly connected L-shaped wall-slab example. To establish a reference solution, we firstly compute the eigenvalues and eigenfunctions on a wall-slab configuration in a conforming discretization in Section 'Results and discussion'. There we already investigate the effect of connecting walls and slabs with different elastomers on the eigenvalues and the eigenfunctions. We then test the new formulation on the same wall-slab configuration in Section 'The new elastomeric coupled mortar formulation'. Section 'Influence of the elastomer thickness' analyses numerically the influence of the elastomer thickness on the new coupling condition. Furthermore a more complex and application relevant example is presented in Section 'A complex example'. In Section 'Conclusions', we give some conclusion according to the numerical results showing the flexibility and robustness of the new mortar method for practical application.

Methods

Problem setting and conforming discretization

In this section, we provide a dimensionally reduced model, resulting in a modified mortar approach. In contrast to the classical mortar setting, we end up with a non-symmetric saddle-point formulation. The surface traction now enters as a spring into the coupling condition.

The eigenvalue problem under investigation is given by

$$-\operatorname{div}\sigma = \omega\rho u, \tag{1}$$

where ρ is the density and the stress tensor σ and the linearized strain tensor ϵ are defined as

$$\sigma = \mathbb{C}\epsilon(u) \qquad \text{and} \qquad \epsilon(u) = \frac{1}{2}\left(\nabla u + \nabla u^T\right). \tag{2}$$

Furthermore $\mathbb{C}$ denotes the Hookes tensor which is given by the material parameters. We have denoted the eigenvalues by ω to avoid a confusion with the Lagrange multiplier λ of the mortar method considered in Section 'Mortar method'. The most important parameters are summarized in Table 1, for convenience. We will assume that the domain $\Omega \subset \mathbb{R}^3$ is bounded and polyhedral. In addition, we enforce Dirichlet boundary conditions on a non-trivial set Γ_D and homogeneous Neumann boundary conditions on Γ_N, where $\Gamma_N \cap \Gamma_D = \emptyset$ and $\Gamma_D \cup \Gamma_N = \partial\Omega$.

The variational formulation of (1) reads: Find the eigenvalues $\omega \in \mathbb{R}$ and the eigenfunctions $u \in V(\Omega) := \left\{u \mid u \in (H^1(\Omega))^3, u\left(\Gamma_D\right) = 0\right\}$ so that

$$\int_\Omega \mathbb{C}\epsilon(u) : \epsilon(v)\, dx = \omega\rho \int_\Omega u \cdot v\, dx. \tag{3}$$

We discretize Equation (3) using conforming finite elements of high order associated with a hexahedral mesh. As basis functions, we use hierarchical shape functions based on integrated Legendre polynomials [20,33].

Mortar method

A mortar method is typically associated with a domain partitioning. Here, the domain Ω is decomposed into two non-overlapping subdomains Ω_m and Ω_s so that

$$\Omega = \Omega_m \cup \Omega_s, \qquad \Omega_m \cap \Omega_s = \emptyset.$$

The indices m and s correspond to the master and slave side, respectively. In our case, the wall is the slave domain and the slab is the mortar domain. We then define a common interface: $\Gamma := \partial\Omega_s \cap \partial\Omega_m$ and the Lagrange multiplier space by $M := \left(H^{-\frac{1}{2}}(\Gamma)\right)^3$ where $H^{-\frac{1}{2}}(\Gamma)$ is the dual space of $H^{\frac{1}{2}}(\Gamma)$. Here, we assume that $\partial\Gamma \cap \overline{\Gamma}_D = \emptyset$ and thus no modifications on $\partial\Gamma$ have to be taken into account. The primal space is defined by $X := V\left(\Omega_m\right) \times V\left(\Omega_s\right)$, where $V(\Omega_i) := \left\{u \mid u \in \left(H^1(\Omega_i)\right)^3, u\left(\Gamma_D \cap \partial\Omega_i\right) = 0\right\}$ with $i \in \{s, m\}$. We can now define our bilinear forms for the mortar method by

$$\begin{aligned} a(u,v) &:= a_{\Omega_m}(u,v) + a_{\Omega_s}(u,v), \quad b(u,\mu) := < u_s - u_m, \mu >^* \\ d(u,v) &:= d_{\Omega_m}(u,v) + d_{\Omega_s}(u,v), \end{aligned}$$

Table 1 Parameter definitions

Parameter	Definition
$\widetilde{\mu}$	Lamé parameter (shear modulus)
$\widetilde{\lambda}$	Lamé parameter
ρ	Density
ω	Eigenvalue
λ	Lagrange multiplier
ν	Poissons ratio
E	Young moduli

where $< \cdot, \cdot >^*$ denotes the duality pairing of $\left(H^{\frac{1}{2}}(\Gamma)\right)^3$ and $\left(H^{-\frac{1}{2}}(\Gamma)\right)^3$ and $a_{\Omega_i}(\cdot, \cdot)$, $d_{\Omega_i}(\cdot, \cdot)$ are defined by

$$a_{\Omega_i}(u, v) := \int_{\Omega_i} \mathbb{C}\epsilon(u) : \epsilon(v)\, dx, \qquad d_{\Omega_i}(u, v) := \rho \int_{\Omega_i} u \cdot v\, dx. \tag{4}$$

The eigenvalue problem (1) can then be written in the following variational form: Find the eigenvalues $\omega \in \mathbb{R}$, the eigenfunctions $u \in X$ and $\lambda \in M$ so that

$$\begin{aligned} a(u, v) + b(v, \lambda) &= \omega d(u, v), \quad v \in X \\ b(u, \mu) &= 0, \qquad\qquad \mu \in M. \end{aligned} \tag{5}$$

Equation (5) now defines the saddle point problem arising from the mortar method. The Lagrange multiplier λ corresponds to the negative surface traction $-\sigma n$ of Ω_s on the interface Γ, where n is the outward unit normal of Ω_s.

For the discretization of the primal variable of (5), we employ hexahedral finite elements of high order on each subdomain Ω_m, Ω_s. The dual space is discretized by the trace space of the discrete primal space on Ω_s. This choice guarantees inf-sup stability [3,34] and the mortar method for solving (5) can be written as

$$A(u, \lambda; v, \mu) = \omega d(u, v)$$

with $A(u, \lambda; v, \mu) := a(u, v) + b(v, \lambda) + b(u, \mu)$.

The bilinearform $A(\cdot, \cdot; \cdot, \cdot)$ fulfills the conditions of Remark 13.4 in [35], and thus the theory given in Section 8 of [35] ensures convergence of the discrete eigenvalues and eigenfunctions.

Modeling of the elastomer

The modeling of an elastomer for vibration isolation has been the subject in [36,37]. These papers take many mechanical properties like strain and damping directly into account. Alternatively, the modal- and spectral-analysis can be realized by the modal superposition. In this case, the eigenmodes of the undamped system are required, and the damping is only taken into account in a postprocessing step. Thus, we neglect the damping. Moreover, the elastomer is modeled in terms of the linear elasticity equations because it is comparatively thin in one space direction [22]. This section will lay out a new modeling approach using a Robin type condition for the coupling, in order to replace an elastomer. This new coupling condition results in a dimensional reduced model which avoids the meshing of the three dimensional subdomain which corresponds to the elastomer. Our new coupling condition still yields a saddle point problem which fits into the implementational framework of mortar methods.

Modified mortar method using a Robin type condition

The goal of this modeling approach is to replace the elements representing the elastomer between the two components by a Robin type condition. The modeling idea is depicted in Figure 1. Because of the very thin elastomer layer, in our case 1.2[cm], we simplify the transversal shear in the elastomer and neglect the mass of the elastomer. We assume the elastomer to act linearly in z-direction on the solution between the slab and the wall. Without loss of generality, we assume the coordinate system of the mortar interface to

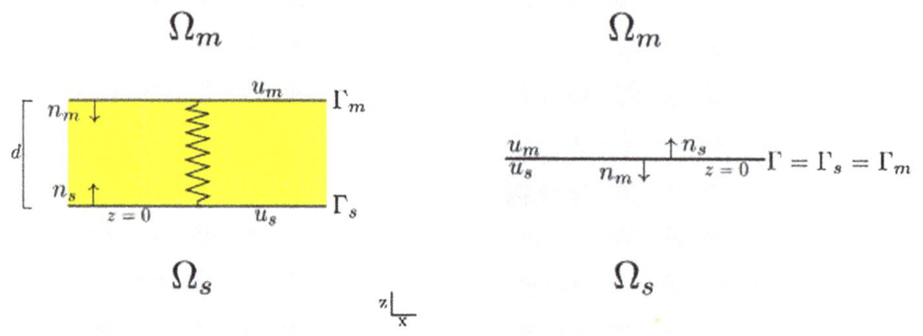

Figure 1 Modeling concept: thin layer (left) and interface formulation (right).

be at $z = 0$. Therefore, we define our displacement in the spirit of a Taylor series with $z \in [0, d]$, where d denotes the thickness of the elastomer as

$$u(x, y, z) = u_s(x, y) + \frac{z}{d}\left(u_m(x, y) - u_s(x, y)\right). \tag{6}$$

With this definition and with $[u] := (u_m(x, y) - u_s(x, y))$, the gradient of the displacement field at $z = 0$ is given by

$$\nabla u|_{z=0} = \begin{pmatrix} (u_s)_{1,x} & (u_s)_{1,y} & \frac{1}{d}[u]_1 \\ (u_s)_{2,x} & (u_s)_{2,y} & \frac{1}{d}[u]_2 \\ (u_s)_{3,x} & (u_s)_{3,y} & \frac{1}{d}[u]_3 \end{pmatrix}.$$

Now the linearized elastic strain reads

$$\epsilon(u|_{z=0}) = \frac{1}{2}\begin{pmatrix} 2(u_s)_{1,x} & (u_s)_{1,y} + (u_s)_{2,x} & \frac{1}{d}[u]_1 + (u_s)_{3,x} \\ (u_s)_{2,x} + (u_s)_{1,y} & 2(u_s)_{2,y} & \frac{1}{d}[u]_2 + (u_s)_{3,y} \\ (u_s)_{3,x} + \frac{1}{d}[u]_1 & (u_s)_{3,y} + \frac{1}{d}[u]_2 & \frac{2}{d}[u]_3 \end{pmatrix}.$$

Further, we assume the following standard linear isotropic stress-strain relationship with the lamé parameters $\widetilde{\mu}$ and $\widetilde{\lambda}$ to hold in the elastomer, i.e.,

$$\sigma = 2\widetilde{\mu}\epsilon + \widetilde{\lambda} tr(\epsilon) Id.$$

As the interface is assumed to be aligned to z=0, the normal vector on Γ directed towards Ω_m is given by $n = [0, 0, 1]^T$. The fluxes are then explicitly given by

$$\begin{aligned} \sigma|_{z=0} n &= \begin{pmatrix} \widetilde{\mu}\left(\frac{1}{d}[u]_1 + (u_s)_{3,x}\right) \\ \widetilde{\mu}\left(\frac{1}{d}[u]_2 + (u_s)_{3,y}\right) \\ 2\widetilde{\mu}\frac{1}{d}[u]_3 \end{pmatrix} + \widetilde{\lambda}\begin{pmatrix} 0 \\ 0 \\ tr(\epsilon) \end{pmatrix} \\ &= \begin{pmatrix} \frac{\widetilde{\mu}}{d}[u]_1 + \widetilde{\mu}(u_s)_{3,x} \\ \frac{\widetilde{\mu}}{d}[u]_2 + \widetilde{\mu}(u_s)_{3,y} \\ \left(\frac{2\widetilde{\mu}}{d} + \frac{\widetilde{\lambda}}{d}\right)[u]_3 + \widetilde{\lambda}\left((u_s)_{1,x} + (u_s)_{2,y}\right) \end{pmatrix}. \end{aligned} \tag{7}$$

Equation (7) is the new coupling condition between displacements and surface traction in the strong form.

Note that in comparison to the standard mortar coupling condition $u_s - u_m = 0$, we additionally obtain dependencies on the derivatives $(u_s)_{3,x}, (u_s)_{3,y}, (u_s)_{1,x}, (u_s)_{2,y}$, and the

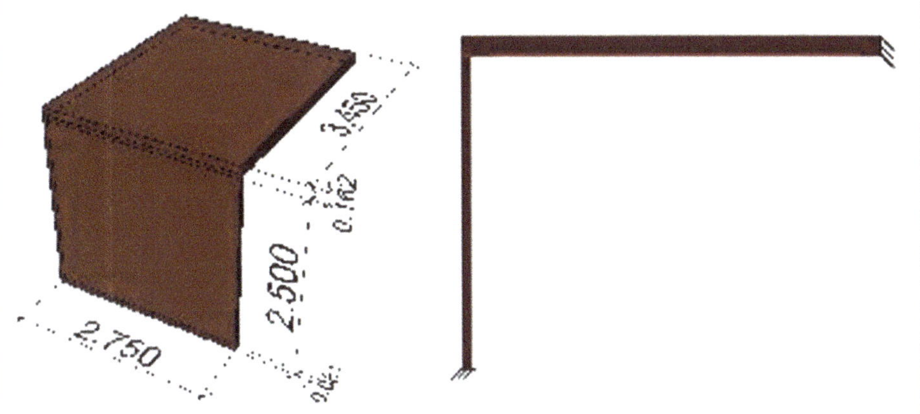

Figure 2 L-shaped connection of a slab to a wall.

surface traction $\lambda_s = -\sigma|_{z=0}n$ interacts as a spring term with the displacement. The corresponding bilinear forms are now given by

$$\widetilde{b}(u,\mu) = \begin{pmatrix} <[u]_1\,,\mu_1>^* + d < (u_s)_{3,x}\,,\mu_1> \\ <[u]_2\,,\mu_2>^* + d < (u_s)_{3,y}\,,\mu_2> \\ <[u]_3\,,\mu_3>^* + \beta\left(< (u_s)_{1,x}\,,\mu_3> + < (u_s)_{2,y}\,,\mu_3>\right) \end{pmatrix},$$

$$c(\lambda,\mu) = <\lambda,\mu>,$$

with $<\cdot,\cdot>$ being the $\left(H^{-\frac{1}{2}}(\Gamma)\right)^3$ scalar product and $\beta = \widetilde{\lambda}d/2\widetilde{\mu} + \widetilde{\lambda}$. We note that this scalar product on the dual space is realized within the discrete setting as a L^2-surface integral. In contrast to the bilinear forms $b(\cdot,\cdot)$ and $\widetilde{b}(\cdot,\cdot)$ no basis functions being defined on different sides of the interface are associated with $c(\cdot,\cdot)$. Both λ and μ are given by the mesh on the slave side, and thus a standard quadrature formula can be easily applied. For given surface tractions λ_i, the force equilibria of both bodies Ω_i reads

$$a_{\Omega_i}(u,v_i) + <v_i,\lambda_i> = \omega d_{\Omega_i}(u,v_i),$$

Neglecting the difference between $-\lambda_s = \sigma|_{z=0}n$ and $\lambda_m = \sigma|_{z=d}n$, setting $\lambda = \lambda_s$ and adding both equations we obtain

$$a(u,v) + b(v,\lambda) = \omega d(u,v). \tag{8}$$

Figure 3 Hexahedral discretization: left conforming, right mortar.

Table 2 Elastomer properties for the simulations

	Timber	Elast 1	Elast 2	Elast 3	Elast 4	Elast 5
Young's-modulus in $[N/m^2]$	$9790 \cdot 10^6$	$1.8 \cdot 10^7$	$8.0 \cdot 10^6$	$3.7 \cdot 10^6$	$1.7 \cdot 10^6$	$8.0 \cdot 10^5$
Poisson v in [−]	0.05	0.4	0.4	0.4	0.4	0.4

The new coupling condition Equation (7) in the weak form and Equation (8) leads to the dimensionally reduced model given by

$$\begin{aligned} a(u,v) \;+\; b(v,\lambda) \;&= \omega d(u,v) \qquad &, v \in X \\ \widetilde{b}(u,\mu) \;-\; \alpha c(\lambda,\mu) \;&= 0 \qquad &, \mu \in M \end{aligned} \tag{9}$$

with the modeling parameter α defined as

$$\alpha := \begin{pmatrix} \frac{d}{\widetilde{\mu}} & 0 & 0 \\ 0 & \frac{d}{\widetilde{\mu}} & 0 \\ 0 & 0 & \frac{d}{2\widetilde{\mu}+\widetilde{\lambda}} \end{pmatrix}.$$

Note that the parameters α and β can be directly computed from the properties of the elastomer. Replacing X by X_h and M by M_h gives the discrete version of Equation (9) yielding approximations ω_h of the eigenvalues.

Results and discussion

Comparison between conforming and mortar discretization

We now consider the example depicted in Figure 2. It resembles a rigidly supported wall connected to a slab on one side and clamped at the other side. The corresponding discretization is depicted in Figure 3. It consists of ten hexahedral elements in the conforming case and eight in the mortar case. At this stage, we do not model an elastomeric coupling yet but assign the material parameters for timber to all hexahedral elements.

Table 3 Comparison of eigenfrequency of the L-shaped wall-slab configuration

	p = 3			p = 7		
EW	Conform	Mortar	%	Conform	Mortar	%
1	50.720	50.852	0.261	50.289	50.298	0.019
2	70.755	72.006	1.768	69.172	69.942	1.113
3	76.534	78.317	2.330	74.456	74.833	0.506
4	90.707	91.976	1.399	87.931	88.491	0.637
5	159.423	168.390	5.624	125.276	126.069	0.632
6	174.712	174.869	0.090	159.311	159.393	0.051
7	179.359	185.147	3.227	172.931	172.966	0.020
	p = 10			p = 15		
EW	Conform	Mortar	%	Conform	Mortar	%
1	50.282	50.288	0.012	50.278	50.281	0.006
2	68.929	69.518	0.854	68.749	69.062	0.455
3	74.304	74.535	0.311	74.220	74.341	0.162
4	87.685	88.101	0.475	87.545	87.768	0.255
5	124.818	125.340	0.418	124.581	124.842	0.210
6	159.264	159.315	0.032	159.237	159.264	0.016
7	172.884	172.911	0.016	172.865	172.882	0.010

No elastomer between the wall and slab. The unit for the eigenfrequencies is given by $[Hz]$.

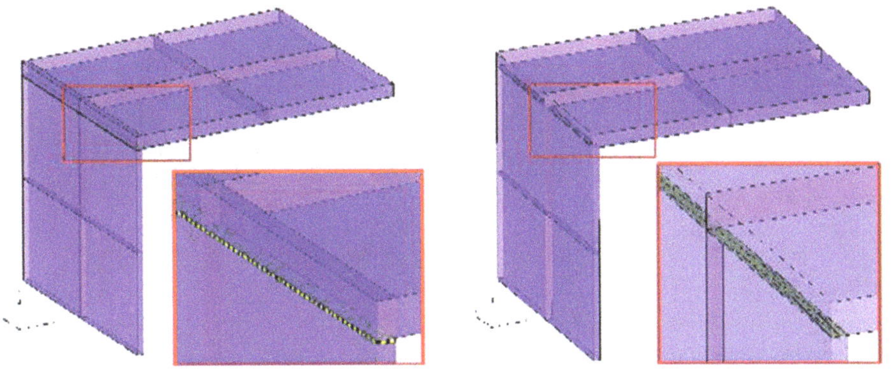

Figure 4 Conforming (left) and mortar discretizations (right) of the structure whose geometry is described in Figure 2. The thin elastomer layer is condensed into the mortar interface.

The material parameters are given in Table 2, where Poisson's ratio and Young's module are denoted by ν and E, respectively.

The eigenvalues for a sequence of p-FEM computations with polynomial degree $p \in \{3, 7, 10, 15\}$ are depicted in Table 3 along with the differences between the conforming and the mortar discretization. The differences decrease for higher orders.

Discrete modeling of the elastomer

In order to obtain a reference solution, the elastomer is discretely represented by a thin layer of hexahedral elements. The discretization is depicted in Figure 4, on the left. The green hexahedral elements in Figure 4 mark the elastomer. The material properties for typical elastomers are given in Table 2, where hard materials are listed first. The specific type of elastomer chosen in a practical application depends on the dead load to be expected on the elastomer. The corresponding eigenvalues of the system wall-elastomer-slab are given in Table 4. Eigenvalues corresponding to a direct connection of wall and slab are provided as well. It is readily apparent that, depending on the mode and the elastomer under contemplation, the eigenvalues of the system with an elastomer layer are about $5 - 35[\%]$ lower than without the elastomer. This is related to the fact that the coupling of the slab to the wall becomes weaker. Figure 5 illustrates the relative decay of each eigenvalue computed from the results depicted in Table 4.

Table 4 Influence of the different elastomers on the eigenfrequencies given in [*Hz*]

EW	No Elast.	Elast. 1	Elast. 2	Elast. 3	Elast. 4	Elast. 5
1	50.282	48.584	47.472	45.933	43.157	38.357
2	68.929	52.437	51.461	50.461	48.676	45.275
3	74.304	64.128	61.773	58.287	52.669	45.588
4	87.685	79.851	77.797	74.245	68.109	59.885
5	124.818	110.669	105.449	98.276	90.290	84.003
6	159.264	149.448	141.577	127.098	106.626	89.151
7	172.884	160.956	154.662	140.762	123.733	105.596
8	178.886	162.633	155.910	145.873	127.320	111.518

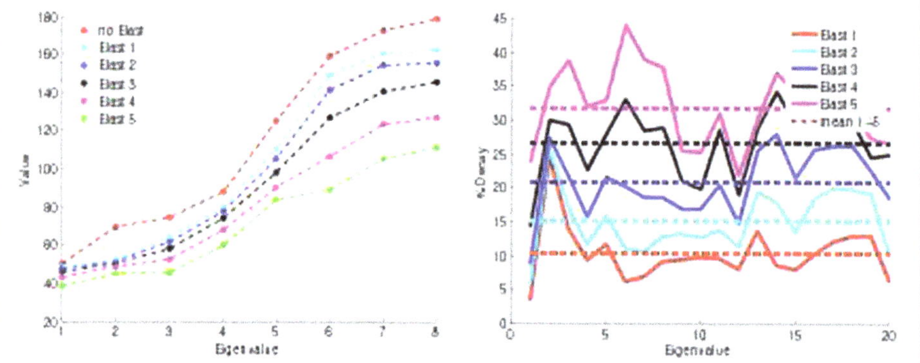

Figure 5 Dependence of the first eight eigenvalues of the elastomer (left) and relative decay with respect to no elastomer for the first 20 eigenvalues (right).

The new elastomeric coupled mortar formulation

We now test the new mortar model given by Equation (9) using the discretization depicted on the right-hand side of Figure 4. The results are compared to the classical, conforming discretization, as depicted on the left-hand side of Figure 4, where the elastomer was modeled explicitly, as described in Section 'Results and discussion'.

Table 5 depicts the first eight eigenvalues obtained by the new mortar model along with the deviation in [%] from the eigenvalues of the explicitly modeled elastic layer whose results were given in Table 4. All computations are carried out with a polynomial degree of $p = 10$. We observe that the new model is able to reproduce the eigenvalues to an accuracy of at least four per cent. Not only the eigenvalues but also the eigenmodes of the two different discretization models have to match closely. Figure 6 shows selected eigenvectors of Elastomer 5. The upper row provides the eigenvectors, as computed by an explicit modeling of the elastomer while the lower row represents the corresponding eigenvectors of the new mortar method. Obviously, different types of modes such as lateral and transversal shear modes as well as pure compression and traction modes are equally well represented. While in the upper row the elastomer undergoes severe deformations, these are approximated by the coupling conditions at the interface between wall and slab in the lower row. Note that the missing elements for the elastomeric layer result from the reduction of the dimension. Moreover, the sequence of the eigenmodes remains the same in both models.

Table 5 Eigenfrequencies given in [*Hz*] for the new modeling approach along with the deviation in percent from the conforming discretization depicted on the left hand side of Figure 4

	Elast 1		Elast 2		Elast 3		Elast 4		Elast 5	
EW	Value	%	Value	%	Value	%	Value	%	Value	%
1	48.664	0.165	47.511	0.082	46.034	0.218	43.206	0.112	38.545	0.490
2	52.678	0.459	51.628	0.325	50.685	0.443	48.997	0.659	45.846	1.262
3	64.315	0.292	61.916	0.231	58.685	0.682	52.891	0.421	46.082	1.083
4	80.059	0.260	78.252	0.585	75.113	1.170	69.159	1.542	61.539	2.763
5	110.912	0.220	105.784	0.317	99.112	0.850	90.606	0.350	84.208	0.243
6	149.371	0.052	141.757	0.127	128.750	1.300	107.303	0.635	89.468	0.355
7	161.365	0.254	155.063	0.259	142.314	1.103	124.127	0.319	109.623	3.814
8	162.967	0.205	157.058	0.737	148.494	1.797	130.530	2.521	111.558	0.036

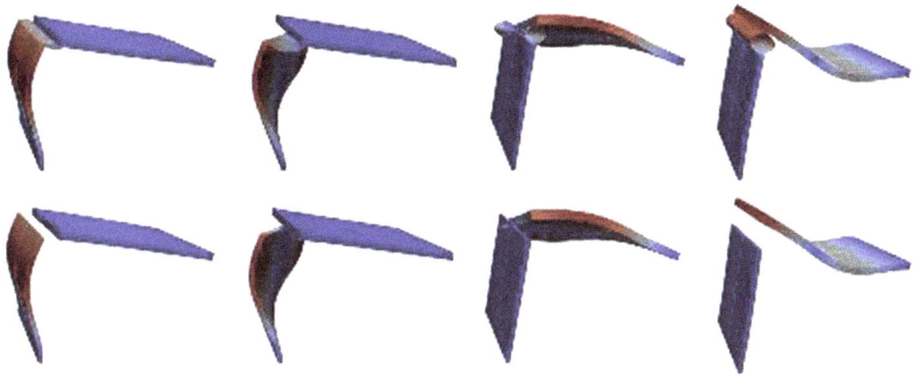

Figure 6 Comparison between eigenmodes 1, 3, 4, 7. Top row: conforming, hexahedral discretization. Bottom row: new mortar method. Note that the greyscale shows the displacement.

We also analyse the eigenmodes by a modal assurance criterion as it is described in [38]. This modal assurance criterion determines the correlation of the eigenmodes. For a good correlation, the resulting matrix should have a diagonal with values greater than 0.9. Values close to 0 mean a poor correlation. The modal assurance criterion matrices show very good results for all practically relevant elastomers investigated in this paper. We show exemplary the modal assurance criterion matrix for elastomer 5 in Table 6. A selection of the eigenmodes are depicted in Figure 6. This confirms the good results for the newly developed coupling condition. Furthermore, it is pointed out that the number of finite elements is reduced by one third even in this small example. Herein, the boundary conforming model required 12 hexahedral elements while only 8 hexahedral elements sufficed for the new mortar approach. However, and most importantly, the mesh generation is simpler using the reduced model in the sense that each wall or slab can now be meshed separately before the discretized components are glued back together.

Influence of the elastomer thickness

A key assumption of the new approach is that the displacement field varies only linearly in the direction perpendicular to the two opposite interface of the elastomer with adjacent structures. In order to investigate the validity of this assumption, we vary the thickness of the elastomer and show its influence on the corresponding eigenvalues.

Table 6 Modal assurance criterion for the modeling of Elastomer 5

		u_1	u_2	u_3	u_4	u_5	u_6	u_7	u_8
	u_1	1.000	0.000	0.004	0.000	0.002	0.000	0.000	0.000
	u_2	0.000	1.000	0.000	0.000	0.000	0.000	0.000	0.000
	u_3	0.003	0.000	0.999	0.000	0.000	0.000	0.003	0.000
$MAC_{E5\_1.2[cm]}=$	u_4	0.000	0.000	0.000	0.999	0.000	0.000	0.000	0.000
	u_5	0.001	0.000	0.000	0.000	1.000	0.000	0.000	0.000
	u_6	0.000	0.000	0.000	0.000	0.000	1.000	0.002	0.000
	u_7	0.000	0.000	0.000	0.000	0.000	0.000	0.998	0.000
	u_8	0.000	0.000	0.000	0.000	0.000	0.000	0.000	1.000

Table 7 Eigenfrequencies given in [*Hz*] for the conform and the new method with the corresponding deviation in [%] for the elastomer thickness 3[cm]

	Elast 1			Elast 3			Elast 5		
EW	Conform method	New method	% diff	Conform method	New method	% diff	Conform method	New method	% diff
1	46.873	46.828	0.096	42.043	42.282	0.568	29.716	30.419	2.367
2	51.223	51.504	0.549	47.783	48.655	1.825	36.608	37.718	3.033
3	60.827	61.081	0.416	50.929	51.658	1.432	37.274	39.155	5.047
4	76.991	78.028	1.347	65.466	68.104	4.031	48.848	51.360	5.143
5	103.857	104.722	0.833	88.359	89.365	1.139	76.357	76.166	0.251
6	138.595	139.275	0.491	101.474	103.915	2.406	79.253	79.477	0.283
7	151.783	152.988	0.794	119.491	121.222	1.449	84.601	88.684	4.826
8	153.680	156.466	1.813	119.571	127.398	6.546	103.182	102.752	0.416

Remark 1. *At this point it is noted that the thickness of the elastomers for typical wall-slab configurations is below* 3[*cm*]. *In practical applications, thicknesses range from* 1[*cm*] *to* 1.5[*cm*].

The reference solution is again computed with the conforming finite element method. We perform our simulation with two further thicknesses of the elastomer. The first thickness is 3[*cm*], which is the maximum relevant thickness and the second thickness is 4[*cm*], which is beyond the typical application range. The results for the investigation for the two elastomer thicknesses are depicted in Table 7 and Table 8 respectively. The tables show the deviation in [%] between the new model and the explicitly modeled elastomer.

While it can be observed that the thicker the elastomer, the bigger the error, the error does not rise above engineering accuracy for practical applications. Table 9 and Table 10 show the model assurance criterion matrices for the eigenmodes for the corresponding 3[*cm*] and 4[*cm*] elastomer simulations.

A complex example

The good performance of the new mortar method carries over to larger examples of engineering relevance even if an orthotropic material law is used for the elastically connected building parts as these changes in the material parameters only have an influence on the

Table 8 Eigenfrequencies given in [*Hz*] for the conform and the new method with the corresponding deviation in [%] for the elastomer thickness 4[cm]

	Elast 1			Elast 3			Elast 5		
EW	Conform method	New method	% diff	Conform method	New method	% diff	Conform method	New method	%diff
1	46.100	46.001	0.215	39.931	40.471	1.354	26.683	27.634	3.562
2	50.780	51.115	0.661	46.091	47.664	3.413	32.869	35.531	8.098
3	59.252	59.736	0.817	47.753	48.835	2.266	35.046	36.449	4.004
4	75.331	76.804	1.956	61.409	65.291	6.323	45.903	48.787	6.283
5	100.661	102.266	1.594	85.356	86.606	1.464	73.943	73.419	0.707
6	132.275	134.570	1.735	93.278	96.202	3.135	78.124	78.372	0.317
7	145.499	148.266	1.902	108.759	115.437	6.140	80.538	84.663	5.122
8	148.833	153.236	2.959	113.584	119.391	5.113	101.426	100.758	0.658

Table 9 Modal assurance criterion for the modeling of Elastomer 5 with thickness 3[cm]

		u_1	u_2	u_3	u_4	u_5	u_6	u_7	u_8
	u_1	1.000	0.000	0.000	0.000	0.000	0.001	0.000	0.000
	u_2	0.000	1.000	0.000	0.000	0.000	0.000	0.000	0.000
	u_3	0.000	0.000	0.996	0.000	0.000	0.000	0.003	0.000
$MAC_{E5\_3cm}=$	u_4	0.000	0.000	0.000	0.999	0.000	0.000	0.000	0.000
	u_5	0.001	0.000	0.000	0.000	0.998	0.000	0.006	0.000
	u_6	0.002	0.000	0.000	0.000	0.002	0.999	0.001	0.000
	u_7	0.000	0.000	0.004	0.000	0.000	0.000	0.994	0.000
	u_8	0.000	0.000	0.000	0.000	0.000	0.000	0.000	1.000

tensor $\mathbb{C}$ for the wood parts in the linear elasticity Equation (3). $\mathbb{C}$ is then given according to [22] by

$$\mathbb{C} = \begin{pmatrix} A_{11} & A_{12} & A_{13} & 0 & 0 & 0 \\ A_{21} & A_{22} & A_{23} & 0 & 0 & 0 \\ A_{31} & A_{32} & A_{33} & 0 & 0 & 0 \\ 0 & 0 & 0 & G_{xy} & 0 & 0 \\ 0 & 0 & 0 & 0 & G_{yz} & 0 \\ 0 & 0 & 0 & 0 & 0 & G_{zx} \end{pmatrix}$$

with

$$A_{11} = \frac{E_x}{D_0}\left(1 - \nu_{yz}^2 \frac{E_z}{E_y}\right), \qquad A_{12} = A_{21} = \frac{E_y}{D_0}\left(\nu_{xy} + \nu_{yz}\nu_{zx}\frac{E_z}{E_y}\right),$$
$$A_{13} = A_{31} = \frac{E_z}{D_0}\left(\nu_{xy}\nu_{yz} + \nu_{zx}\right), \qquad A_{22} = \frac{E_y}{D_0}\left(1 - \nu_{zx}^2 \frac{E_z}{E_x}\right),$$
$$A_{23} = A_{32} = \frac{E_z}{D_0}\left(\nu_{yz} + \nu_{xy}\nu_{zx}\frac{E_y}{E_x}\right), \qquad A_{33} = \frac{E_z}{D_0}\left(1 - \nu_{xy}^2 \frac{E_y}{E_x}\right)$$

where

$$D_0 = 1 - \nu_{yz}^2 \frac{E_z}{E_y} - \nu_{xy}^2 \frac{E_y}{E_x} - \nu_{xy}\nu_{yz}\nu_{zx}\frac{E_z}{E_x} - \nu_{zx}^2 \frac{E_z}{E_x}.$$

Figure 7 depicts a floor plan of a timber building along with a 2 1/2D submodel consisting of three rooms.

This model forms the basis of the three-dimensional computational solid model comprising all conforming hexahedral elements depicted in Figure 8. Note that walls and slabs consist of several layers of wood, as depicted in Figure 9. The thickness of the

Table 10 Modal assurance criterion for the modeling of Elastomer 5 with thickness 4[cm]

		u_1	u_2	u_3	u_4	u_5	u_6	u_7	u_8
	u_1	1.000	0.000	0.000	0.000	0.000	0.001	0.000	0.000
	u_2	0.000	1.000	0.000	0.000	0.000	0.000	0.000	0.000
	u_3	0.000	0.000	0.994	0.000	0.000	0.000	0.003	0.000
$MAC_{E5\_4cm}=$	u_4	0.000	0.000	0.000	0.999	0.000	0.000	0.000	0.000
	u_5	0.002	0.000	0.000	0.000	0.998	0.000	0.008	0.000
	u_6	0.002	0.000	0.000	0.000	0.002	0.999	0.001	0.000
	u_7	0.000	0.000	0.008	0.000	0.000	0.000	0.990	0.000
	u_8	0.000	0.000	0.000	0.000	0.000	0.000	0.000	1.000

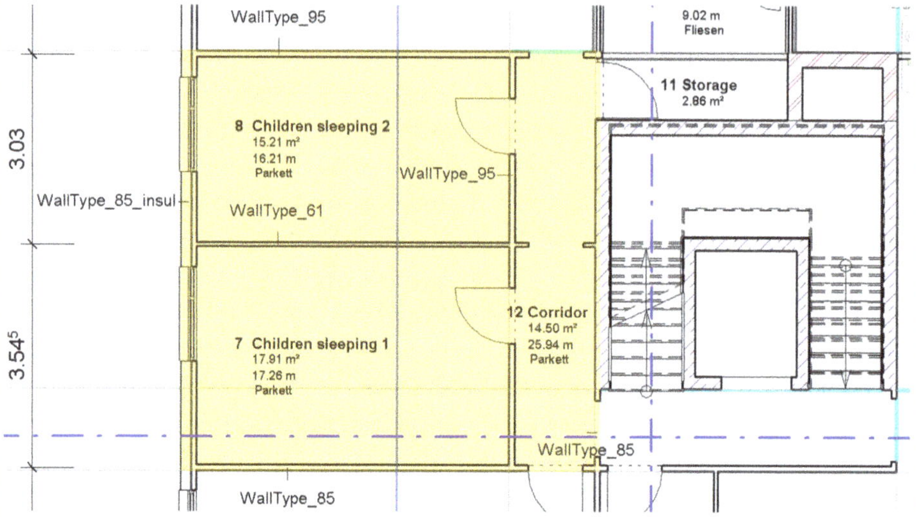

Figure 7 Detail of the ground floor plan considered for acoustical analysis.

layers is given in Table 11. Each layer is explicitly modeled with the characteristic, orthotropic material parameters of timber. We set the Young's moduli in fiber direction $E_x = 137 \times 10^6 \left[N/m^2\right]$, in-plane orthogonal $E_y = 1424 \times 10^6 \left[N/m^2\right]$, and perpendicular to the plane $E_z = 10211 \times 10^6 \left[N/m^2\right]$. The Poisson's ratios are $\nu_{zx} = 0.035$, $\nu_{yz} = 0.045$, $\nu_{xy} = 0.037$. In addition, we apply the shear moduli $G_{zx} = 459 \times 10^6 \left[N/m^2\right]$, $G_{yz} = 102 \times 10^6 \left[N/m^2\right]$ and $G_{xy} = 171 \times 10^6 \left[N/m^2\right]$. The density is assumed to be $\rho = 450 \left[kg/m^3\right]$ for all layers. Although the individual layers have the same material properties, their fiber orientation in plane is orthogonal in adjacent layers in such a way that the orientation is equal on every other layer only. This situation is accurately resolved by the finite element mesh. The elastomer is situated only at the interface where the slab rests on the walls and possesses the isotropic material properties of Elastomer 5, as given in Table 2. The conforming model is depicted in Figure 8. In total, the mesh consists of 7578 hexahedral elements.

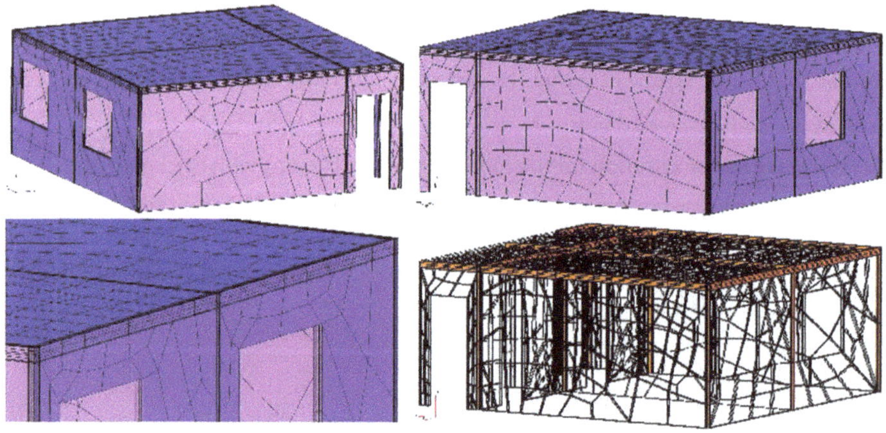

Figure 8 Conforming hexahedral discretization.

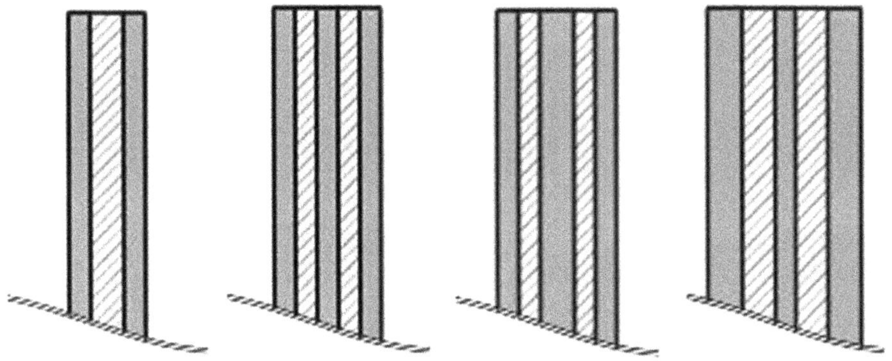

Figure 9 Wall types from left to right: Wall type 61, 85, 95 and slab 125.

The computational mesh for the mortar method is depicted in Figure 10. It consists of only 2475 hexahedral elements. It is evident how the components wall and slab were meshed independently of one another and are non-conforming at their interface. Not only does this greatly simplify the mesh generation process itself, it also avoids the generation of hexahedral elements due to continuity constraints at the interfaces of walls and/or slabs. A further reduction of hexahedral elements is possible by choosing mesh densities individually for all involved components. Also note that local refinements do not branch out to other walls. The elastomer where the slab rests on the walls is now modeled using the new mortar method given in Equation (9).

Table 12 also summarizes the comparison for the first eight eigenvalues and then selected higher eigenvalues up to one hundred. Note that the modeling error introduced by the new mortar approach remains below one per cent for all investigated eigenvalues. The error (in comparison to the conforming method) obtained when using the mortar method with the new coupling condition is comparable to the error obtained when using the standard mortar method. The upper row of Figure 11 depicts selected eigenvectors resulting from the conforming discretization given in Figure 8, while the lower half depicts the corresponding eigenvectors of the mortar discretization of Figure 10. All eigenvectors match within an accuracy which is considered sufficient for engineering applications.

Conclusions

The aim of this contribution was to model the behavior of eigenvalue problems of elastomerically supported, cross-laminated timber structures by means of an extended mortar method.

To this end, we first evaluated the applicability of the mortar method to the p-version of the finite element method of an eigenvalue problem for three-dimensional shell and

Table 11 Layer thicknesses of walls and slab 125

Type	Layering [*mm*]
61	17*–27–17*
85	17*–17–17*–17–17*
95	17*–17–27*–17–17*
125	27*–27–17*–27–27*

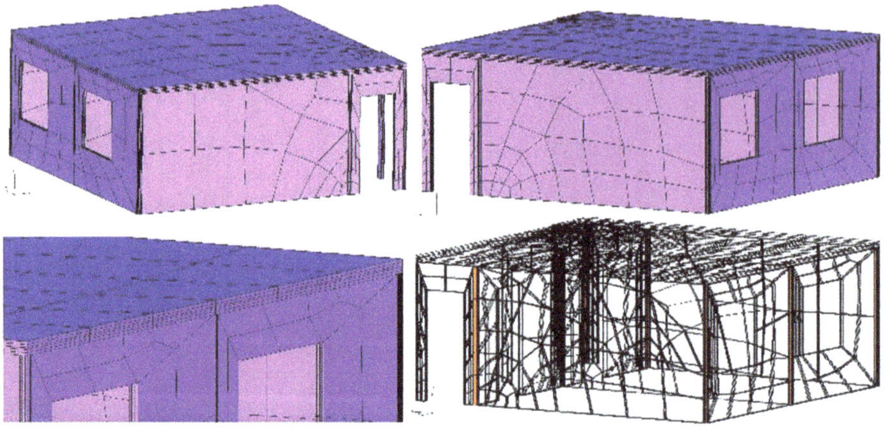

Figure 10 Non-conforming hexahedral discretization.

plate-like structures. The deviation from a conformingly discretized, stiffly coupled wall-slab configuration for higher order p is below 1[%] for all investigated eigenvalues. The eigenmodes likewise provided an excellent match within the required engineering tolerance. Secondly we derived a new coupling condition for the mortar method which is able to replace an explicit resolution of an elastomer. This new transmission condition

Table 12 Computed eigenfrequencies given in [*Hz*] for the building example

EW	Conform no elast	Mortar no elast	% diff	Conform elast	New coupling elast	% diff
1	11.357	11.471	1.007	9.883	9.960	0.779
2	13.738	13.861	0.899	12.439	12.496	0.459
3	14.347	14.425	0.547	13.302	13.346	0.330
4	15.807	15.947	0.884	13.938	14.067	0.926
5	16.988	17.133	0.856	14.980	15.134	1.030
6	21.070	21.329	1.227	19.256	19.398	0.737
7	21.832	21.988	0.715	20.765	20.833	0.325
8	24.038	24.265	0.947	21.072	21.165	0.437
...	...	...	...	...	...	...
20	36.868	37.071	0.552	34.033	34.437	1.189
...	...	...	...	...	...	...
30	48.414	48.769	0.732	43.329	43.850	1.202
...	...	...	...	...	...	...
40	61.815	62.479	1.073	53.238	53.574	0.631
...	...	...	...	...	...	...
50	69.224	70.028	1.162	60.897	61.468	0.938
...	...	...	...	...	...	...
60	77.711	78.402	0.889	66.702	67.982	1.919
...	...	...	...	...	...	...
70	86.225	86.443	0.253	76.123	76.488	0.479
...	...	...	...	...	...	...
80	93.425	93.893	0.501	83.881	84.382	0.597
...	...	...	...	...	...	...
90	101.063	101.673	0.603	88.875	89.558	0.769
...	...	...	...	...	...	...
100	108.871	109.382	0.469	94.814	95.145	0.349

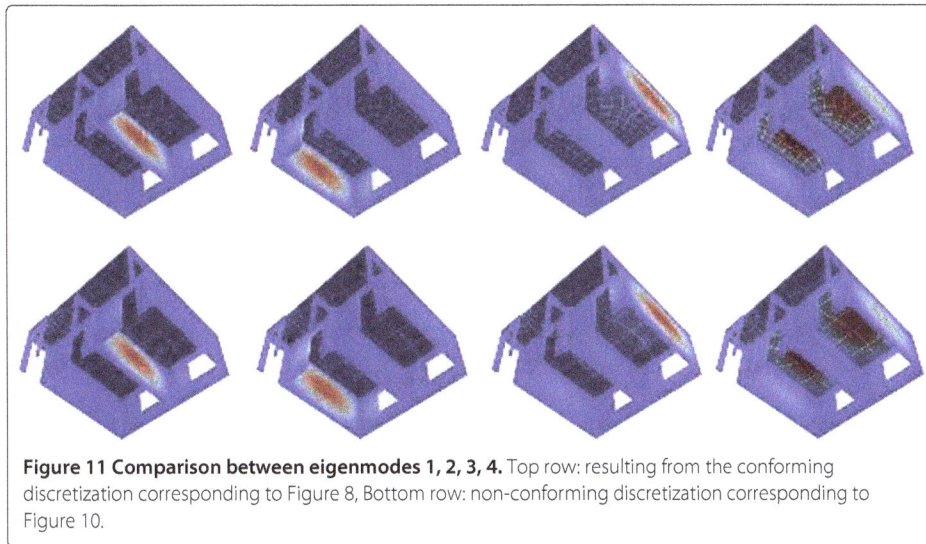

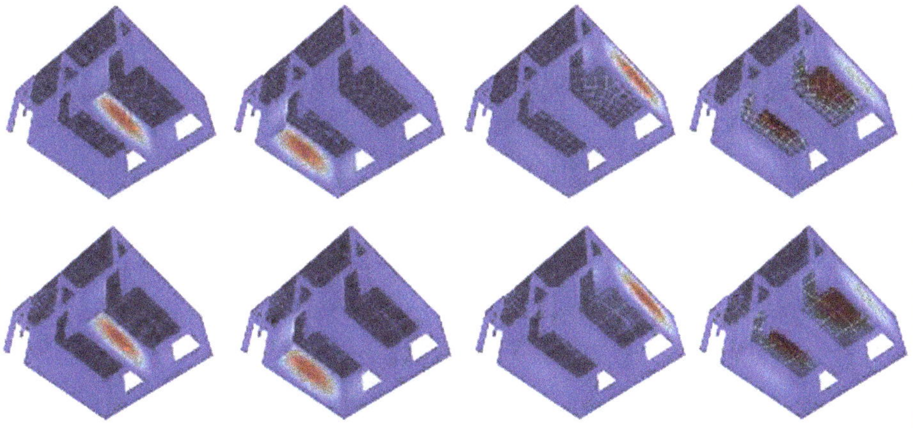

Figure 11 Comparison between eigenmodes 1, 2, 3, 4. Top row: resulting from the conforming discretization corresponding to Figure 8, Bottom row: non-conforming discretization corresponding to Figure 10.

is obtained from a dimension reduction. We then compared the eigenvalues and eigenmodes computed within this approach to the conformingly discretized wall-slab example, the wall now being connected to the slab by means of an elastomer. The resulting lowest eight eigenvalues of the two models correspond within a tolerance of less than 1[%]. This accuracy is sufficient for the application at hand. We finally demonstrate that the good results obtained by the newly developed mortar variant also extend to larger examples of engineering relevance.

The practical motivation of using the new mortar method was to greatly simplify both the engineering modeling effort and the meshing process by dispensing with the need for a conformal element coupling between construction components like slabs and walls. An interesting side effect, however, was that it was also possible to significantly reduce the overall computational workload. The conforming model of the engineering example resulted in 7578 hexahedral elements while only 2475 hexahedral elements were needed for the mortar model. This reduction is due to the facts that: a) a componentwise mesh generation naturally introduces the possibility to choose local mesh densities, b) necessary refinements in other building components do not need to be respected and, accordingly, do not spread across interfaces, and c) at the interfaces of orthogonally coupled, laminated structures it was possible to avoid unnecessary hexahedral elements naturally due to the relaxed topological constraints, and d) it is not required to resolve the geometrically thin elastomer layer.

Competing interests

The authors declare that they have no competing interests.

Authors' contributions

All authors have prepared the manuscript. All authors have read and approved the final manuscript.

Acknowledgements

We would like to gratefully acknowledge the funds provided by the "Deutsche Forschungsgemeinschaft" under the contract/grant numbers: RA-624/21-1 and WO-671/13-1.

Author details

[1]Institute for Numerical Mathematics, Technische Universität München, Boltzmannstraße 3, 85748 Garching b, München, Germany. [2]Chair of Computation in Engineering, Technische Universität München, Arcisstraße 21, 80290 München, Germany.

References

1. Bernardi C, Maday Y, Patera AT (1994) A new non conforming approach to domain decomposition: the mortar element method. In: Brezis H, Lions J-L (eds). Collège de France Seminar. Pitman, Paris, France, XI. pp 13–51
2. Ben Belgacem F (1999) The mortar finite element method with Lagrange multipliers. Numerische Mathematik 84:173–197
3. Wohlmuth BI (2001) Discretization methods and iterative solvers based on domain decomposition. In: Lecture Notes in Computational Science and Engineering. Springer, Berlin, New York Vol. 17
4. Hauret P, Tallec P (2007) A discontinuous stabilized mortar method for general 3d elastic problems. Comput Methods Appl Mech Eng 196(49-52):4881–4900
5. Puso MA (2004) A 3d mortar method for solid mechanics. Int J Numeric Methods Eng 59:315–336
6. Wohlmuth BI, Popp A, Gee MW, Wall WA (2012) An abstract framework for a priori estimates for contact problems in 3D with quadratic finite elements. Comput Mech 49(6):735–747
7. Sitzmann S, Willner K, Wohlmuth BI (2014) A dual lagrange method for contact problems with regularized contact conditions. Int J Numeric Methods Eng 99:221–238
8. Casadei F, Gabellini E, Fotia G, Maggio F, Quarteroni A (2002) A mortar spectral/finite element method for complex 2D and 3D elastodynamic problems. Comput Methods Appl Mech Eng 191(45):5119–5148
9. Flemisch B, Wohlmuth B (2007) Nonconforming methods for nonlinear elasticity problems. In: Widlund O, Keyes D (eds). Domain Decomposition Methods in Science and Engineering XVI. Lect. Notes Comput. Sci. Eng. Springer, Berlin, Germany. pp 65–76
10. Hauret P, Tallec P (2007) Two-scale Dirichlet-Neumann preconditioners for elastic problems with boundary refinements. Comput Methods Appl Mech Eng 196(8):1574–1588
11. Klöppel T, Popp A, Küttler U, Wall W (2011) Fluid-structure interaction for non-conforming interfaces based on a dual mortar formulation. Comput Methods Appl Mech Eng 200(45-46):3111–3126
12. Peszynska M, Wheeler M, Yotov I (2002) Mortar upscaling for multiphase flow in porous media. Comput Geoscience 6(1):73–100
13. Peszynska M (2005) Mortar adaptivity in mixed methods for flow in porous media. Int J Numerical Anal Models 2(3):241–282
14. Triebenbacher S, Kaltenbacher M, Wohlmuth B, Flemisch B (2010) Applications of the mortar finite element method in vibroacoustics and flow induced noise computations. Acta Acustica United Acustica 96(3):536–553
15. Flemisch B, Kaltenbacher M, Wohlmuth B (2006) Elasto-acoustic and acoustic-acoustic coupling on non-matching grids. Int J Numeric Methods Eng 67(13):1791–1810
16. Buffa A, Perugia I, Warburton T (2009) The mortar-discontinuous Galerkin method for the 2D Maxwell eigenproblem. J Sci Comput 40:86–114
17. Lamichhane BP, Wohlmuth B (2005) Mortar finite elements with dual Lagrange multipliers: some application. Lect Notes Comput Sci Eng 40:319–326
18. Seshaiyer P, Suri M (1999) Uniform hp convergence results for the mortar finite element method. Math Comput 69:521–546
19. Belgacem B, Seshaiyer P, Suri M (2000) Optimal convergence rates of hp mortar finite element methods for second-order elliptic problems. ESAIM: Math Model Numerical Anal 34:591–608
20. Szabó BA, Düster A, Rank E (2004) The p-version of the Finite Element Method. In: Stein E, de Borst R, Hughes TJR (eds). Encyclopedia of Computational Mechanics. John Wiley & Sons, Hoboken, New Jersey, USA Vol. 1, Chap. 5. pp 119–139
21. Rank E, Düster A, Nübel V, Preusch K, Bruhns OT (2005) High order finite elements for shells. Comput Methods Appl Mech Eng 194:2494–2512
22. Rabold A (2010) Anwendung der finite element Methode auf die Trittschallberechnung. Dissertation, Chair for Computation in Engineering, Fakultät für Bauingenieur- und Vermessungswesen, Technische Universität München
23. Sorger C, Frischmann F, Kollmannsberger S, Rank E (2014) TUM.GeoFrame: Automated high-order hexahedral mesh generation for shell-like structures. Eng Comput 30(1):41–56
24. Rüberg T, Martin Schanz M (2008) Coupling finite and boundary element methods for static and dynamic elastic problems with non-conforming interfaces. Comput Methods Appl Mech Eng 198:449–458
25. Wang L, Hou S, Shi L (2013) A numerical method for solving 3d elasticity equations with sharp-edged interfaces. Int J Partial Differential Equations 2013:1–10
26. Hauret P, Ortiz M (2006) Bv estimates for mortar methods in linear elasticity. Comput Methods Appl Mech Engrg 195:4783–4793
27. XIE H, LI Z, QIAO Z (2011) A finite element method for elasticity interface problems with locally modified triangulations. Int J Numer Anal Model 8(2):189–200
28. Nwofor TC (2012) Finite element modeling of brick-mortar interface stresses. Int J Civil Environ Eng 12:48–67
29. Boström A, Bövik P, Olsson P (1992) A comparison of exact first order and spring boundary conditions for scattering by thin layers. J Nondestructive Eval 11:175–184
30. Bare DZ, Orlik J, Panasenko G (2014) Asymptotic dimension reduction of a Robin-type elasticity boundary value problem in thin beams. Appl Anal 93(6):1217–1238
31. Wassouf Z (2010) The mortar method for the finite element method of high order. PhD thesis, Technische Universitä, t München
32. Gander MJ, Japhet C, Maday Y, Nataf F (2004) A new cement to glue nonconforming grids with Robin interface conditions : the finite element case. Lect Notes Comput Sci Eng 40:259–266. Springer, Berlin Heidelberg
33. Szabó BA, Babuška I (1991) Finite Element Analysis. John Wiley & Sons, Hoboken, New Jersey, USA

34. Belgacem FB, Maday Y (1997) The mortar element method for three dimensional finite elements. ESAIM: Mathematical Modelling and Numerical Analysis - Modé, lisation Mathématique et Analyse Numérique 31(2):289–302
35. Babuška I, Osborn J (1991) Eigenvalue problems. Handbook Numerical Anal II 2:642–787
36. Banks HT, Lybeck N (1997) Modeling methodology for elastomer dynamics. Syst Control: Foundations Appl 22:37–50
37. Castellani A, Kajon G, Panzeri P, Pezzoli P (1998) Elastomeric materials used for vibration isolation of railway lines. J Eng Mech 124:614–621
38. Pastor M, Binda M, Harcarik T (2012) Modal assurance criterion. In: Procedia enginering MMaMS, 48. pp 543–548

6

Real time simulation for computational surgery: a review

Elías Cueto[1*] and Francisco Chinesta[2,3]

*Correspondence:
ecueto@unizar.es
[1] Aragón Institute of Engineering Research, Universidad de Zaragoza, Zaragoza, Spain
Full list of author information is available at the end of the article

Abstract

In this paper a non-exhaustive review is made on the existing literature for real-time simulation in the field of computational surgery. Many methods have been proposed so far to deal with the very astringent assumption of real-time response in the field, specially for simulators equipped with haptic peripherals. A special emphasis is made on techniques that respond to the so-called second generation of surgery simulators, that able to adequately model the mechanics of the problem. Techniques employing supercomputing facilities, notably those base upon parallel implementations on GPUs will be covered, while special attention will be paid to techniques based upon model order reduction, a promising technique in the field. Finally, some review is made on techniques able to give some insight in the so-called third generation of surgery simulators, i.e., that able to include physiological details into the simulation.

Keywords: Numerical simulation; Computational surgery; Real time; Model order reduction

Introduction

Computational surgery has been defined recently [1] as "the application of mathematics and algorithm design, enabling imaging, robotics, informatics, and simulation technologies, incorporating biological and physical principles, to improve surgery". It has been since the early times of development of computers that their promising use in the field of medicine has been investigated [2,3]. With the irruption of endoscopical and minimally invasive procedures, on one side, and robot-operated surgery, on the other, training of surgeons has becoming a task in which computer simulation has acquired a preeminent role in recent years [4]. In essence, surgeons have begun to access organs during surgery in an indirect way, through a screen, and this needs for a period of intensive training to avoid costly errors.

R. Satava [5] proposed some fifteen years ago a taxonomy of virtual anatomy that he divided into five different generations. The first generation was composed by systems representing accurately the geometry of the organs at a macroscopic level. The second generation would include an accurate description of the physical dynamics of the body. While it is still hard, more than a decade after, to find a real-time surgical simulator that incorporates accurate, state-of-the-art models for soft tissues at a continuum level, this taxonomy included three more generations. From the third to the fifth one, these virtual descriptions of the patient should include, respectively, accurate descriptions of physiology, microscopic anatomy (at a neurovascular level, for instance) and, finally, biochemical systems.

Many computational methods can be found in the biomechanics literature that deal with different levels of description, see for instance [4,6-11] among many others. The last three, however, have not been yet fully incorporated into virtual reality simulators due to the impressive computational cost that they involve, since they incorporate descriptions at the molecular level.

To understand the challenge of including state-of-the-art constitutive equations in the simulator, let us consider the structure of a laparoscopic surgery simulator for training purposes [4]. In essence, a virtual surgery simulator is composed by, see Figure 1, a haptic peripheral that communicates with a simulation engine (see also Figure 2). This simulation core is responsible of detecting contact between the virtual tool and the organ and, by integrating the stresses produced by such a contact, give back to the user hand a realistic sensation of touch. This is precisely one of the bottlenecks of such systems: to provide the user with a realistic sensation of touch some 500 Hz-1 kHz of feedback response (which coincides with the free hand gesture frequency) should be achieved [4]. For a detailed explanation on haptics the reader can consult [12,13]. It can be easily noticed how this feedback rate becomes a true challenge when we deal, as in the case of living soft tissues, with large deformation settings, frequently in a hyperelastic framework [14,15]. This type of simulations take hours in today's computers, while we are now requiring to be solved from 500 to one thousand times per second.

Sometimes real-time constraints are less astringent, and require lower feedback rates but, on the contrary, are expected to provide long-time responses such as, for instance, a simulated outcome of surgery. This is typical in surgery planning simulators [16,17], that should provide a forecast of the results of a given decision taken in the operating room. In augmented reality applications, for instance, only visual continuity is required, and then feedback at some 25 Hz are enough to perceive this sensation of continuous movement [18].

In this review paper a chronological order is pursued, that coincides roughly with the generations of surgical simulators defined in [5]. It can be said that a true example of second-generation simulator, that which includes state-of-the-art constitutive modeling for soft living tissues, has never been reported in the literature. However, some attempts

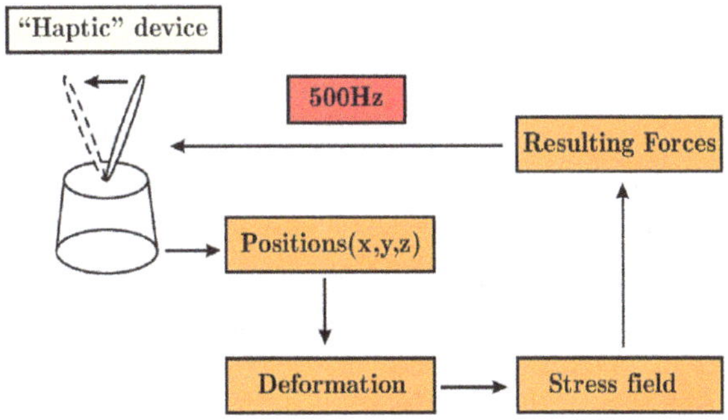

Figure 1 Schematic architecture of a real-time simulator for surgery.

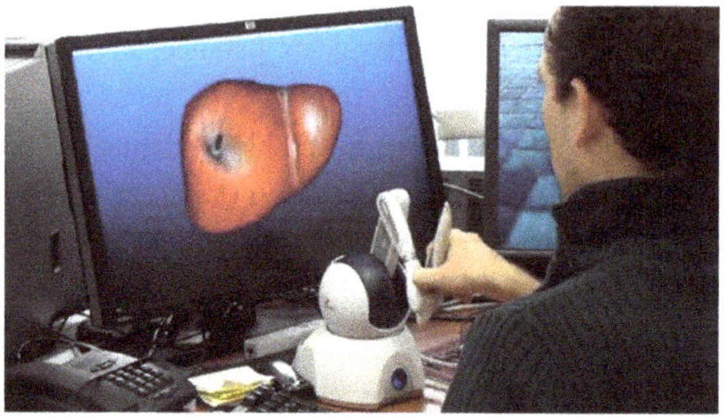

Figure 2 Example of a surgery simulator equipped with a haptic peripheral.

have been made to develop simulators including accurate descriptions of large deformations. These will be reviewed in Section 'Refining the appearance: the second generation of surgery simulation.' In Section 'Surgery simulation by model order reduction' particular emphasis is made on methods based upon some form of model reduction, which has revealed as one promising alternative for this challenging problem. In Section 'Adding physiological details: third to fifth generations of surgery simulators' a review is made of those works aiming at incorporating physiological, microscopic or biochemical details into the simulation.

Review

Early times of surgery simulation: the first generation

The origins of the so-called "deformable models" in the world of computer imaging date back to the mid eighties [19,20]. Basically, they deal with the problem of deforming a solid following the laws of continuum mechanics. However, restrictions posed by real-time requirements did not allow, at that time, to perform these free-form deformations in real-time yet.

In 2005, a review by Monserrat *et al.* [21] divided existing deformable models for surgery simulation into *heuristic, hybrid* and *continuum mechanical* approaches, nearly all belonging by that time to the first generation of models. Under the heuristic category, the most relevant approaches are that of spring-masses. Since real-time finite element descriptions of continuum mechanics were not achievable by that time, most approaches tried to simplify (sometimes over-simplify) the equations of linear elasticity [22] by substituting the continuum by a series of springs and masses obeying Newton's second law. In general, these approaches did not obtain much success, since the deformations they produced were far from realistic.

Much more interesting are the approaches based upon continuum mechanics [23]. By that time, most approaches were still based upon linear elasticity. Particularly noteworthy was the attempt to solve linear elasticity by employing boundary element methods, thus avoiding to mesh the interior of the domains [24]. In [25], the authors, on the contrary, computed the equations of motion for a linearly elastic material using FEM and then by using coarse volumetric meshes to do simulations at interactive rates.

Simulation of surgical cutting deserves a special comment. From the early times of cutting spring-mass systems [26], continuum-based simulations have therefore preferred the use of X-FEM techniques [27,28] or GPU implementations [29].

In general, this first generation of surgery simulation lacks today of interest, since some prototypes are now present that incorporate, at least, non-linear strain measures into the simulation. These are reviewed in the next section.

Refining the appearance: the second generation of surgery simulation

During the fast developments in the first generation of surgery simulators, it was commonly accepted that existing simulators did not provide with a realistic sensation [4]. This had an origin obviously on the poor constitutive models that real-time constraints allowed to solve by that time. At a visual level it was soon realized that linear elasticity gave raise to a poor perception of deformation, due to the apparent gain in volume that it produces when employed to solve large deformations. In Figure 3 it can be noticed how when solving a beam bending problem undergoing large strains, if linear elasticity is employed, a large gain of volume is perceived.

Thus, it is of utmost importance to consider, at least, Saint Venant-Kirchhoff (SVK) models (linear elasticity under large deformation settings) [30,31]. Although SVK models are known to be unstable under compression, for instance [32], they have been considered in a number of recent works as a means to introduce at least a good visual description of large deformations, see [33]. Methods other than finite elements, for instance meshfree methods have also been used for this purpose [34]. Under this same rationale, De and coworkers applied the point collocation-based method of finite spheres (PCMFS) technique to simulate tissue deformations that are geometrically nonlinear [13,35]. They also developed a technique coined as Point-Associated Finite Field (PAFF) [36] that includes a local approximation to the large deformation field near the surgical tool. Even a system based upon the use of neural networks has been presented in [37]. In it, the system is trained with a large set of possible load states in order to achieve real-time performance in the execution loop. A recent survey on the field could also provide the reader with some more insight on the field [38].

An even more surprising approach to the problem of including non-linear strain measures can be found in the work of Delingette [39]. In it, a non-linear (quadratic) spring-mass model is developed so as to be able to reproduce SVK models. The resulting formulation, in addition, eliminates the typical instabilities in compression of this constitutive model.

However, it is known that soft tissues are often composed of collagen fibers and include considerable amount of water [14], and can therefore be modeled as quasi-incompressible. This complex composition cannot be well simulated using only geometrically non-linear models and other constitutive laws should be used for a better approximation of the behavior of these tissues. Among them the so-called Holzapfel model [14] and neo-Hookean strain energy functions [32] are very common, see also [15]. Also in [40] a comparison of six different constitutive models was made for the human liver, concluding that an Ogden viscohyperelastic model provided the best result for the registered deformation of a liver on a CT scanner. Furthermore, in [41] a viscoelastic model under large deformations was considered. It can be considered as a particular implementation of a series of Lagrangian codes coined as Total Lagrangian Explicit Dynamics (TLED), developed for that purpose [42]. Basically, it consists of an implementation of a traditional explicit, lumped-mass, finite element code for structural dynamics, in which the constitutive law is evaluated also explicitly.

Briefly speaking, TLED approaches consider the equations of solid dynamics (including non-linear strain measures) that, after finite element semi-discretisation give rise to a system of equations

$$M\ddot{U} + D\dot{U} + K(U)U = F$$

where M is a lumped mass matrix and $D = \alpha M$ is a dumping matrix, also diagonal. This allows for a computation of the equilibrium equations at an element level, without the

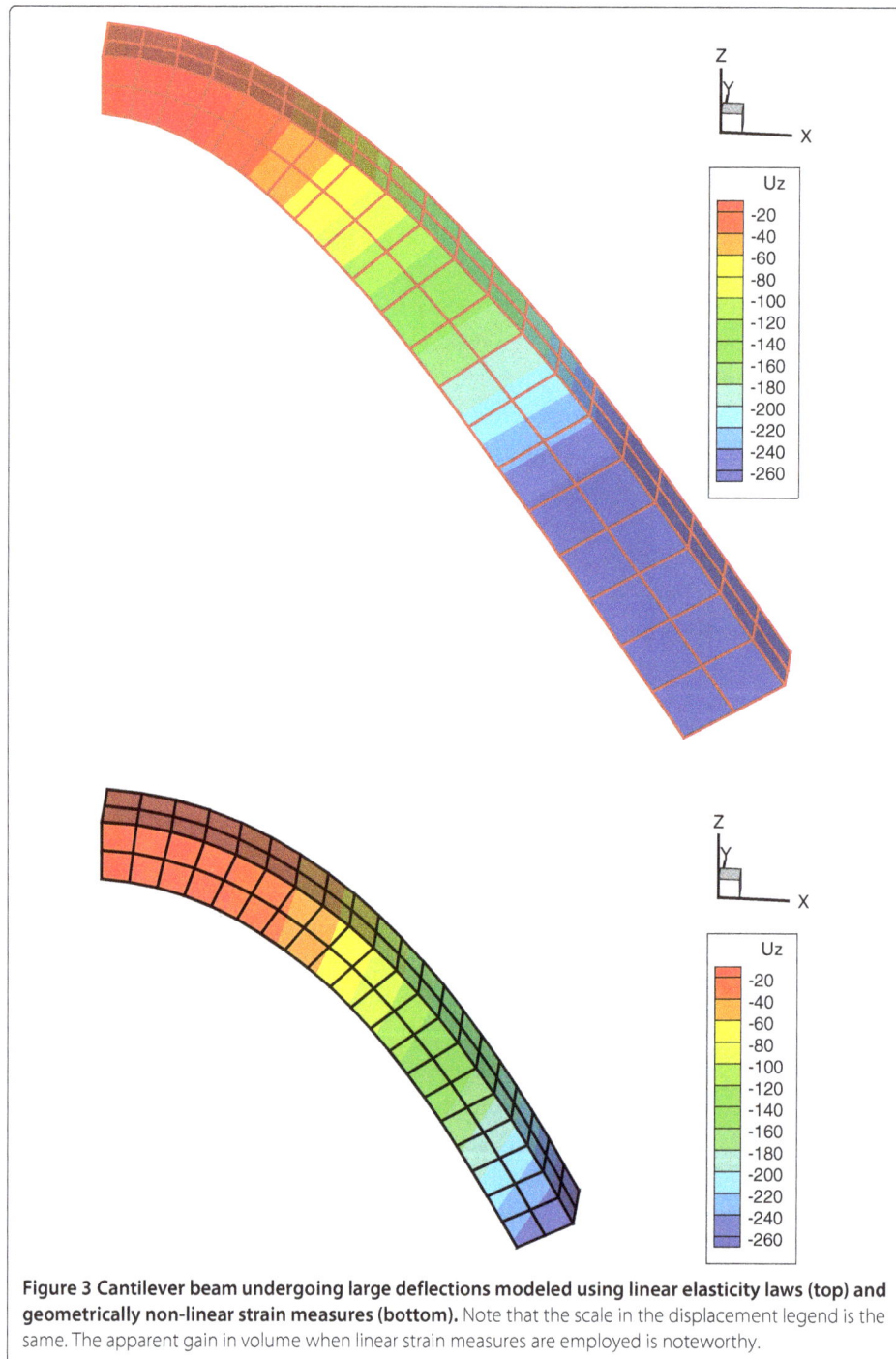

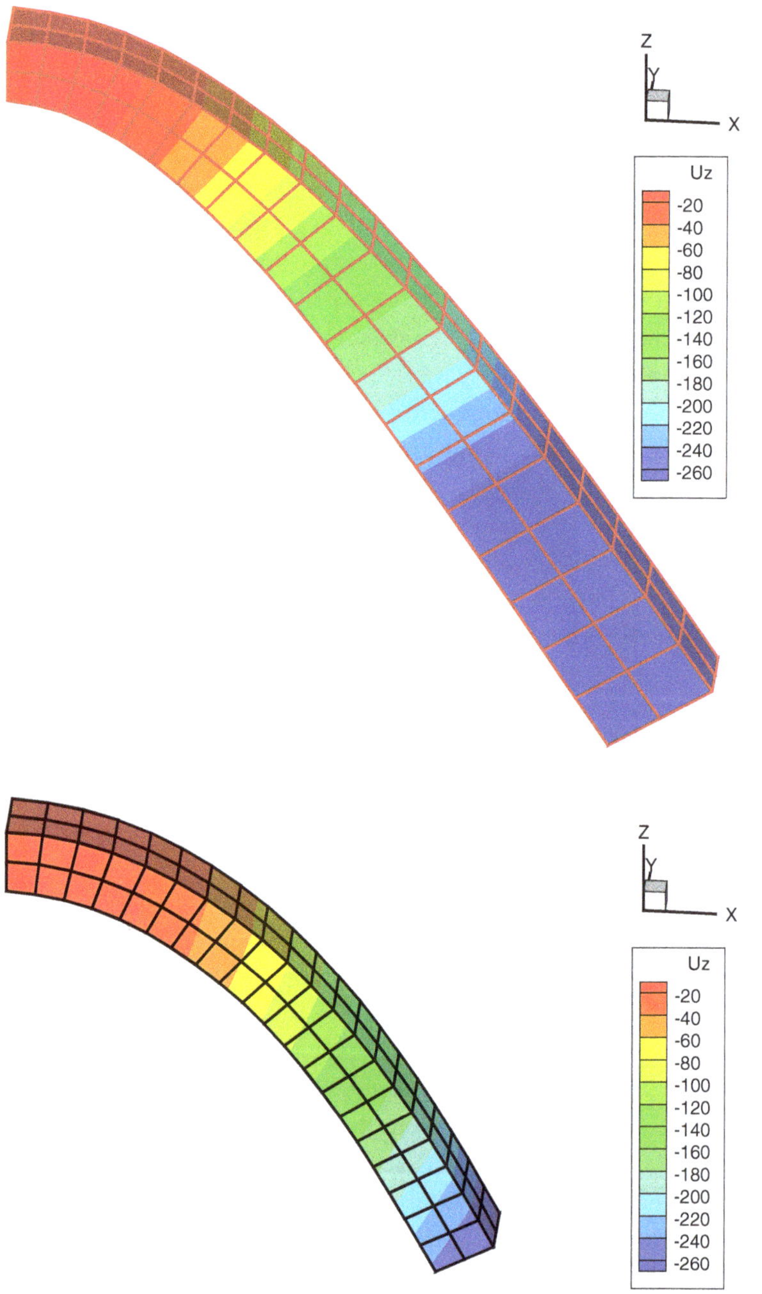

Figure 3 Cantilever beam undergoing large deflections modeled using linear elasticity laws (top) and geometrically non-linear strain measures (bottom). Note that the scale in the displacement legend is the same. The apparent gain in volume when linear strain measures are employed is noteworthy.

need for an assembly procedure. This fact is specially important in a GPU implementation, since each elemental contribution to the equation could be computed separately and independently.

In general, these methods do not work well for large integration times, as is well known for classical explicit finite elements. With these techniques, medium-sized models (some 16000 tetrahedra) can be solved at real-time (500 Hz) rates.

Of particular interest are also methods based on the use of implicit time integration. Particularly, we highlight works based upon asynchronous preconditioning on CPUs and GPUs [43].

Surgery simulation by model order reduction

In view of the above, it seems that the only possibility to date to consider constitutive laws other than SVK was to implement them in Lagrangian finite element codes in which the chosen constitutive law is evaluated explicitly. The main limitation of this approach comes from the fact that the dynamics of the organ is evaluated explicitly during large time intervals (in the order of hours of real user time), therefore lacking of stability in energy conservation, for instance. The possibility of employing implicit energy and momentum conserving integration schemes for the dynamics, along with a consistent evaluation of tangent stiffness matrices in a Newton-Raphson scheme, for instance, seems to be out of reach for the moment.

One appealing possibility to solve these limitations comes from model order reduction (MOR) techniques. Roughly speaking, MOR provides the simulation with an "optimal" (in some sense) set of global basis functions, specific to the considered problem, so as to perform simulations with a minimum of degrees of freedom. MOR techniques have been discovered and re-discovered in almost all scientific disciplines in the last century, giving rise to a plethora of different names for essentially similar techniques [44-48].

The first to apply a technique similar to model order reduction in the field of visualization of solid deformations seems to be James [33]. In this work, SVK models were considered, with an heuristic choice of the global (Ritz) basis functions to be employed.

In general, in real-time simulation, [8], "... We do not care about the time taken for one-time pre-calculation such as setting up equations, inverting matrices, etc.". This means that the possibility exists of employing as much time as needed in off-line computations whose results are to be used on-line. This paves the way for a consistent off-line determination of the optimal set of basis functions for a given problem.

Proper orthogonal decomposition

In Proper Orthogonal Decomposition (POD) techniques, it is assumed that the evolution of the unknown field (here assumed scalar, for simplicity of the exposition), $u(\boldsymbol{x}, t)$ is known. In practical applications (assume that some numerical simulations have been performed off-line), this field is expressed in a discrete form which is known at the nodes of a spatial mesh and for some instants of time t^m. Thus, we consider that $u(\boldsymbol{x}_i, t^m) = u^m(\boldsymbol{x}_i) \equiv u_i^m$ $(t^m = m \times \Delta t)$ are known. We can also write $\boldsymbol{u}^m$ for the vector containing the nodal degrees of freedom at time t^m. The main idea of the Karhunen-Loève (KL) decomposition is to obtain the most typical or characteristic structure $\phi(\boldsymbol{x})$ among these $u^m(\boldsymbol{x})$, $\forall m$. This is equivalent to obtain a function that maximizes the functional α:

$$\alpha = \frac{\sum_{m=1}^{m=M} \left[\sum_{i=1}^{i=N} \phi\left(\boldsymbol{x}_i\right) u^m\left(\boldsymbol{x}_i\right) \right]^2}{\sum_{i=1}^{i=N} \left(\phi\left(\boldsymbol{x}_i\right)\right)^2}, \tag{1}$$

where N represents the number of nodes of the complete model and M the number of computed time steps. The maximization leads to:

$$\sum_{m=1}^{m=M} \left[\left(\sum_{i=1}^{i=N} \tilde{\phi}\left(\boldsymbol{x}_i\right) u^m\left(\boldsymbol{x}_i\right) \right) \left(\sum_{j=1}^{j=N} \phi\left(\boldsymbol{x}_j\right) u^m\left(\boldsymbol{x}_j\right) \right) \right] = \alpha \sum_{i=1}^{i=N} \tilde{\phi}\left(\boldsymbol{x}_i\right) \phi\left(\boldsymbol{x}_i\right); \ \forall \tilde{\phi},$$

which can be rewritten in the form

$$\sum_{i=1}^{i=N} \left\{ \sum_{j=1}^{j=N} \left[\sum_{m=1}^{m=M} u^m\left(\boldsymbol{x}_i\right) u^m\left(\boldsymbol{x}_j\right) \phi\left(\boldsymbol{x}_j\right) \right] \tilde{\phi}\left(\boldsymbol{x}_i\right) \right\} = \alpha \sum_{i=1}^{i=N} \tilde{\phi}\left(\boldsymbol{x}_i\right) \phi\left(\boldsymbol{x}_i\right); \ \forall \tilde{\phi}. \tag{2}$$

Defining the vector ϕ such that its i-th component is $\phi(\boldsymbol{x}_i)$, Eq. (2) takes the following matrix form

$$\tilde{\phi}^T c\, \phi = \alpha \tilde{\phi}^T \phi; \ \forall \tilde{\phi} \Rightarrow c\, \phi = \alpha \phi, \tag{3}$$

where the two-point correlation matrix is given by

$$c_{ij} = \sum_{m=1}^{m=M} u^m\left(\boldsymbol{x}_i\right) u^m\left(\boldsymbol{x}_j\right) \Leftrightarrow \boldsymbol{c} = \sum_{m=1}^{m=M} \boldsymbol{u}^m \left(\boldsymbol{u}^m\right)^T,$$

which is symmetric and positive definite. If we define the matrix $\boldsymbol{Q}$ containing the discrete field history:

$$\boldsymbol{Q} = \begin{pmatrix} u_1^1 & u_1^2 & \cdots & u_1^M \\ u_2^1 & u_2^2 & \cdots & u_2^M \\ \vdots & \vdots & \ddots & \vdots \\ u_N^1 & u_N^2 & \cdots & u_N^M \end{pmatrix}, \tag{4}$$

then it is easy to verify that the matrix $\boldsymbol{c}$ in Eq. (3) results in

$$\boldsymbol{c} = \boldsymbol{Q}\,\boldsymbol{Q}^T.$$

Assume now that some direct simulations have been carried out (off-line, this process may take as much time as needed). From them, we can determine u_i^m, $\forall i \in [1, \cdots, N]$ and $\forall m \in [1, \cdots, M]$, and from these solutions the n eigenvectors related to the n-highest eigenvalues that are expected to contain the most important information (at least, statistically speaking) about the problem solution. For this purpose the eigenvalue problem defined by Eq. (3) is solved and all the eigenvalues ϕ_k belonging to some pre-defined interval are stored. In practice n is much lower than N, and this constitutes the main advantage of the technique.

POD techniques use therefore these n eigenfunctions ϕ_k as Ritz-like (thus, globally supported) basis functions for approximating the solution of a problem *slightly* different to the one that has served to define u_i^m. For this purpose we need to define the matrix $\boldsymbol{A} = [\phi_1 \cdots \phi_n]$

$$\boldsymbol{A} = \begin{pmatrix} \phi_1\left(\boldsymbol{x}_1\right) & \phi_2\left(\boldsymbol{x}_1\right) & \cdots & \phi_n\left(\boldsymbol{x}_1\right) \\ \phi_1\left(\boldsymbol{x}_2\right) & \phi_2\left(\boldsymbol{x}_2\right) & \cdots & \phi_n\left(\boldsymbol{x}_2\right) \\ \vdots & \vdots & \ddots & \vdots \\ \phi_1\left(\boldsymbol{x}_N\right) & \phi_2\left(\boldsymbol{x}_N\right) & \cdots & \phi_n\left(\boldsymbol{x}_N\right) \end{pmatrix}.$$

Consider now the linear system of equations coming from the discretization of a generic problem, in the form:

$$K\,u^{m} \;=\; F^{m-1}, \tag{5}$$

where the superscript refers to the time step. By assuming that the unknown vector contains the nodal degrees of freedom, it can be expressed as:

$$u^{m} = \sum_{i=1}^{i=n} \zeta_i^{m}\,\phi_i \;=\; A\,\zeta^{m},$$

from which Eq. (5) results

$$K\,u^{m} \;=\; F^{m-1} \Rightarrow K\,A\,\zeta^{m} \;=\; H^{m-1},$$

and by multiplying both terms by A^T

$$A^T K\,A\,\zeta^{m} \;=\; A^T F^{m-1},$$

is obtained, which proves that the final system of equations is of low order, i.e. the dimension of $A^T G\,A$ is $n \times n$, with $n \ll N$.

It is well known that reduced-order models allow for larger time steps in structural dynamics simulations [49]. This is the approach followed in [50,51], where POD reduced models were employed trying to enlarge admissible time steps in explicit dynamics solved by finite element methods. In [52] a local/global approach to the problem of model reduction, where a reduced model is employed far from the "conflict" zones, i.e., those with crack sharp gradients, and a more classical finite element implementation is done where the crack influence is bigger. A similar approach (a coupling between POD and FEM) was previously accomplished in [53]. The reader interested in dynamical aspects of model order reduction could also consult an excellent paper by Matthies [54].

In sum, from the results of complete-order models solved previously and off-line, it is possible to obtain a basis which is optimal (in the sense that it minimizes the functional in Eq. (1)) to be used in problems *slightly* different to the original ones. This is essentially the approach followed in [55-57]. As can be noticed from Figure 4, where an example of palpation of a human cornea is shown, there are clear differences between both results, even if the reduced model, that employs only six degrees of freedom, could be refined.

The origin of these discrepancies comes from the fact that a non-linear, hyperelastic, constitutive equation was used for the corneal tissue [15], while the strong requirements

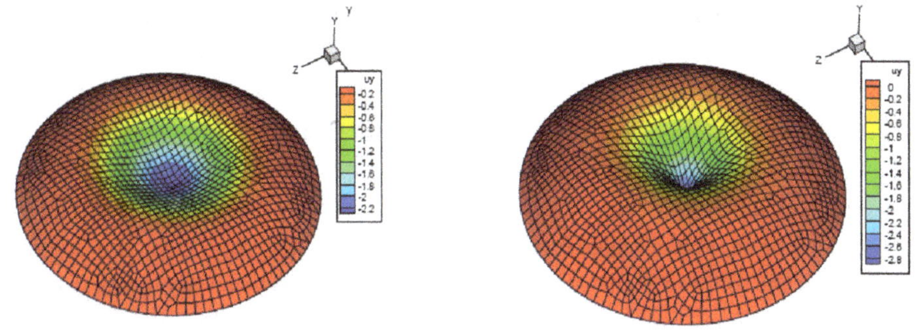

Figure 4 Vertical displacement field for a particular load position in the palpation of a cornea. Complete model (left) vs. reduced model (six d.o.f., right).

of real-time feedback rates prevent from computing a consistently linearized tangent stiffness matrix. Thus, the simulation is actually linear, and the tangent stiffness matrix at the origin is actually being employed at any time in the simulation.

Other source of problems, which is classical in POD-based model order reduction, is the choice of the complete models from which to extract the snapshots matrix, Eq. (4). More precisely, it is of utmost importance how to interpolate between reduced models when they are obtained after complete models reproducing conditions (load positions, for instance), which are not exactly reproduced in the on-line phase of the method. In [58] it was pointed out that the set of empirical eigenfunctions given by Eq. (3) for a given model forms the so-called Grassman manifold $\mathcal{G}(n, N)$. Therefore, in order to interpolate the set of basis functions, that consequently do not form a vector space, we must move to the tangent plane at a point of the manifold, which is a "flat" space, interpolate there, and project back to the manifold.

A less rigorous, but much simpler, method to interpolate among previously computed reduced models was established in [59] and coined as POD with interpolation (PODI). Although in the standard PODI technique the POD procedure is applied to the complete set of snapshots (for different load positions, say) of the system to obtain an orthonormal basis $\boldsymbol{A} = [\phi_1 \cdots \phi_n]$, in [55] POD was just applied to each complete model (i.e., to each load position). Thus, an orthonormal basis for each system's parameter value is obtained. Basis were then interpolated for intermediate positions of the load. Although it is clear that the interpolation of orthogonal sets of functions does not yield, in general, new orthogonal basis functions, the technique works well if the "distance" of reduced models in the Grassman manifold is not too large (an aspect always very difficult to determine and covered very elegantly in [60]).

Further developments in the real-time simulation based upon POD techniques include the employ of asymptotic numerical methods (ANM) [61-64] in combination with POD [65-67]. By means of an asymptotic expansion of the variables of interest, POD-ANM techniques allow to solve non-linear problems without the burden associated to the update and inversion of tangent stiffness matrices. Instead, a series of linear problems, all with the same tangent operator, are solved, one for each level of expansion. These methods will be detailed in Section 'Methods based on proper generalized decomposition' below.

Another way to deal with non-linear problems in a reduced order context is to consider the Empirical Interpolation Method (EIM) [68] or its discrete counterpart, the DEIM [69].

Some problems, in general, presents special difficulties for their reduction. These are problems in which advancing fronts are present, in general, that lead to hardly separable models that need for many modes (or degrees of freedom). A very interesting problem that poses difficulties in this sense for simulation in general, but very specially for real-time simulation, is that of electrophysiology (see [70,71] and references therein, for instance). The main cause of these difficulties comes from the fact that electrophysiology is governed by anisotropic reaction-diffusion equations with a very rapidly varying reaction term. Accurate schemes need for very fine meshes and extremely small time step size. The fact that the solution of the problem is roughly composed by an advancing front (very similar to a soliton wave) makes this problem specially difficult for model reduction methods, that do not capture properly the solution. In this framework, the work by Gerbeau and coworkers [72-74] is noteworthy. They use mainly POD techniques to that end, although very recently they employed the concept of Lax pairs to avoid the mentioned problems, characteristic of electrophysiology [74].

Methods based on proper generalized decomposition

Model reduction methods based on POD or related techniques are known as "a posteriori" methods, since they need for a number of complete model simulations in order to obtain the snapshots. It is possible, however, to develop "a priori" methods in which there is no need for a complete-model solution. Proper Generalized Decomposition (PGD) methods belong to this class of methods. Although initially designed for a space-time decomposition of the solution within the LATIN method [75] or, independently, for high dimensional problems in statistical mechanics [76,77], PGD methods provide a very appealing alternative to solve in real time complex computational problems.

In fact, PGD approaches to the problem at hand constitute a change of paradigm, since they are based upon the off-line computation of a general (high dimensional) solution to the problem, that is after evaluated (rather than simulated) in real time in the on-line phase [78,79]. So to speak, it generates a sort of *computational vademecum* [80] that provides the solution for any possible situation during the on-line phase of the simulation. Of course, obtaining such a general solution and storing it efficiently is not an easy task. PGD methods tackle these difficulties by means of the following ingredients:

1. The vademecum solution is in fact a high dimensional solution to the problem, depending on many different parameters such as initial or boundary conditions, material parameters, etc. To deal with the so-called curse of dimensionality (the number of degrees of freedom of the solution grows exponentially with the number of dimensions in traditional mesh-based methods), PGD proposes a *separated representation*. Consider, for the sake of simplicity in the exposition, that the solution of an organ palpation depends solely on the physical coordinates and the load position:

$$u_j^n(\boldsymbol{x}, \boldsymbol{s}) = \sum_{k=1}^{n} X_j^k(\boldsymbol{x}) \cdot Y_j^k(\boldsymbol{s}), \tag{6}$$

   where u_j represents the j-th component of the displacement field, depending on the physical position $\boldsymbol{x}$ and the mentioned position of the load exerted by the scalpel, $\boldsymbol{s}$. Dependence of the solution on further parameters can be treated under this same rationale without difficulties. Problems defined in spaces up to 100 dimensions have been solved successfully in the literature, see [78] and references therein.
2. Superscript n in Eq. (6) indicates that PGD proceeds actually in an iterative way. Improvements in the approximation are obtained by applying a greedy algorithm, i.e.,

$$u_j^{n+1}(\boldsymbol{x}, \boldsymbol{s}) = u_j^n(\boldsymbol{x}, \boldsymbol{s}) + R_j(\boldsymbol{x}) \cdot S_j(\boldsymbol{s}),$$

   where $\boldsymbol{R}(\boldsymbol{x})$ and $\boldsymbol{S}(\boldsymbol{s})$ are the sought functions that improve the approximation. To obtain these functions, consider first the admissible variation of the displacement, given by

$$u_j^*(\boldsymbol{x}, \boldsymbol{s}) = R_j^*(\boldsymbol{x}) \cdot S_j(\boldsymbol{s}) + R_j(\boldsymbol{x}) \cdot S_j^*(\boldsymbol{s}). \tag{7}$$

Several options are at hand so as to determine the new pair of functions $\boldsymbol{R}$ and $\boldsymbol{S}$. The most frequently used, due to both its ease of implementation and good convergence properties, in general, is a fixed-point algorithm in which functions $\boldsymbol{R}$ and $\boldsymbol{S}$ are sought iteratively so as to minimize the residual. See the extensive bibliography on PGD for more details.

3. PGD methods are in fact a model order reduction technique, since the order of the approximation, n, can be fixed so as to have a minimum of degrees of freedom in the model. To this end, error indicators are indispensable [81,82]. Functions $\boldsymbol{X}^k$ and $\boldsymbol{Y}^k$ in Eq. 6 can actually be seen as the PGD counterpart of the POD modes ϕ_k. However, as opposed to POD, PGD does not always provide with optimal modes. Analytical proofs of optimality exist for elliptic problems with symmetric operators [83].

Another difficult problem, common to virtually all model reduction methods, is that of the non-linearity of the problem. Since most living tissues are inherently non-linear, these problems must be solved within the PGD framework. In [84,85] an explicit approach and one based upon ANM are presented, respectively. Both approaches can lead to similar levels of accuracy, although obtained with very different modes, see Figure 5.

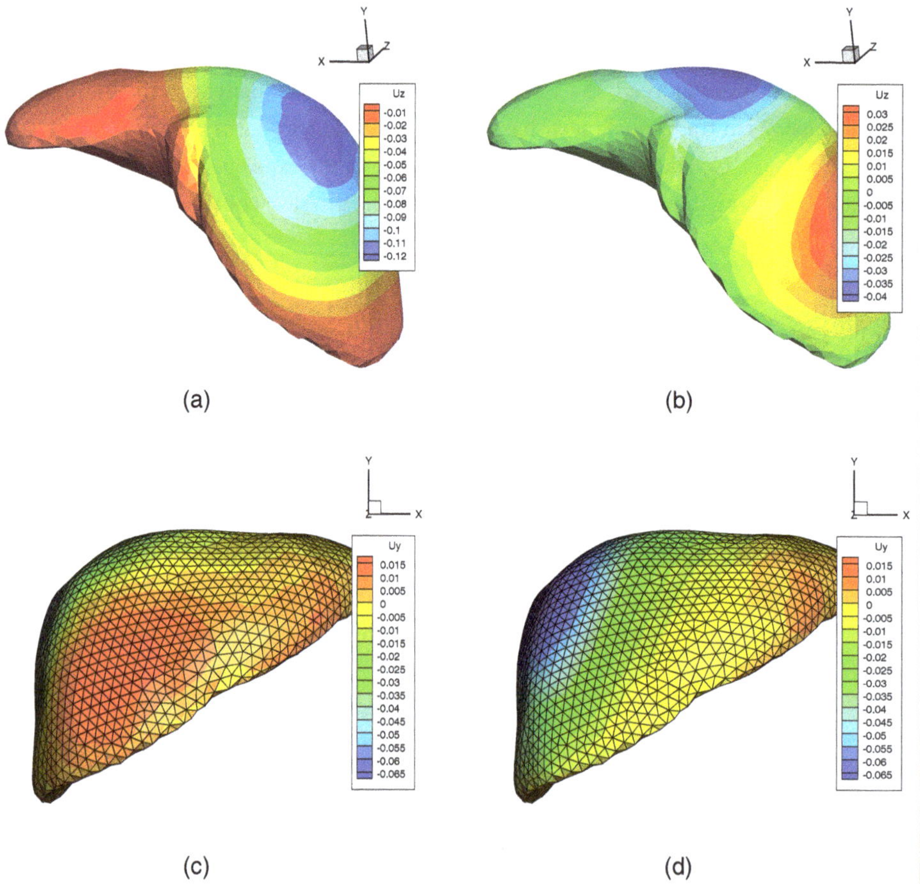

Figure 5 Two first modes $X^{1,2}(x)$ in the simulation of liver palpation. Top: explicit version in [84]. **(a)** first mode, **(b)** second mode. Bottom: PGD-ANM approach in [85]. **(c)** first mode, **(d)** second mode.

Since, in principle, it is only necessary to store these modes in the form of vectors (and, in the case of functions $Y^k(s)$ they are only defined on the boundary of the organ $\Gamma = \partial\Omega$), the storage in memory of these results is very efficient. As an example, this method allows for an interative implementation on deployed, handheld devices such as smartphones and tables (see Figure 6) or web-based demonstrators (see Figure 7).

Under the model reduction rationale, aspects such as cutting are much less investigated, but in [67], for instance, an approach based upon a combination of X-FEM and POD is developed. Regarding PGD applications, a similar approach could be envisaged, based on a PGD-XFEM coupling strategy [86].

Adding physiological details: third to fifth generations of surgery simulators

Understanding and quantitatively simulating physiology of the human has been a challenge since some fifteen years ago. The Physiome project [87-89] is a clear paradigm of this, opening the era of computational physiology. But an effective, real-time simulation of physiology is nowadays even further than a successful, purely mechanical, real-time simulation at a continuum level of living tissues, as discussed in Section 'Refining the appearance: the second generation of surgery simulation' before.

While introducing physiology in the simulation is not strictly necessary for many applications (most minimally invasive surgery training procedures, for instance), simulation based medical or surgical planning [90] very often needs it. For instance, in [91] a novel haptic simulation system is developed that includes not only force but thermal feedback, useful for the location of tumors, for instance.

In other problems, including physiology in the simulation is simply unavoidable. For instance, predicting the outcome of vein graft surgery is one of these problems that combines the need for simulation at a macroscopic level, but also at a gene regulatory network level [1]. It is hypothesized that blood shearing forces modulates a specific gene regulatory network determining the adaptive response of the vein wall. But simulation at a gene

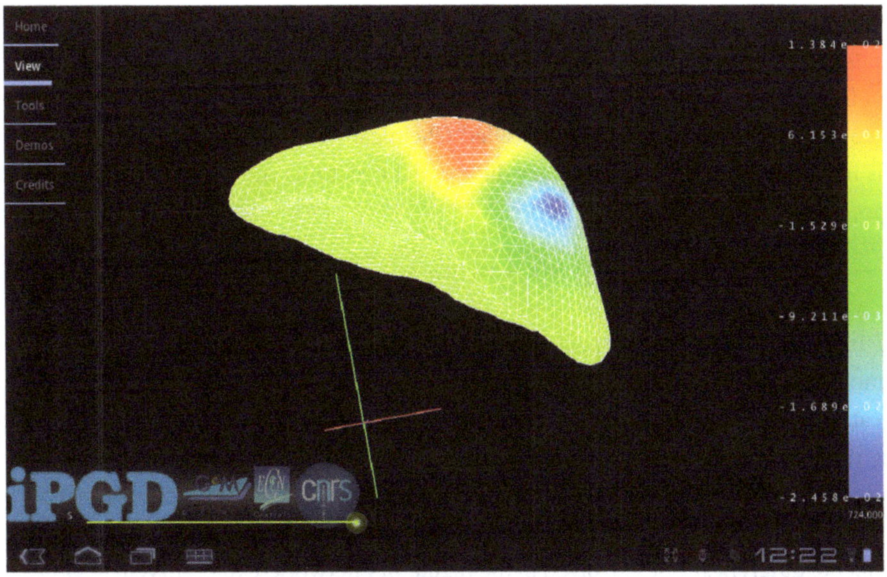

Figure 6 Implementation of the PGD approach to the liver palpation problem on a Motorola Xoom tablet. It provides continuous visual feedback without problems.

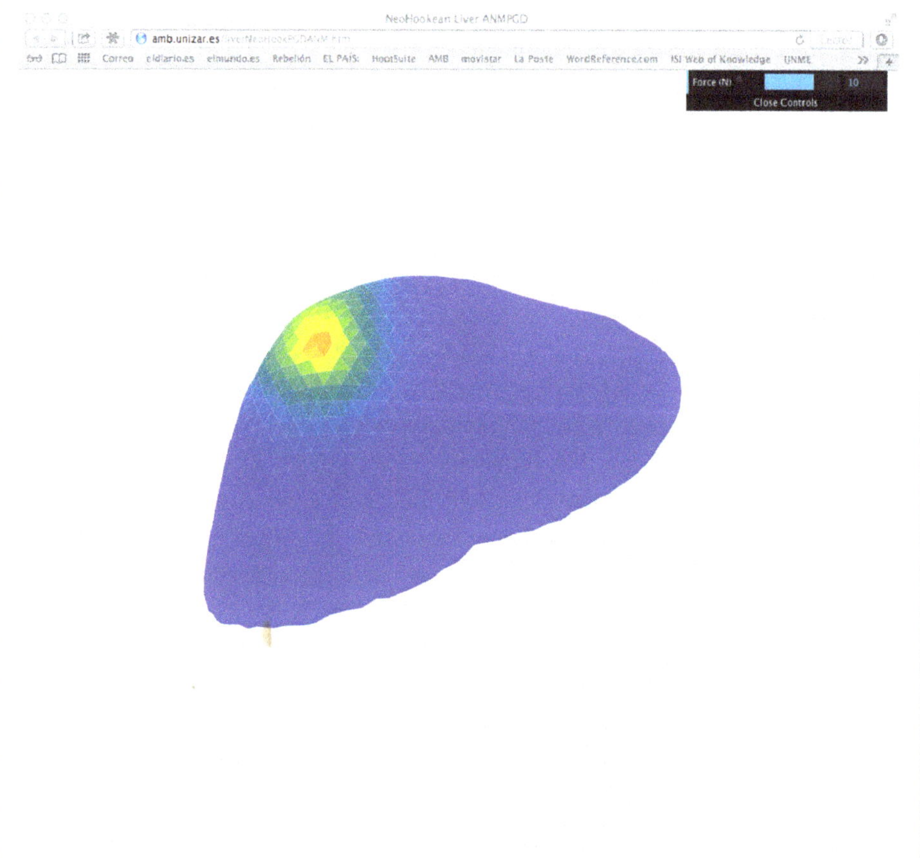

Figure 7 Implementation of the PGD approach to the liver palpation problem on a web page using javascript.

regulatory network level is a challenge in itself. This is so since at this level a stochastic description of the reactions occurring within the cell must be done, since it is nonsense to speak of concentration at this level, where only few specimens of each component are present [9,92,93]. It is commonly accepted that under some weak hypothesis (well stirred mixture, fixed volume and temperature), the system can be considered as Markovian, and can be consequently modeled by the so-called Chemical Master Equation (CME), [94], which is in fact no more than an ordinary differential equations stating the conservation of the probability density function P in time:

$$\frac{\partial P(\boldsymbol{z}, t|\boldsymbol{z}_0, t_0)}{\partial t} = \sum_j \left[a_j \left(\boldsymbol{z} - \boldsymbol{v}_j\right) P\left(\boldsymbol{z} - \boldsymbol{v}_j, t|\boldsymbol{z}_0, t_0\right) - a_j(\boldsymbol{z}) P\left(\boldsymbol{z}, t|\boldsymbol{z}_0, t_0\right)\right], \tag{8}$$

where $P\left(\boldsymbol{z}, t|\boldsymbol{z}_0, t_0\right)$ represents the probability of being at a state in which there are a number of molecules of each species stored in the vector $\boldsymbol{z}$ at time t when we started from a state $\boldsymbol{z}_0$ at time t_0. a_j represents the *propensity* (i.e., the probability) of reaction j to occur, while $\boldsymbol{v}_j$ represents the change in the number of molecules of each species if reaction j takes place. This change is given, of course, by the stoichiometry of the reaction at hand.

This set of equations are defined in a state space which possess as many dimensions as the number of different species involved in the regulatory network. Under this challenging framework, if we consider N different species, present at a number n of copies, the number of different possible states of the system is n^N. This number can take the astronomical value of 10^{6000} if we consider some types of proteins, for instance [94].This phenomenon is known as the *curse of dimensionality* and has been already mentioned in this survey, see Section 'Refining the appearance: the second generation of surgery simulation'. For instance, Nobel prize winner R. B. Laughlin said, when talking about this problem [95] that "No computer existing, or that will ever exist, can break this barrier because it is a catastrophe of dimension".

Monte Carlo-like algorithms (the so-called stochastic simulation algorithm, SSA [94,96,97]) are often the preferred approach to these problems. But Monte Carlo techniques need for as many as possible individual realizations of the problem that compromise its simple application in inverse identification, leading to excessive time consuming simulations, together with great variance in the results.

In [98], however, a PGD technique is used to avoid precisely the curse of dimensionality. Again, an off-line/on-line strategy is developed that allows to obtain vademecum solutions easy to evaluate on-line under very astringent real-time restrictions, for example when considering the propensities as extra-coordinates. As commented before, PGD approximates the variable of interest, the probability P, as a finite sum of separable functions:

$$P^N(\boldsymbol{z}, t) = \sum_{j=1}^{N} F_1^j(z_1) \cdot F_2^j(z_2) \cdot \ldots \cdot F_D^j(z_D) \cdot F_t^j(t),$$

where, as mentioned before, the variables z_i represent the number of molecules of species i present at a given time instant. Note the presence of a function depending solely on time, $F_t^j(t)$. This means that the algorithm is not incremental. Instead, it solves for the whole time history of the chemical species at each iteration of the method, so as to be available to be evaluated at any point in the multiscale simulation pipeline. With this technique, in [98] cascades of up to 20 terms were successfully simulated giving results in good accordance with the best ones available in the literature. This approach allows to compute the steady state of a system, impossible when considering stochastic approaches.

Generally speaking, adding physiological details to real-time simulations for surgery planning is still out of reach in many cases. Additional efforts of research are needed to successfully incorporate many aspects such as bleeding, microscopical details such as glandular anatomy or, finally, biochemical aspects such as those just commented.

Conclusions

Real-time simulation in the field of computational surgery has evolved a lot in the last fifteen years. Although not exhaustive, in this review paper we have presented a survey of some of the most prominent techniques nowadays. Particular attention has been paid to techniques based on model order reduction, a framework that has somewhat revolutionize the field and that seems the only way to incorporate to the field detailed physiological descriptions, with microscopic details and micro-macro approaches to the always challenging work of simulating the human body.

Other aspects, such as patient-specific modeling, deserve additional attention. Due to their inherent complexity and the infancy of their state of development, they have not been covered in this review.

Competing interests
The authors declare that they have no competing interests.

Authors' contributions

Both authors, EC and FC, participated equally in the manuscript writing. Both read and approved the final manuscript.

Acknowledgments

This work has been partially supported by the Spanish Ministry of Economy and Competitiveness, through grant number CICYT-DPI2011-27778-C02-01.

Author details

[1]Aragón Institute of Engineering Research, Universidad de Zaragoza, Zaragoza, Spain. [2]Ecole Centrale de Nantes, Nantes, France. [3]Institut Universitaire de France, Paris, France.

References

1. Garbey M, Bass B, Berceli S (2012) Multiscale mechanobiology modeling for surgery assessment. Acta Mechanica Sinica 28:1186–1202. http://dx.doi.org/10.1007/s10409-012-0133-4, doi:10.1007/s10409-012-0133-4
2. Champion HR, Gallagher AG (2003) Surgical simulation: a good idea whose time has come. Br J Surg 90(7):767–768. doi:10.1002/bjs.4187, http://dx.doi.org/10.1002/bjs.4187
3. Winslow RL, Trayanova N, Geman D, Miller MI (2012) Computational medicine: translating models to clinical care. Sci Transl Med 4(158). doi:10.1126/scitranslmed.3003528
4. Delingette H, Ayache N (2004) Soft tissue modeling for surgery simulation. In: Ayache N (ed) Computational models for the human body, handbook of numerical analysis (Ph. Ciarlet, Ed.) Elsevier, Idem, pp 453–550
5. Satava RM (1996) Medical virtual reality: the current status of the future. In: Sieburg H, Weghorst S, Morgan K (eds) Healthcare in the information age. IOS Press
6. Delingette H, Ayache N (2005) Hepatic surgery simulation. Comm ACM 48:31–36. doi:http://doi.acm.org/10.1145/1042091.1042116, http://doi.acm.org/10.1145/1042091.1042116
7. Cotin S, Delingette H, Ayache N (1999) Real-time elastic deformations of soft tissues for surgery simulation. In: Hagen H (ed) IEEE transactions on visualization and computer graphics, vol 5 (1). IEEE Computer Society, pp 62–73. citeseer.ist.psu.edu/cotin98realtime.html
8. Bro-Nielsen M, Cotin S (1996) Real-time volumetric deformable models for surgery simulation using finite elements and condensation. Comput Graph Forum 15(3):57–66
9. Hasty J, McMillen D, Isaacs F, Collins JJ (2001) Computational studies of gene regulatory networks: in numero molecular biology. Nat Rev Genet 2:268–279
10. Hegland M, Burden C, Santoso L, MacNamara S, Boothm H (2007) A solver for the stochastic master equation applied to gene regulatory networks. J Comput Appl Math 205:708–724
11. Sasai M, Wolynes PG (2003) Stochastic gene expression as a many-body problem. Proc Natl Acad Sci 100(5): 2374–2379
12. Kim J, De S, Srinivasan MA (2002) Proceedings 10th symposium on haptic interfaces for virtual environment and teleoperator systems. haptics 2002 In: Haptic Interfaces for Virtual Environment and Teleoperator Systems, 2002. HAPTICS 2002. Proceedings. 10th Symposium on, pp 51–57. doi:10.1109/HAPTIC.2002.998933
13. Basdogan C, De S, Kim J, Muniyandi M, Kim H, Srinivasan M, 2004 Haptics in minimally invasive surgical simulation and training. IEEE Comput Graph Appl Mag 24(2):56–64. doi:10.1109/MCG.2004.1274062
14. Holzapfel GA, Gasser TC (2000) A new constitutive framework for arterial wall mechanics and a comparative study of material models. J Elasticity 61:1–48
15. Alastrue V, Calvo B, Pena E, Doblare M (2006) Biomechanical modeling of refractive corneal surgery. J Biomecha Eng-Trans ASME 128:150–160
16. Berceli SA, Tran-Son-Tay R, Garbey M, Jiang Z (2009) Hemodynamically driven vein graft remodeling: A systems biology approach. Vascular 17(suppl 1):S2–S9. doi:10.2310/6670.2008.00083, http://vas.sagepub.com/content/17/suppl_1/S2.abstract, http://vas.sagepub.com/content/17/suppl_1/S2.full.pdf+html
17. Vilanova G, Colominas I, Gomez H (2013) Capillary networks in tumor angiogenesis: From discrete endothelial cells to phase-field averaged descriptions via isogeometric analysis. Int J Numer Methods Biomed Eng. doi:10.1002/cnm.2552, http://dx.doi.org/10.1002/cnm.2552
18. Azuma R (1997) A survey of augmented reality. Presence-Teleoperators Virtual Environ 6(4):355–385
19. Sederberg TW, Parry SR (1986) Free-form deformation of solid geometric models In: Proceedings of the 13th annual conference on Computer graphics and interactive techniques, SIGGRAPH '86. New York, pp 151–160. doi:10.1145/15922.15903, http://doi.acm.org/10.1145/15922.15903
20. Terzopoulos D, Fleischer K (1988) Deformable models. Vis Comput 4(6):306–331. doi:10.1007/BF01908877, http://dx.doi.org/10.1007/BF01908877
21. Meier U, Lopez O, Monserrat C, Juan M, Alcaniz M (2005) Real-time deformable models for surgery simulation: a survey. Comput Meth Programs Biomed 77(3):183–197. doi:10.1016/j.cmpb.2004.11.002, http://www.sciencedirect.com/science/article/pii/S0169260704002093
22. Lee Y, Terzopoulos D, Waters K (1995) Realistic modeling for facial animation In: Proceedings of the 22nd annual conference on computer graphics and interactive techniques, SIGGRAPH '95. ACM, New York, pp 55–62. doi:10.1145/218380.218407, http://doi.acm.org/10.1145/218380.218407
23. Famaey N, Stolen J (2008) Soft tissue modeling for applications in virtual surgery and surgical robotics. Comput Methods Biomech Biomed Engin 11(4):351–366
24. Monserrat C, Meier U, Alcaniz M, Chinesta F, Juan M (2001) A new approach for the real-time simulation of tissue deformations in surgery simulation. Comput Methods Programs Biomed 64(2):77–85. doi:10.1016/S0169-2607(00)00093-6, http://www.sciencedirect.com/science/article/pii/S0169260700000936

25. Capell S, Green S, Curless B, Duchamp T, Popovic Z (2002) Interactive skeleton-driven dynamic deformations In: ACM SIGGRAPH. ACM, New York
26. Cotin S, Delingette H, Ayache N (2000) A hybrid elastic model for real-time cutting, deformations, and force feedback for surgery training and simulation. Vis Comput 16(8):437–452. doi:10.1007/PL00007215, http://dx.doi.org/10.1007/PL00007215
27. Vigneron L, Verly J, Warfield S On extended finite element method (xfem) for modelling of organ deformations associated with surgical cuts. In: Cotin S, Metaxas D (eds) Medical Simulation, Lecture Notes in Computer Science, vol 3078. Springer Berlin Heidelberg, pp 134–143. doi:10.1007/978-3-540-25968-8_15, http://dx.doi.org/10.1007/978-3-540-25968-8_15
28. Jeřábková L, Kuhlen T (2009) Stable cutting of deformable objects in virtual environments using xfem. IEEE Comput Graph Appl 29(2):61–71. doi:10.1109/MCG.2009.32, http://dx.doi.org/10.1109/MCG.2009.32
29. Courtecuisse H, Jung H, Allard J, Duriez C, Lee DY, Cotin S (2010) GPU-based real-time soft tissue deformation with cutting and haptic feedback. Prog Biophys Mol Biol 103(2–3):159–168. doi: http://dx.doi.org/10.1016/j.pbiomolbio.2010.09.016, http://www.sciencedirect.com/science/article/pii/S0079610710000842, special Issue on Biomechanical Modelling of Soft Tissue Motion
30. Zhong H, Peters T (2007) A real time hyperelastic tissue model. Comput Methods Biomech Biomed Engin 10(3): 185–193. doi:10.1080/10255840701292732, http://www.tandfonline.com/doi/abs/10.1080/10255840701292732, pMID: 17558647, http://www.tandfonline.com/doi/pdf/10.1080/10255840701292732
31. Zhong H, Wachowiak MP, Peters TM (2005) A real time finite element based tissue simulation method incorporating nonlinear elastic behavior. Comput Methods Biomech Biomed Engin 8(3):177–189. doi:10.1080/10255840500295852, http://www.tandfonline.com/doi/abs/10.1080/10255840500295852, pMID: 16214712, http://www.tandfonline.com/doi/pdf/10.1080/10255840500295852
32. Bonet J, Wood RD (2008) Nonlinear continuum mechanics for finite element analysis. Cambridge University Press, Cambridge
33. Barbič J, James DL (2005) Real-time subspace integration for St. Venant-Kirchhoff deformable models. ACM Trans Graph 24(3):982–990
34. Horton A, Wittek A, Joldes GR, Miller K (2010) A meshless total lagrangian explicit dynamics algorithm for surgical simulation. Int J Numer Methods Biomed Eng 26(8):977–998. doi:10.1002/cnm.1374, http://dx.doi.org/10.1002/cnm.1374
35. Lim YJ, De S (2007) Real time simulation of nonlinear tissue response in virtual surgery using the point collocation-based method of finite spheres. Comput Methods Appl Mech Eng 196:3011–3024
36. De S, Lim YJ, Manivannan M, Srinivasan MA (2006) Physically realistic virtual surgery using the point-associated finite field (paff) approach. Presence: Teleoperators Virtual Environ 15(3):294–308. doi:10.1162/pres.15.3.294, http://dx.doi.org/10.1162/pres.15.3.294
37. Deo D, De S (2009) Phyness: a physics-driven neural networks-based surgery simulation system with force feedback In: Third joint eurohaptics conference and symposium on haptic interfaces for virtual environment and teleoperator systems, Salt Lake City, UT, USA
38. Misra S, Ramesh KT, Okamura AM (2008) Modeling of tool-tissue interactions for computer-based surgical simulation: a literature review. Presence: Teleoper Virtual Environ 17(5):463–491. doi:10.1162/pres.17.5.463, http://dx.doi.org/10.1162/pres.17.5.463
39. Delingette H (2008) Biquadratic and quadratic springs for modeling st venant kirchhoff materials In: Proceedings of the 4th international symposium on biomedical simulation, ISBMS '08. Springer-Verlag, Berlin, Heidelberg, pp 40–48. doi:10.1007/978-3-540-70521-5_5, http://dx.doi.org/10.1007/978-3-540-70521-5_5
40. Martinez-Martinez F, Ruperez M, Martin-Guerrero J, Monserrat C, Lago M, Pareja E, Brugger S, Lopez-Andujar R (2013) Estimation of the elastic parameters of human liver biomechanical models by means of medical images and evolutionary computation. Comput Methods Programs Biomed 111(3):537–549. doi:http://dx.doi.org/10.1016/j.cmpb.2013.05.005, http://www.sciencedirect.com/science/article/pii/S0169260713001491
41. Comas O, Taylor ZA, Allard J, Ourselin S, Cotin S, Passenger J (2008) Efficient nonlinear FEM for soft tissue modelling and its GPU implementation within the open source framework SOFA In: Proceedings of the 4th international symposium on biomedical simulation, ISBMS '08. Springer-Verlag, Berlin, Heidelberg, pp 28–39. doi:10.1007/978-3-540-70521-5_4, http://dx.doi.org/10.1007/978-3-540-70521-5_4
42. Miller K, Joldes G, Lance D, Wittek A (2007) Total lagrangian explicit dynamics finite element algorithm for computing soft tissue deformation. Comm Numer Meth Eng 23(2):121–134. doi:10.1002/cnm.887, http://dx.doi.org/10.1002/cnm.887
43. Courtecuisse H, Allard J, Kerfriden P, Bordas SP, Cotin S, Duriez C (2014) Real-time simulation of contact and cutting of heterogeneous soft-tissues. Med Image Anal 18(2):394–410. doi:http://dx.doi.org/10.1016/j.media.2013.11.001, http://www.sciencedirect.com/science/article/pii/S1361841513001692
44. Karhunen KK (1947) Über lineare Methoden in der Wahrscheinlichkeitsrechnung. Ann Acad Sci Fennicae Ser A I Math-Phys 37:1–79
45. Loève MM (1963) Probability theory. The University Series in Higher Mathematics. 3rd ed. Van Nostrand, Princeton
46. Lorenz EN (1956). MIT, Departement of Meteorology, Scientific Report Number 1, Statistical Forecasting Project
47. Park HM, Cho DH (1996) The use of the Karhunen-Loève decomposition for the modeling of distributed parameter systems. Chem Eng Sci 51(1):81–98
48. Sirovich L (1987) Turbulence and the dynamics of coherent structures part I: coherent structures. Q Appl Math XLV: 561–571
49. Krysl P, Lall S, Marsden J (2001) Dimensional model reduction in non-linear finite element dynamics of solids and structures. Int J Numer Meth in Engng 51:479–504
50. Taylor Z, Crozier S, Ourselin S (2011) A reduced order explicit dynamic finite element algorithm for surgical simulation. IEEE Trans Med Imaging 30(9):1713–1721. doi:10.1109/TMI.2011.2143723

51. Taylor Z, Ourselin S, Crozier S (2010) A reduced order finite element algorithm for surgical simulation In: Engineering in Medicine and Biology Society (EMBC). 2010 Annual International Conference of the IEEE, pp 239–242. doi:10.1109/IEMBS.2010.5627720
52. Kerfriden P, Passieux JC, Bordas SPA (2012) Local/global model order reduction strategy for the simulation of quasi-brittle fracture. Int J Numer Methods Eng 89(2):154–179. doi:10.1002/nme.3234, http://dx.doi.org/10.1002/nme.3234
53. Ammar A, Pruliere E, Ferec J, Chinesta F, Cueto E (2009) Coupling finite elements and reduced approximation bases. Eur J Comput Mech 18(5–6):445–463
54. Meyer M, Matthies HG (2003) Efficient model reduction in non-linear dynamics using the Karhunen-Loève expansion and dual-weighted-residual methods. Comput Mech 31(1–2):179–191. doi:10.1007/s00466-002-0404-1, http://dx.doi.org/10.1007/s00466-002-0404-1
55. Niroomandi S, Alfaro I, Cueto E, Chinesta F (2008) Real-time deformable models of non-linear tissues by model reduction techniques. Comput Methods Programs Biomed 91(3):223–231. doi:10.1016/j.cmpb.2008.04.008, http://www.sciencedirect.com/science/article/B6T5J-4SNPPVY-2/2/8a417e7f1371768b4c928d1f12fc7a0f
56. Dogan F, Celebi MS (2011) Real-time deformation simulation of non-linear viscoelastic soft tissues. Simulation 87(3): 179–187
57. Radermacher A, Reese S (2013) Proper orthogonal decomposition-based model reduction for nonlinear biomechanical analysis. Int J Mater Eng Innovat 4(4):149–165. doi:10.1504/IJMATEI.2013.054393
58. Amsallem D, Farhat C (2008) An interpolation method for adapting reduced-order models and application to aeroelasticity. AIAA J 46:1803–1813
59. Ly HV, Tran HT (2005) Modeling and control of physical processes using proper orthogonal decomposition. Math Comput Model 33:223–236
60. Son NT (2013) A real time procedure for affinely dependent parametric model order reduction using interpolation on grassmann manifolds. Int J Numer Meth Eng 93(8):818–833. doi:10.1002/nme.4408, http://dx.doi.org/10.1002/nme.4408
61. Cochelin B, Damil N, Potier-Ferry M (1994) Asymptotic-numerical methods and Padéapproximants for non-linear elastic structures. Int J Numer Methods Eng 37:1187–1213
62. Abichou H, Zahrouni H, Potier-Ferry M (2002) Asymptotic numerical method for problems coupling several nonlinearities. Comput Meth Appl Mech Eng 191(51–52):5795–5810
63. Cao HL, Potier-Ferry M (1999) An improved iterative method for large strain viscoplastic problems. Int J Numer Methods Eng 44:155–176
64. Cochelin B, Damil N, Potier-Ferry M (1994) The asymptotic numerical method: an efficient perturbation technique for nonlinear structural mechanics. Revue Europeenne des Elements Finis 3:281–297
65. Niroomandi S, Alfaro I, Cueto E, Chinesta F (2010) Model order reduction for hyperelastic materials. J Numer Meth Eng 81(9):1180–1206. doi:10.1002/nme.2733, http://dx.doi.org/10.1002/nme.2733
66. Niroomandi S, Alfaro I, Cueto E, Chinesta F (2012) Accounting for large deformations in real-time simulations of soft tissues based on reduced-order models. Comput Methods Programs Biomed 105(1):1–12. doi:DOI: 10.1016/j.cmpb.2010.06.012, http://www.sciencedirect.com/science/article/B6T5J-50VGHDD-1/2/1201566766c0d280af9195bf07bfaf91
67. Niroomandi S, Alfaro I, Gonzalez D, Cueto E, Chinesta F (2012) Real-time simulation of surgery by reduced-order modeling and x-fem techniques. Int J Numer Methods Biomed Eng 28(5):574–588. doi:10.1002/cnm.1491, http://dx.doi.org/10.1002/cnm.1491
68. Barrault M, Maday Y, Nguyen N, Patera A (2004) An 'empirical interpolation' method: application to efficient reduced-basis discretization of partial differential equations. Comptes Rendus Mathematique 339(9):667–672. doi:10.1016/j.crma.2004.08.006
69. Chaturantabut S, Sorensen DC (2010) Nonlinear model reduction via discrete empirical interpolation. SIAM J Sci Comput 32:2737–2764. doi:10.1137/090766498, http://dx.doi.org/10.1137/090766498
70. Boulakia M, Cazeau S, Fernandez M, Gerbeau JF, Zemzemi N (2010) Mathematical modeling of electrocardiograms: A numerical study. Ann Biomed Eng 38(3):1071–1097. doi:10.1007/s10439-009-9873-0, http://dx.doi.org/10.1007/s10439-009-9873-0
71. Heidenreich EA, Ferrero JM, Doblare M, Rodriguez JF (2010) Adaptive macro finite elements for the numerical solution of monodomain equations in cardiac electrophysiology. Ann Biomed Eng 38(7):2331–2345. doi:10.1007/s10439-010-9997-2, http://dx.doi.org/10.1007/s10439-010-9997-2
72. Boulakia M, Schenone E, Gerbeau JF (2012) Reduced-order modeling for cardiac electrophysiology. application to parameter identification. Int J Numer Methods Biomed Eng 28(6–7):727–744. doi:10.1002/cnm.2465, http://dx.doi.org/10.1002/cnm.2465
73. McLeod K, Caiazzo A, Fernandez MA, Mansi T, Vignon-Clementel IE, Sermesant M, Pennec X, Boudjemline Y, Gerbeau JF (2010) Atlas-based reduced models of blood flows for fast patient-specific simulations. In: Camara O, Pop M, Rhode K, Sermesant M, Smith N, Young A (eds) Statistical Atlases and Computational Models of the Heart, Lecture Notes in Computer Science, vol 6364. Springer Berlin Heidelberg, pp 95–104. doi:10.1007/978-3-642-15835-3_10, http://dx.doi.org/10.1007/978-3-642-15835-3_10
74. Gerbeau JF, Lombardi D (2012) Reduced-order modeling based on approximated lax pairs. J Comput Phys 265(2014):246–269
75. Ladeveze P (1999) Nonlinear computational structural mechanics. Springer, New York
76. Ammar A, Mokdad B, Chinesta F, Keunings R (2006) A new family of solvers for some classes of multidimensional partial differential equations encountered in kinetic theory modeling of complex fluids. J Non-Newtonian Fluid Mech 139:153–176

77. Ammar A, Mokdad B, Chinesta F, Keunings R (2007) A new family of solvers for some classes of multidimensional partial differential equations encountered in kinetic theory modeling of complex fluids. part ii: transient simulation using space-time separated representations. J Non-Newtonian Fluid Mech 144:98–121
78. Chinesta F, Ammar A, Cueto E (2010) Recent advances in the use of the proper generalized decomposition for solving multidimensional models. Arch Comput Meth Eng 17(4):327–350
79. Chinesta F, Ladeveze P, Cueto E (2011) A short review on model order reduction based on proper generalized decomposition. Arch Comput Meth Eng 18:395–404
80. Chinesta F, Leygue A, Bordeu F, Aguado J, Cueto E, Gonzalez D, Alfaro I, Ammar A, Huerta A (2013) Pgd-based computational vademecum for efficient design, optimization and control. Arch Comput Meth Eng 20(1):31–59. doi:10.1007/s11831-013-9080-x, http://dx.doi.org/10.1007/s11831-013-9080-x
81. Ammar A, Chinesta F, Diez P, Huerta A (2010) An error estimator for separated representations of highly multidimensional models. Comput Meth Appl Mech Eng 199(25–28):1872–1880. doi:http://dx.doi.org/10.1016/j.cma.2010.02.012, http://www.sciencedirect.com/science/article/pii/S0045782510000708
82. Ladeveze P, Chamoin L (2011) On the verification of model reduction methods based on the proper generalized decomposition. Comput Methods Appl Mech Eng 200(23–24):2032–2047. doi:http://dx.doi.org/10.1016/j.cma.2011.02.019, http://www.sciencedirect.com/science/article/pii/S0045782511001058
83. Nouy A (2010) A priori model reduction through proper generalized decomposition for solving time-dependent partial differential equations. Comput Meth Appl Mech Eng 199(23–24):1603–1626. doi:http://dx.doi.org/10.1016/j.cma.2010.01.009, http://www.sciencedirect.com/science/article/pii/S0045782510000186
84. Niroomandi S, Gonzalez D, Alfaro I, Bordeu F, Leygue A, Cueto E, Chinesta F (2013b) Real-time simulation of biological soft tissues: a pgd approach. Int J Numer Methods Biomed Eng 29(5):586–600. doi:10.1002/cnm.2544, http://dx.doi.org/10.1002/cnm.2544
85. Niroomandi S, Alfaro I, Gonzalez D, Cueto E, Chinesta F (2013) Model order reduction in hyperelasticity: a proper generalized decomposition approach. Int J Numer Meth Eng. doi:10.1002/nme.4531, http://dx.doi.org/10.1002/nme.4531
86. Ammar A, Chinesta F, Cueto E (2011) Coupling finite elements and proper generalized decompositions. Int J Multiscale Comput Eng 9(1):17–33
87. Hunter P, Nielsen P (2005) A strategy for integrative computational physiology. Physiology 20(5): 316–325. doi:10.1152/physiol.00022.2005, http://physiologyonline.physiology.org/content/20/5/316.abstract, http://physiologyonline.physiology.org/content/20/5/316.full.pdf+html
88. Hunter P (2006) Modeling human physiology: The iups/embs physiome project. Proc IEEE 94(4):678–691. doi:10.1109/JPROC.2006.871767
89. Bassingthwaighte J, Hunter P, Noble D (2009) The cardiac physiome: perspectives for the future. Exp Physiol 94(5): 597–605. doi:10.1113/expphysiol.2008.044099, http://ep.physoc.org/content/94/5/597.abstract, http://ep.physoc.org/content/94/5/597.full.pdf+html
90. Wilson N, Wang K, Dutton RW, Taylor C (2001) A software framework for creating patient specific geometric models from medical imaging data for simulation based medical planning of vascular surgery. In: Niessen W, Viergever M (eds) Medical Image Computing and Computer-Assisted Intervention, MICCAI 2001, Lecture Notes in Computer Science, vol 2208. Springer Berlin Heidelberg, pp 449–456. doi:10.1007/3-540-45468-3_54, http://dx.doi.org/10.1007/3-540-45468-3_54
91. Guiatni M, Riboulet V, Duriez C, Kheddar A, Cotin S (2013) A combined force and thermal feedback interface for minimally invasive procedures simulation. IEEE ASME Trans Mechatron 18(3):1170–1181. doi:10.1109/TMECH.2012.2197862
92. Turner TE, Schnell S, Burrage K (2004) Stochastic approaches for modelling in vivo reactions. Comput Biol Chem 28: 165–178
93. Sreenath SN, Cho KH, Wellstead P (2008) Modelling the dynamics of signalling pathways. Essays Biochem 45:1–28
94. Munsky B, Khammash M (2006) The finite state projection algorithm for the solution of the chemical master equation. J Chem Phys 124(4):044104. doi:10.1063/1.2145882, http://link.aip.org/link/?JCP/124/044104/1
95. Laughlin RB, Pines D (2000) The theory of everything. Proc Natl Acad Sci 97(1): 28–31. doi:10.1073/pnas.97.1.28, http://www.pnas.org/content/97/1/28.abstract, http://www.pnas.org/content/97/1/28.full.pdf+html
96. Gillespie DT (1977) Exact stochastic simulation of coupled chemical reactions. J Phys Chem 81(25):2340–2361. doi:10.1021/j100540a008, http://pubs.acs.org/doi/abs/10.1021/j100540a008, http://pubs.acs.org/doi/pdf/10.1021/j100540a008
97. Gillespie DT (2001) Approximate accelerated stochastic simulation of chemically reacting systems. J Chem Phys 115: 1716–1733
98. Ammar A, Cueto E, Chinesta F (2012) Reduction of the chemical master equation for gene regulatory networks using proper generalized decompositions. Int J Numer Methods Biomed Eng 28(9):960–973

A port-reduced static condensation reduced basis element method for large component-synthesized structures: approximation and *A Posteriori* error estimation

Jens L Eftang[1,2*] and Anthony T Patera[1]

*Correspondence: jleftang@gmail.com
[1] Department of Mechanical Engineering, Massachusetts Institute of Technology, 77 Massachusetts Avenue, Cambridge, MA-02139, USA
[2] Current address: DNV GL - Software, P.O.Box 300, NO-1322, Høvik, Norway

Abstract

Background: We consider a static condensation reduced basis element framework for efficient approximation of parameter-dependent linear elliptic partial differential equations in large three-dimensional component-based domains. The approach features an offline computational stage in which a library of interoperable parametrized components is prepared; and an online computational stage in which these component archetypes may be instantiated and connected through predefined ports to form a global synthesized system. Thanks to the component-interior reduced basis approximations, the online computation time is often relatively small compared to a classical finite element calculation.

Methods: In addition to reduced basis approximation in the component interiors, we employ in this paper port reduction with empirical port modes to reduce the number of degrees of freedom on the ports and thus the size of the Schur complement system. The framework is equipped with efficiently computable *a posteriori* error estimators that provide asymptotically rigorous bounds on the error in the approximation with respect to the underlying finite element discretization. We extend our earlier approach for two-dimensional scalar problems to the more demanding three-dimensional vector-field case.

Results and Conclusions: This paper focuses on linear elasticity analysis for large structures with tens of millions of finite element degrees of freedom. Through our procedure we effectively reduce the number of degrees of freedom to a few thousand, and we demonstrate through extensive numerical results for a microtruss structure that our approach provides an accurate, rapid, and *a posteriori* verifiable approximation for relevant large-scale engineering problems.

Keywords: Static condensation; Reduced basis element method; Component synthesis; Domain decomposition; Port reduction; Interface reduction; *A posteriori* error estimation; Non-conforming methods; Structural analysis; Large-scale simulation

Background

For several decades the finite element (FE) method has been a popular and important tool in engineering design and analysis of systems modelled by partial differential equations (PDEs). In particular, in fields such as structural analysis and strength assessment, the FE method is in widespread use in industry through a variety of commercial software packages. Many of the structures that are subject to industrial FE analysis are composed of a large number of components — consider for example a truss bridge, a space satellite [1], or a building or vehicle frame. Such large and at first sight complicated structures pose challenges both in terms of initial manual labor related to domain modelling and meshing, and in terms of subsequent computational cost.

Component-based structures which contain many identical or similar components are often analyzed through substructuring or superelement techniques [2], which mitigate some of these issues. Mathematically, superelement techniques are based on static condensation of all FE degrees of freedom that are interior to components, and hence the size of the global but condensed linear-algebraic (Schur complement) system is equal to the number of degrees of freedom associated with component interfaces, henceforth in this paper referred to as ports. The static condensation step necessitates a large number of component-interior FE "bubble" solves — one FE solve for each degree of freedom on each port of each component — and is for this reason rather expensive; however this step is embarrassingly parallel, and is furthermore required only once for each unique component instantiation.

Model order reduction techniques can be applied to substructuring or superelement procedures in order to further reduce the computational cost. A well-known approach is the classical component mode synthesis (CMS) [3,4], which replaces the original FE spaces for the component-interior bubble solves with spaces spanned by a few component-interior eigenmodes. As a result, the cost associated with each bubble calculation is reduced, and the formation of the global Schur complement system is consequently much less expensive.

A more recent approach, which is relevant in the context of parameter-dependent PDEs and which we for this reason consider here in this paper, is the static condensation reduced basis element method (SCRBE) introduced in [5]. Rather than the eigenmodal expansion typically used in the CMS, the SCRBE employs the reduced basis method (RB) [6] for the bubble function approximations. Each RB approximation space is specifically tailored to a particular bubble and the associated parameter dependence defined by the PDE within each component; the SCRBE thus accommodates parametric variations for example related to component geometry, loads, material properties, or boundary conditions. Furthermore, thanks to the typically very rapid (often exponential) convergence of the RB approximation [7,8], these RB spaces are low-dimensional and thus bubble function approximation is computationally inexpensive.

In addition to enabling parametric variations, the SCRBE features a strict offline-online computational decoupling. In the offline stage, the RB spaces and associated datasets for each component *archetype* in a component library is computed and stored. This stage requires FE solves and may thus be relatively expensive, but is carried out only once as a library preprocessing step. In the subsequent online stage, the user may instantiate any of the interoperable library archetypes, and assign to each component instantiation the desired parameter values; the RB bubble function approximations are then computed, and the Schur complement system is assembled and solved. This online step is much less expensive and in particular does never invoke the underlying FE discretization.

However, common to all these static-condensation-based approaches — including the SCRBE — is a global Schur complement linear-algebraic system of size equal to the total number of degrees of freedom associated with ports. For large systems with many components and ports, and in particular for problems with three-dimensional vector-valued field variables — such as in linear elasticity — the size of this system is considerable and thus clearly prohibits the fast response required in, say, an interactive design or optimization context. To overcome this limitation various port reduction techniques may be used. For example, for the CMS approaches an eigenmode expansion (with subsequent truncation) for the port degrees of freedom is considered in [9,10], and an adaptive procedure based on *a posteriori* error estimators for the port reduction is considered in [11]. For the SCRBE, we introduce in [12] port reduction with *empirical modes*; in this case the port approximation spaces are informed by snapshots of relevant port-restricted solutions which are obtained through an offline pairwise empirical training algorithm.

Unique to the SCRBE is a certification framework that allows efficient computation of *a posteriori* bounds or estimators for the error in the SCRBE approximation with respect to the underlying FE "truth" discretization. This framework invokes classical residual arguments on the (RB) bubble level [6], a non-conforming approximation to the error-residual equation at the port level, and finally matrix perturbation at the system level in order to bound (under an eigenvalue proximity assumption) the error contributions from both RB approximation [5] and port reduction [12]. In actual practice, we may reduce online computational cost by consideration of a plausible and asymptotically rigorous error estimator rather than a rigorous error bound.

In this paper, we extend our earlier work for two-dimensional scalar problems in [12] to the more demanding three-dimensional vector-field case. We focus here on applications in linear elasticity, but we note that the component synthesis and indeed RB and port approximations can be readily extended to problems in heat transfer or (frequency domain) acoustics, or any phenomenon described by a linear elliptic or parabolic [13] PDE.[a] Through our procedure we effectively reduce the number of degrees of freedom from tens of millions (in the underlying FE discretization) to only a few thousand (in the port-reduced SCRBE approximation); the associated computation time is thus reduced from minutes or hours to only a few seconds.

Our approach here features several important innovations. First, as we consider here larger global systems with a much larger number of instantiated components we introduce a new non-symmetric SCRBE approximation, which reduces both offline and online cost and memory footprint; the corresponding linear-algebraic system is subsequently symmetrized in order to (say) accommodate efficient linear solvers. We also demonstrate that our central theoretical results in particular related to *a posteriori* error estimation survive intact for this more efficient revision of our earlier formulations in [12]. Second, we provide a precise formulation for general geometric mappings and port space compatibility, and we demonstrate that (in the isotropic linear-elastic case) rigid-body parameters related to "docking" of component instantiations in a system do not affect the associated bilinear forms and thus do not impact offline — thanks to smaller RB space dimensions — or online — thanks to treatment of differently oriented component instantiations as effectively identical — computational cost. Third, we introduce a new functional interpretation of our algebraic *a posteriori* error estimation framework in [12], which may serve to extend our approach here to larger classes of problems. And finally, we consider multi-reference parameter bound conditioners [14] for sharper error estimation.

The remainder of the paper is organized as follows. We start with a brief presentation of a general parametrized component static condensation framework for d-dimensional vector-valued linear elliptic partial differential equations; we focus on the concepts relevant in the SCRBE framework and we formulate the port compatibility requirements. Next, we discuss the RB and port reduction strategies for the computational cost reduction associated with component interiors and component interfaces, respectively. Then, we introduce our *a posteriori* error estimation framework. Finally, we present extensive results for a three-dimensional microtruss application, and provide some conclusive remarks.

We include with this manuscript [Additional file 1]. This short movie presents the main ingredients of the port-reduced SCRBE method, and sums up the key numerical results reported in this paper.

Component-based static condensation

Concepts: library components and system

We now introduce the key concepts for our SCRBE approximation: a *library* of parametrized and interoperable *archetype components*, which is prepared in the offline stage; and a *system* of *component instantiations* connected at *ports*, which is assembled and solved (and, if desired, visualized) in the online stage.

In the context of structural analysis, an archetype component typically (but not necessarily) corresponds to a physical construction unit, such as a beam, a plate, or a connector; in physical d-dimensional space ($d = 1, 2, 3$) we denote by $\hat{\Omega}_m \subset \mathbb{R}^d$ the reference domain associated with archetype component m, $1 \leq m \leq M$, where M is the number of archetypes in the library. The boundary of this domain, $\partial\hat{\Omega}_m$, has a set of n_m^γ disjoint *local ports*, denoted as $\hat{\gamma}_{m,j} \subseteq \partial\hat{\Omega}_m$, $1 \leq j \leq n_m^\gamma$; these ports enable the components to connect to other components. Note we shall assume that all ports on an archetype component are mutually separated by (at least) a non-port, non-Dirichlet boundary segment. If this is not the case, modifications to our procedures below must be considered [10].

The physical behavior of each archetype component is governed by a vector-valued (we consider d field components) parametrized linear elliptic partial differential equation. We thus introduce for $1 \leq m \leq M$ the continuous (and here, in this paper, symmetric) archetype bilinear form $\hat{a}_m(\cdot,\cdot;\hat{\mu}_m) : (H^1(\hat{\Omega}_m))^d \times (H^1(\hat{\Omega}_m))^d \to \mathbb{R}$, and the bounded archetype linear functional $\hat{f}_m(\cdot,\cdot;\hat{\mu}_m) : (H^1(\hat{\Omega}_m))^d \to \mathbb{R}$. Here, $\hat{\mu}_m \in \hat{\mathcal{D}}^m \subset \mathbb{R}^{\hat{P}_m}$ is a vector of $\hat{P}_m$ scalar parameters that describe (say) the component geometry, boundary conditions, loads, or material properies, and $(H^1(\hat{\Omega}_m))^d$ is the usual (d-tensorized) first-order Sobolev space over $\hat{\Omega}_m$. We shall assume that $\hat{a}_m$ and $\hat{f}_m$ admit affine expansions as

$$\hat{a}_m(\cdot,\cdot;\hat{\mu}_m) = \sum_{q=1}^{\hat{Q}_m^a} \hat{a}^q(\cdot,\cdot)\Theta_a^q(\hat{\mu}_m), \qquad \hat{f}_m(\cdot;\hat{\mu}_m) = \sum_{q=1}^{\hat{Q}_m^f} \hat{f}^q(\cdot)\Theta_f^q(\hat{\mu}_m), \tag{1}$$

where the $\hat{a}^q$ and $\hat{f}^q$ are parameter-independent forms and the Θ_a^q and Θ_f^q are parameter-dependent functions; for computational efficiency of the SCRBE evaluation stage it is critical that $\hat{Q}_m^a$ and $\hat{Q}_m^f$ are relatively small.

We next introduce the discrete archetype component spaces $\hat{X}_m^h \subset (H^1(\hat{\Omega}_m))^d$, $1 \leq m \leq M$, which correspond to standard FE discretizations [15] of $(H^1(\hat{\Omega}_m))^d$; and we introduce the discrete *port spaces*, the restrictions

$$\hat{P}_{m,j} \equiv \hat{X}_m^h|_{\hat{\gamma}_{m,j}} \tag{2}$$

of dimension $\mathcal{N}_m^\gamma \equiv \dim(\hat{P}_{m,j})$. We denote the bases for these port spaces by $\{\hat{\chi}_{m,j,k}\}_{k=1}^{\mathcal{N}_{m,j}^\gamma}$ such that

$$\hat{P}_{m,j} = \operatorname{span}\{\hat{\chi}_{m,j,1}, \ldots, \hat{\chi}_{m,j,\mathcal{N}_{m,j}^\gamma}\}. \tag{3}$$

For simplicity of presentation here we shall assume that Dirichlet conditions are enforced only on ports and thus not through the archetype component discrete spaces $\hat{X}_m^h$ (this is the case for our numerical results later).

The library component archetypes may be *instantiated* and connected at ports to form a global system. To this end we introduce a mapping $\mathcal{M} : \{1, \ldots, I\} \to \{1, \ldots, M\}$ from any of the I instantiations in the system to exactly one of the M archetypes in the library. For instantiated component i, we introduce the parameter vector $\mu_i \in \mathcal{D}_i$, where $\mathcal{D}_i \subseteq \hat{\mathcal{D}}_{\mathcal{M}(i)}$. We then introduce a (parameter-dependent) geometric mapping $\mathcal{T}_i : \hat{\Omega}_i \to \Omega_i$ from archetype (reference) to system (physical) coordinates; thus $\Omega_i = \mathcal{T}_i(\hat{\Omega}_{\mathcal{M}(i)})$ is the instantiated component domain and $\gamma_{i,j} = \mathcal{T}_i(\hat{\gamma}_{\mathcal{M}(i),j})$, $1 \leq j \leq n_{\mathcal{M}(i)}^\gamma$, are the instantiated ports. We consider for each of our mappings $\mathcal{T}_i$ application of a deformation $\mathcal{T}_i^{\text{def}}$ and then a rotation $\mathcal{T}_i^{\text{rot}}$ such that $\mathcal{T}_i \equiv \mathcal{T}_i^{\text{rot}}\mathcal{T}_i^{\text{def}}$. In this paper, we consider for $\mathcal{T}_i^{\text{def}}$ only dilation *and* translation, and we further assume that $\mathcal{T}_i^{\text{def}}$, when applied to a port, is pure translation (such that $\gamma_{i,j} = \mathcal{T}_i(\hat{\gamma}_{\mathcal{M}(i),j})$ corresponds to a rigid-body transformation). We illustrate the situation (for $d = 2$) in Figure 1 and Figure 2: in Figure 1 we show a single

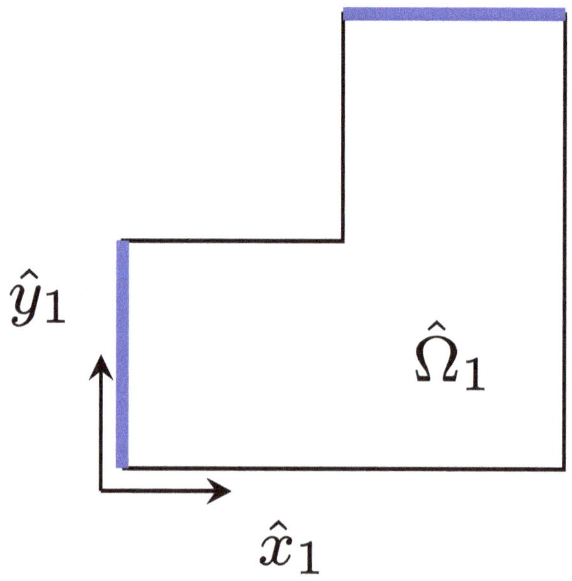

Figure 1 An archetype component in coordinates $(\hat{x}_m, \hat{y}_m)$.

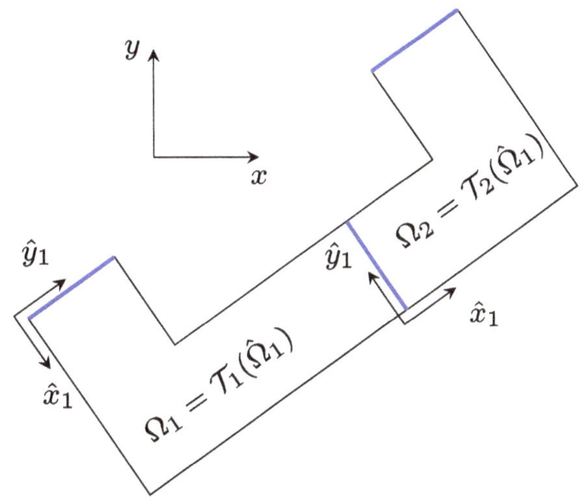

Figure 2 Two component instantiations form a system in coordinates (x, y).

archetype library component; in Figure 2 we instantiate two components of the same archetype subject to different mappings $\mathcal{T}_1$ and $\mathcal{T}_2$, the first of which has a non-trivial (piecewise dilation) $\mathcal{T}_1^{\text{def}}$.

We also introduce a mapped discrete component-local space

$$X_i^h = \text{span}\{\mathcal{T}_i^{\text{rot}}(v \circ \mathcal{T}_i^{-1}), \quad v \in \hat{X}_{\mathcal{M}(i)}^h\}; \tag{4}$$

and further, with

$$\chi_{i,j,k} \equiv \mathcal{T}_i^{\text{rot}}(\hat{\chi}_{\mathcal{M}(i),j,k} \circ \mathcal{T}_i^{-1}), \tag{5}$$

we introduce, for $1 \leq j \leq n_{\mathcal{M}(i)}^{\gamma}$, $1 \leq i \leq I$, the mapped discrete port spaces

$$P_{i,j}^h = \text{span}\{\chi_{i,j,k}, \quad 1 \leq k \leq \mathcal{N}_{\mathcal{M}(i),j}^{\gamma}\}. \tag{6}$$

Note that here and in the following the notation $[\cdot] \circ \mathcal{T}_i^{-1}$ denotes the usual composition,[b] and the notation $\mathcal{T}_i^{\text{rot}}(\cdot)$ denotes pointwise application of $\mathcal{T}_i^{\text{rot}}$ to the (vector-valued) argument;[c] we apply $\mathcal{T}_i^{\text{rot}}$ to the dependent variables to eliminate parameters related to spatial orientation of components from the bilinear forms, and to accommodate compatibility of basis functions on instantiated ports.

We may now introduce the synthesized system domain Ω as $\overline{\Omega} = \cup_{i=1}^{I} \overline{\Omega}_i$, the system parameter domain $\mathcal{D} = \oplus_{i=1}^{I} \mathcal{D}_i$, and the system parameter vector $\mu = (\mu_1, \ldots, \mu_I)$; we denote the total number of system parameters by P.

When an instantiated component becomes part of a system, its local ports are associated to global ports. Each global port Γ_p, $1 \leq p \leq n_0^{\Gamma}$, in the system is either a coincidence of two local ports and hence in the interior of Ω, or a single local port on the boundary $\partial\Omega$. We define the connectivity of the system through global-to-local index sets π_p, $1 \leq p \leq n_0^{\Gamma}$: an interior global port is associated to two local ports $\gamma_{i,j}$ and $\gamma_{i',j'}$, and we thus set $\pi_p = \{(i,j), (i',j')\}$; a boundary global port is associated to a single local port $\gamma_{i,j}$, and we thus set $\pi_p = \{(i,j)\}$. We also introduce for instantiated component i, $1 \leq i \leq I$, a local-to-global map $\mathcal{G}_i$ such that for local port j, $1 \leq j \leq n_{\mathcal{M}(i)}^{\gamma}$, we have $\mathcal{G}_i(j) = p$ if $(i,j) \in \pi_p$. Note that on any global port Γ_p, $1 \leq p \leq n_0^{\Gamma}$, we may elect to impose Dirichlet boundary conditions; we denote by $n^\Gamma \leq n_0^\Gamma$ the number of global ports on which we do *not* impose Dirichlet boundary conditions.

To ensure global continuity of the solution we must require conforming port spaces and bases in the sense that for any shared (that is, interior) global port $\pi_p = \{(i,j),(i',j')\}$ we must have

$$\chi_{i,j,k} = \chi_{i',j',k}; \tag{7}$$

we discuss this port compatibility requirement further in the "Port compatibility" subsection below.

We may now introduce for any $w, v \in (H^1(\Omega))^d$ and any $\mu \in \mathcal{D}$ the system-level symmetric, continuous bilinear form as

$$a(w,v;\mu) = \sum_{i=1}^{I} \hat{a}_{\mathcal{M}(i)}((\mathcal{T}_i^{\mathrm{rot}})^{-1}(w|_{\Omega_i} \circ \mathcal{T}_i), (\mathcal{T}_i^{\mathrm{rot}})^{-1}(v|_{\Omega_i} \circ \mathcal{T}_i); \mu_i), \tag{8}$$

and the system-level bounded linear functional

$$f(v;\mu) = \sum_{i=1}^{I} \hat{f}_{\mathcal{M}(i)}((\mathcal{T}_i^{\mathrm{rot}})^{-1}(v|_{\Omega_i} \circ \mathcal{T}_i); \mu_i); \tag{9}$$

note that the effect of the mapping $\mathcal{T}_i$ to each archetype bilinear and linear form (defined over the archetype reference domain) is reflected through the parameter μ_i.

In the case that $\mathcal{T}_i$ is a pure rigid-body transformation (that is, $\mathcal{T}_i$ is a rotation and a translation) and the material properties of the component do not depend on spatial orientation — such as in isotropic linear elasticity — the application of $\mathcal{T}^{\mathrm{rot}}$ to the dependent variables results in cancellation of the mapping Jacobians, and thus the archetype bilinear form does not reflect the associated mapping parameters. Similarly, when $\mathcal{T}_i$ is a combination of a rigid-body map and (say) dilation, only the latter must be parametrized through the archetype bilinear form. We explicitly demonstrate this cancellation for the case of isotropic linear elasticity in the "Microtruss beam application" section, and we comment on the computational implications in the "Model reduction" section.

We now introduce a global space $X(\Omega) \subset (H^1(\Omega))^d$ such that $X(\Omega)$ is equal to $(H^1(\Omega))^d$ except for restrictions to enforce port (and in general also non-port) Dirichlet boundary conditions; we assume that sufficient boundary conditions are enforced such that $a(\cdot,\cdot;\mu)$ is coercive over $X(\Omega)$. The well-posed system-level variational problem then reads as follows. For any $\mu \in \mathcal{D}$, find $u(\mu) \in X(\Omega)$ such that

$$a(u(\mu), v; \mu) = f(v;\mu), \quad \forall v \in X(\Omega); \tag{10}$$

we also introduce a compliance output as $s(\mu) = f(u(\mu);\mu)$. (Note that, as discussed in [5], restrictions apply to the geometric maps $\mathcal{T}_i$ to maintain well-posedness of (10).)

Similarly, we introduce a global FE discretization $X^h(\Omega) \subset X(\Omega)$ as $X^h(\Omega) = \oplus_{i=1}^I X_i^h(\Omega) \cap X(\Omega)$; hence $X^h(\Omega)$ inherits the boundary conditions as well as the global continuity enforced by $X(\Omega)$. The FE discretization of (10) now reads as follows. For any $\mu \in \mathcal{D}$, find $u^h(\mu) \in X^h(\Omega)$ such that

$$a(u^h(\mu), v; \mu) = f(v;\mu), \quad \forall v \in X^h(\Omega); \tag{11}$$

we also introduce the FE compliance output $s^h(\mu) = f(u^h(\mu);\mu)$.

Mathematical formulation: static condensation

To formulate the static condensation procedure we decompose our discrete global space $X^h(\Omega)$ into bubble spaces associated with component-interior degrees of freedom and a skeleton space associated with port degrees of freedom.

To this end we introduce on archetype component m, $1 \le m \le M$, the bubble space

$$\hat{B}^h_{m;0} = \{v \in \hat{X}^h_m : v|_{\hat{\gamma}_{m,j}} = 0, \quad 1 \le j \le n^{\gamma}_m\}; \tag{12}$$

note that members of $\hat{B}^h_{m;0}$ vanish on local ports. We next introduce the coupling modes $\hat{\psi}_{m,j,k} \in \hat{X}^h_m$ such that

$$\int_{\hat{\Omega}_m} \nabla \hat{\psi}_{m,j,k} \cdot \nabla v = 0, \quad \forall v \in \hat{B}^h_{m;0}, \tag{13}$$

$$\hat{\psi}_{m,j,k} = \begin{cases} \hat{\chi}_{m,j,k}, & \text{on } \hat{\gamma}_{m,j}, \\ 0, & \text{on } \hat{\gamma}_{m,j'} \text{ for } j' \neq j, \end{cases} \tag{14}$$

for $1 \le k \le \mathcal{N}^{\gamma}_{m,j}$, $1 \le j \le n^{\gamma}_m$; we define, on instantiated component i, $1 \le i \le I$, $\psi_{i,j,k} \equiv \mathcal{T}^{\text{rot}}_i(\hat{\psi}_{\mathcal{M}(i),j,k} \circ \mathcal{T}^{-1}_i)$, and we introduce the global functions $\Psi_{p,k} \in X^h(\Omega)$ such that, for $\pi_p = \{(i,j),(i',j')\}$,

$$\Psi_{p,k} \equiv \begin{cases} \psi_{i',j',k}, & \text{in } \Omega_{i'}, \\ \psi_{i,j,k}, & \text{in } \Omega_i, \\ 0, & \text{in } \Omega \setminus (\Omega_{i'} \cup \Omega_i). \end{cases} \tag{15}$$

We may then introduce the global skeleton space

$$\mathcal{S}(\Omega) \equiv \text{span}\{\Psi_{p,k}, \quad 1 \le k \le \mathcal{N}^{\Gamma}_p, 1 \le p \le n^{\Gamma}\} \tag{16}$$

of dimension

$$n_{\text{SC}} \equiv \sum_{p=1}^{n^{\Gamma}} \mathcal{N}^{\Gamma}_p. \tag{17}$$

Note that $\mathcal{S}(\Omega)$ is a continuous space thanks to the port compatibility requirement (7). Also note that in the definition of $\mathcal{S}(\Omega)$ we include only the $n^{\Gamma} \le n^{\Gamma}_0$ ports on which we do not impose Dirichlet boundary conditions (we assume without loss of generality that we enforce Dirichlet boundary conditions on global ports $\Gamma_{n^{\Gamma}+1}, \ldots, \Gamma_{n^{\Gamma}_0}$).

Given the bubble spaces and the coupling modes, we now first introduce, for $1 \le i \le I$, the source bubble $\hat{b}^{f;h}_i(\mu_i) \in \hat{B}^h_{\mathcal{M}(i);0}$, which satisfies

$$\hat{a}_{\mathcal{M}(i)}(\hat{b}^{f;h}_i(\mu_i), v; \mu_i) = \hat{f}_{\mathcal{M}(i)}(v; \mu_i), \quad \forall v \in \hat{B}^h_{\mathcal{M}(i);0}; \tag{18}$$

we define $b^{f;h}_i(\mu_i) \equiv \mathcal{T}^{\text{rot}}_i(\hat{b}^{f;h}_i(\mu_i) \circ \mathcal{T}^{-1}_i)$. Note that $b^{f;h}_i(\mu)$ is a component-local particular solution to our global equation. We next introduce fundamental solutions $\hat{\phi}_{i,j,k}(\mu_i) \equiv \hat{b}^h_{i,j,k}(\mu_i) + \hat{\psi}_{i,j,k}$ associated with each coupling mode $\hat{\psi}_{i,j,k}$ and bubble $\hat{b}^h_{i,j,k}(\mu_i) \in B^h_{\mathcal{M}(i);0}$ such that $\hat{\phi}_{i,j,k}(\mu_i)$ satisfies

$$\hat{a}_{\mathcal{M}(i)}(\hat{\phi}^h_{i,j,k}(\mu_i), v; \mu_i) = 0, \quad \forall v \in \hat{B}^h_{\mathcal{M}(i);0} \tag{19}$$

(note (19) is an equation for $\hat{b}^h_{i,j,k}(\mu_i)$ given the known $\hat{\psi}_{i,j,k}$); we define, on instantiated component i, $1 \le i \le I$, $\phi_{i,j,k}(\mu_i) \equiv \mathcal{T}_i^{\mathrm{rot}}(\hat{\phi}_{i,j,k}(\mu_i) \circ \mathcal{T}_i^{-1})$ and we introduce the global functions $\Phi_{p,k}(\mu) \in X^h(\Omega)$ such that, for $\pi_p = \{(i,j),(i',j')\}$,

$$\Phi_{p,k}(\mu) \equiv \begin{cases} \phi_{i',j,k}(\mu_{i'}), & \text{in } \Omega_{i'}, \\ \phi_{i,j,k}(\mu_i), & \text{in } \Omega_i, \\ 0, & \text{in } \Omega \setminus (\Omega_{i'} \cup \Omega_i). \end{cases} \tag{20}$$

Note that each $\Phi_{\mathcal{G}_j(i),k}(\mu)$ is the fundamental solution (local to a component pair) of our (homogeneous) global equation associated with the particular port mode $\chi_{i,j,k}$. Also note that $\Phi_{\mathcal{G}_j(i),k}(\mu)$ and $b_i^{f;h}(\mu_i)$ scale linearly with certain "free" parameters, such as component-wide thermal conductivity or Young's modulus, which enter outside the bilinear form in (18) and (19); this will have important cost-saving implications in the context of RB approximation.

For each instantiated component we introduce a global function $u_i^h(\mu_i) \in X^h(\Omega)$ which represents the local solution on component i in terms of the source bubbles $b_i^{f;h}(\mu_i)$ and the fundamental solutions $\phi_{i,j,k}(\mu_i)$ as

$$u_i^h(\mu_i) = \begin{cases} \sum\limits_{i=1}^{I} b_i^{f;h}(\mu_i) + \sum\limits_{j=1}^{n^\gamma_{\mathcal{M}(i)}} \sum\limits_{k=1}^{\mathcal{N}^\gamma_{\mathcal{M}(i),j}} \mathbb{U}_{\mathcal{G}_i(j),k}(\mu)\phi_{i,j,k}(\mu_i), & \text{in } \Omega_i, \\ 0, & \text{in } \Omega \setminus \Omega_i, \end{cases} \tag{21}$$

where the coefficient vector $\mathbb{U}_{\mathcal{G}_i(j),k}(\mu)$ contains global unknowns to be determined below. To couple the solutions in neighboring components we require weak flux continuity across global ports:[d] we write

$$u^h(\mu) = \sum_{i=1}^{I} u_i^h(\mu_i) = \sum_{i=1}^{I} b_i^{f;h}(\mu_i) + \sum_{p=1}^{n^\Gamma} \sum_{k=1}^{\mathcal{N}_p^\Gamma} \mathbb{U}_{p,k}(\mu)\Phi_{p,k}(\mu); \tag{22}$$

we then test on all $v \in \mathcal{S}(\Omega)$ such that $u^h(\mu) \in X^h(\Omega)$ satisfies

$$a(u^h(\mu), v; \mu) = f(v; \mu), \quad \forall v \in \mathcal{S}(\Omega); \tag{23}$$

as before, our FE compliance output is $s^h(\mu) = f(u^h(\mu); \mu)$.

We emphasize that (23) is, thanks to Galerkin orthogonality of the fundamental solutions in (19) with respect to the associated bubble space, equivalent to (11). For this same reason we may further define an alternative skeleton space

$$\mathcal{S}_{\mathrm{symm}} \equiv \mathrm{span}\{\Phi_{p,k}(\mu), \quad 1 \le k \le \mathcal{N}_p^\Gamma, 1 \le p \le n^\Gamma\}, \tag{24}$$

such that $u^h(\mu) \in X^h(\Omega)$ satisfies

$$a(u^h(\mu), v; \mu) = f(v; \mu), \quad \forall v \in \mathcal{S}_{\mathrm{symm}}(\Omega). \tag{25}$$

There is no distinction between (23) and (25) in the FE static condensation context; however in the context of the SCRBE, direct approximation of (23) leads to a non-symmetric Schur complement system, while direct approximation of (25) leads to a symmetric Schur complement system. In this paper we shall pursue the former with subsequent Schur complement *symmetrization* as the latter implies significantly larger online computational cost.

The formulation (23) is equivalent to the *symmetric* linear-algebraic Schur complement system

$$\mathbb{A}(\mu)\mathbb{U}(\mu) = \mathbb{F}(\mu) \tag{26}$$

of size $n_{\rm SC}$, in which

$$\mathbb{A}_{(p,k),(p',k')}(\mu) = a(\Phi_{p',k'}(\mu), \Psi_{p,k}; \mu), \tag{27}$$

$$\mathbb{F}_{(p,k)}(\mu) = f(\Psi_{p,k}; \mu) - \sum_{i=1}^{I} a(b_i^{f;h}(\mu_i), \Psi_{p,k}; \mu). \tag{28}$$

We may readily demonstrate the symmetry: by (19) and symmetry of $\hat{a}_m(\cdot, \cdot; \mu_i)$ we obtain

$$\hat{a}_{\mathcal{M}(i)}(\hat{\phi}_{i,j,k}(\mu_i), \hat{\psi}_{i,j',k'}; \mu) = \hat{a}_{\mathcal{M}(i)}(\hat{b}^h_{i,j,k}(\mu_i) + \hat{\psi}_{i,j,k}, \hat{b}^h_{i,j',k'}(\mu_i) + \psi_{i,j',k'}; \mu_i) \tag{29}$$

$$= \hat{a}_{\mathcal{M}(i)}(\hat{b}^h_{i,j',k'}(\mu_i) + \hat{\psi}_{i,j',k'}, \hat{b}^h_{i,j,k}(\mu_i) + \psi_{i,j,k}; \mu_i) \tag{30}$$

$$= \hat{a}_{\mathcal{M}(i)}(\hat{b}^h_{i,j',k'}(\mu_i) + \hat{\psi}_{i,j',k'}, \hat{\psi}_{i,j,k}; \mu_i) \tag{31}$$

$$= \hat{a}_{\mathcal{M}(i)}(\hat{\phi}_{i,j',k'}(\mu_i), \hat{\psi}_{i,j,k}; \mu_i), \tag{32}$$

and as a result

$$a(\Phi_{p',k'}(\mu), \Psi_{p,k}; \mu) = a(\Phi_{p,k}(\mu), \Psi_{p',k'}; \mu). \tag{33}$$

The matrix $\mathbb{A}(\mu)$ is thus symmetric and in particular may be rewritten as

$$\mathbb{A}_{(p,k),(p',k')}(\mu) = \frac{1}{2}a(\Phi_{p',k'}(\mu), \Psi_{p,k}; \mu) + \frac{1}{2}a(\Phi_{p,k}(\mu), \Psi_{p',k'}; \mu) \tag{34}$$

We shall invoke the interpretation (34) of $\mathbb{A}(\mu)$ to symmetrize the SCRBE Schur complement system below.

Port compatibility

The port compatibility requirement (7) between port basis functions associated with ports which may interconnect in a system — port of the same *type* — ensures solution continuity across shared global ports. We recall the archetype port basis functions $\hat{\chi}_{\mathcal{M}(i),j,k}$ introduced in (3), and we recall the associated physical (instantiated) port space basis functions $\chi_{i,j,k}$ introduced in (6). To honor (7), it is clear that the basis functions $\hat{\chi}_{m,j,k}$ on different archetype ports of the same port type must be defined differently according to the archetype port orientation.

To render this more precise we introduce for each unique port type a reference port domain $\beta \subset \mathbb{R}^{d-1}$; we assume for simplicity of exposition that there is only a single port type and thus β needs no subscript. We then consider, on archetype component m, each archetype port domain $\hat{\gamma}_{m,j}$ as the image of β under a rigid-body map $\mathcal{R}_{m,j} = \mathcal{R}^{\rm rot}_{m,j}\mathcal{R}^{\rm tran}_{m,j}$, where $\mathcal{R}^{\rm rot}_{m,j}$ corresponds to rotation and $\mathcal{R}^{\rm tran}_{m,j}$ corresponds to translation, such that

$$\hat{\gamma}_{m,j} = \mathcal{R}_{m,j}(\beta); \tag{35}$$

this map is the key to honor the port compatibility requirement (7).[e]

We then introduce, on the reference port domain β, a set of reference port modes $\hat{\chi}^\beta_k$, $1 \le k \le \mathcal{N}^\beta$, and an associated reference port space

$$P^\beta = \text{span}\{\hat{\chi}^\beta_k, \quad 1 \le k \le \mathcal{N}^\beta\} \tag{36}$$

of dimension $\mathcal{N}^\beta$. We then define, on archetype port domain $\hat{\gamma}_{m,j}$ of type β, the archetype port space basis functions $\hat{\chi}_{m,j,k}$, $1 \leq k \leq \mathcal{N}_m^\gamma$, $1 \leq j \leq n_m^\gamma$, as

$$\hat{\chi}_{m,j,k} \equiv \mathcal{R}_{m,j}^{\text{rot}}(\hat{\chi}_k^\beta \circ \mathcal{R}_{m,j}^{-1}), \tag{37}$$

Note that $\mathcal{N}_m^\gamma = \mathcal{N}^\beta$.

We now consider two instantiated port domains

$$\gamma_{i,j} = \mathcal{T}_i(\hat{\gamma}_{\mathcal{M}(i),j}), \quad \gamma_{i',j'} = \mathcal{T}_{i'}(\hat{\gamma}_{\mathcal{M}(i'),j'}), \tag{38}$$

on different instantiated components i and i'. From (5) we have for the associated port space basis functions

$$\chi_{i,j,k} = \mathcal{T}_i^{\text{rot}}(\hat{\chi}_{\mathcal{M}(i),j,k} \circ \mathcal{T}_i^{-1}), \tag{39}$$

$$\chi_{i',jv,k} = \mathcal{T}_{i'}^{\text{rot}}(\hat{\chi}_{\mathcal{M}(i'),j',k} \circ \mathcal{T}_{i'}^{-1}), \tag{40}$$

and so, with (37),

$$\chi_{i,j,k} = \mathcal{T}_i^{\text{rot}}(\mathcal{R}_{\mathcal{M}(i),j}^{\text{rot}}(\hat{\chi}_k^\beta \circ \mathcal{R}_{\mathcal{M}(i),j}^{-1}) \circ \mathcal{T}_i^{-1}), \tag{41}$$

$$\chi_{i',j',k} = \mathcal{T}_{i'}^{\text{rot}}(\mathcal{R}_{\mathcal{M}(i'),j'}^{\text{rot}}(\hat{\chi}_k^\beta \circ \mathcal{R}_{\mathcal{M}(i'),j'}^{-1}) \circ \mathcal{T}_{i'}^{-1}). \tag{42}$$

Now, suppose that $\pi_p = \{(i,j),(i',j')\}$ for a shared global port Γ_p such that $\gamma_{i,j} = \gamma_{i',j'}$. In this case, from (35) and (38), we obtain

$$\mathcal{T}_i(\mathcal{R}_{\mathcal{M}(i),j}(\beta)) = \mathcal{T}_{i'}(\mathcal{R}_{\mathcal{M}(i'),j'}(\beta)). \tag{43}$$

We recall that $\mathcal{T}_i^{\text{def}}$ (for $1 \leq i \leq I$) when applied to a port corresponds to pure translation. As a result, application of the port mapping $\mathcal{T}_i\mathcal{R}_{\mathcal{M}(i),j}$ corresponds only to translation and rotation. We now recall that the rotation applied to β on each side of (43) is unique, and we may thus conclude from (43) that $\mathcal{T}_i^{\text{rot}}\mathcal{R}_{\mathcal{M}(i),j}^{\text{rot}} = \mathcal{T}_{i'}^{\text{rot}}\mathcal{R}_{\mathcal{M}(i'),j'}^{\text{rot}}$. With (41) and (42), we then obtain $\chi_{i,j,k} = \chi_{i',j',k}$, and we thus honor our port compatibility requirement (7).

Model reduction

The computational efficacy of our port-reduced SCRBE approach is realized through two separate model reduction techniques. As in the standard SCRBE approach [5] we consider component-interior model reduction through RB approximation [6] of the source bubbles (18) and of the fundamental solutions (19) to reduce the cost of each of the many component-interior linear solves required to form the Schur complement system. In addition to RB approximation in the component interiors, we employ port reduction [12] with empirical port modes to reduce the number of degrees of freedom on the ports and thus the size of the Schur complement system. We now discuss each of these techniques in more detail.

Component-interior reduction

For the component-interior model reduction we employ RB approximations

$$\tilde{b}_i^f(\mu_i) \approx b_i^{f;h}(\mu_i), \tag{44}$$

$$\tilde{\phi}_{i,j,k}(\mu_i) \approx \phi_{i,j,k}^h(\mu_i), \tag{45}$$

and thus $\tilde{\Phi}_{p,k}(\mu) \approx \Phi_{p,k}(\mu)$. The purpose of these RB approximations is to allow for efficient formation of an *approximation to* the Schur complement system (54): each RB approximation $\tilde{b}_i^f(\mu_i)$ or $\tilde{\phi}_{i,j,k}(\mu_i)$ is associated with a rapidly convergent [7] RB space specifically tailored to the particular bubble and to the parameter dependence defined by the corresponding (archetype domain) PDE (18) or (19). All RB bubble spaces are thus different, and furthermore each space is typically of much lower dimension than the original FE spaces $\hat{B}^h_{m;0}$. As a consequence, the RB approximations to the solutions of (18) and (19) are obtained at significantly reduced computational cost with minimal compromise to solution accuracy. The RB method is now considered standard, and we refer the reader to [6] for all technical details relevant to the particular class of problems (linear elliptic) that we consider here.

We now introduce the SCRBE approximation $\tilde{u}^*(\mu) \approx u^h(\mu)$ as

$$\tilde{u}^*(\mu) = \sum_{i=1}^{I} \tilde{b}_i^f(\mu_i) + \sum_{p=1}^{n^\Gamma} \sum_{k=1}^{\mathcal{N}_p^\Gamma} \tilde{\mathbb{U}}^*_{p,k}(\mu) \tilde{\Phi}_{p,k}(\mu), \tag{46}$$

and we again choose $\mathcal{S}(\Omega)$ as the test space such that $\tilde{u}^*(\mu) \in X^h(\Omega)$ satisfies

$$a(\tilde{u}^*(\mu), v; \mu) = f(v; \mu), \quad \forall v \in \mathcal{S}(\Omega); \tag{47}$$

the equivalent linear-algebraic system is

$$\tilde{\mathbb{A}}^*(\mu)\tilde{\mathbb{U}}^*(\mu) = \tilde{\mathbb{F}}(\mu) \tag{48}$$

where

$$\tilde{\mathbb{A}}^*_{(p,k),(p',k')}(\mu) = a(\tilde{\Phi}_{p',k'}(\mu), \tilde{\Psi}_{p,k}; \mu), \tag{49}$$

$$\tilde{\mathbb{F}}_{(p,k)}(\mu) = f(\tilde{\Psi}_{p,k}; \mu) - \sum_{i=1}^{I} a(b_i^{f;h}(\mu_i), \tilde{\Psi}_{p,k}; \mu), \tag{50}$$

for $1 \le k \le \mathcal{N}_p^\Gamma$, $1 \le k \le \mathcal{N}_{p'}^\Gamma$, $1 \le p, p' \le n^\Gamma$. Note that $\tilde{\mathbb{A}}^*(\mu)$ in (49) is non-symmetric because each RB approximation $\tilde{\phi}_{i,j,k}(\mu_i)$ (mapped to the respective archetype domain) satisfies (19) only with respect to the associated RB bubble subspace. These RB approximations are thus not Galerkin orthogonal with respect to other bubble spaces; recall that this Galerkin orthogonality (together with symmetry of $\hat{a}_m(\cdot, \cdot; \mu_i)$) is the key to the symmetry of $\mathbb{A}(\mu)$ as demonstrated in (29). To recover symmetry we have two options: we may either, as in [5,12], test on a space

$$\tilde{\mathcal{S}}_{\text{symm}}(\Omega) = \text{span}\{\tilde{\Phi}_{p,k}(\mu), \quad 1 \le k \le \mathcal{N}_p^\Gamma, 1 \le p \le n^\Gamma\} \tag{51}$$

in (47), or we may explicitly symmetrize $\tilde{\mathbb{A}}^*(\mu)$ by algebraic manipulation. The former option necessitates larger offline and online computational cost and storage, in fact, when compared to the latter, by a multiplicative factor equal to the number of RB basis functions.

We thus elect to recover symmetry by algebraic manipulation: we exploit the interpretation (34) of $\mathbb{A}(\mu)$ and we define $\tilde{\mathbb{A}}(\mu) \approx \mathbb{A}(\mu)$ as

$$\tilde{\mathbb{A}}(\mu) = \frac{1}{2}\tilde{\mathbb{A}}^*(\mu) + \frac{1}{2}\tilde{\mathbb{A}}^*(\mu)^{\mathrm{T}}, \tag{52}$$

such that

$$\tilde{\mathbb{A}}_{(p,k),(p',k')}(\mu) = \frac{1}{2}a(\tilde{\Phi}_{p',k'}(\mu), \Psi_{p,k}; \mu) + \frac{1}{2}a(\tilde{\Phi}_{p,k}(\mu), \Psi_{p',k'}; \mu). \tag{53}$$

We may then finally introduce our symmetric SCRBE linear-algebraic system as

$$\tilde{\mathbb{A}}(\mu)\tilde{\mathbb{U}}(\mu) = \tilde{\mathbb{F}}(\mu), \tag{54}$$

and we define the SCRBE field approximation $\tilde{u}(\mu) \approx u^h(\mu)$ as

$$\tilde{u}(\mu) = \sum_{i=1}^{I} \tilde{b}_i^f(\mu_i) + \sum_{p=1}^{n^\Gamma} \sum_{k=1}^{\mathcal{N}_p^\Gamma} \tilde{\mathbb{U}}_{p,k}(\mu) \tilde{\Phi}_{p,k}(\mu). \tag{55}$$

The associated SCRBE compliance output approximation is $\tilde{s}(\mu) = f(\tilde{u}(\mu); \mu)$.

Note that in actual practice, we assemble (54) through a direct-stiffness procedure from component-local matrix and vector blocks associated with and assembled for each of the I component instantiations; the procedure is described in detail in [5,12]. The assembly of these component-local quantities constitutes the majority of online computational cost. However, we need only perform the assembly for each *unique* component instantiation, as identical (or "cloned") components may share local matrices and vectors. We thus realize significant computational savings for systems which consist of instantiations of many component clones, such that we need only consider $I_{\text{eff}} \ll I$ *effective* component instantiations for this assembly proceedure.

There are two particularly important situations in which *different* component instantiations are effectively clones in the sense that the component-local matrix and vector blocks may still be re-used: First, matrix and vector blocks computed for component instantiations which differ only in spatial orientation are (in the case that material properties do not depend on spatial orientation, such as in isotropic linear elasticity) identical thanks to cancellation of the mapping Jacobians in the archetype domain bilinear form; second, "free" parameters such as component-wide thermal conductivity or Young's modulus enter outside the bilinear forms in (18) and (19), and thus the associated matrix blocks will only differ by a scaling factor. As a result, we often obtain $I_{\text{eff}} \ll I$ in practice. We discuss this situation further under "Computational procedures" later in this section.

Port reduction

Framework

While the RB approximation is concerned with component-interior model reduction, we apply port reduction to reduce the number of degrees of freedom associated with component interfaces. For the port reduction procedure we shall consider on each global port Γ_p only $n_{\text{A},p}^\Gamma \leq \mathcal{N}_p^\Gamma$ port modes as "Active" and thus contributing to the approximation; for substantial computational savings we require $n_{\text{A},p}^\Gamma \ll \mathcal{N}_p^\Gamma$. We consider in this subsection the generic port reduction framework and in the next subsection our particular choice of port space basis functions which realizes $n_{\text{A},p}^\Gamma \ll \mathcal{N}_p^\Gamma$.

Based on the $n_{\text{A},p}^\Gamma$ active modes associated with each global port we introduce a port-reduced skeleton space $\mathcal{S}^{\text{PR}}(\Omega) \subseteq \mathcal{S}(\Omega)$ as

$$\mathcal{S}^{\text{PR}}(\Omega) \equiv \text{span}\{\Psi_{p,k}, \quad 1 \leq k \leq n_{\text{A},p}^\Gamma, 1 \leq p \leq n^\Gamma\} \tag{56}$$

of dimension

$$n_{\text{A}} \equiv \sum_{p=1}^{n^\Gamma} n_{\text{A},p}^\Gamma \leq n_{\text{SC}}. \tag{57}$$

We further introduce a port-reduced approximation $\tilde{u}^{\text{PR},*}(\mu) \approx u^h(\mu)$ as

$$\tilde{u}^{\text{PR},*}(\mu) = \sum_{i=1}^{I} \tilde{b}_i^f(\mu_i) + \sum_{p=1}^{n^\Gamma} \sum_{k=1}^{n_{\text{A},p}^\Gamma} \tilde{\mathbb{U}}_{p,k}^*(\mu) \tilde{\Phi}_{p,k}(\mu). \tag{58}$$

We now choose $\mathcal{S}^{\mathrm{PR}}(\Omega)$ as our test space such that

$$a(\tilde{u}^{\mathrm{PR},*}(\mu), v; \mu) = f(v; \mu), \quad \forall v \in \mathcal{S}^{\mathrm{PR}}(\Omega), \tag{59}$$

which leads to the linear-algebraic system

$$\tilde{\mathbb{A}}^{\mathrm{PR},*}(\mu)\tilde{\mathbb{U}}^{\mathrm{PR},*}(\mu) = \tilde{\mathbb{F}}^{\mathrm{PR}}(\mu) \tag{60}$$

of size n_{A}, where

$$\tilde{\mathbb{A}}^{\mathrm{PR},*}_{(p,k),(p',k')}(\mu) = a(\tilde{\Phi}_{p',k'}(\mu), \Psi_{p,k}; \mu), \tag{61}$$

$$\tilde{\mathbb{F}}^{\mathrm{PR}}_{(p,k)}(\mu) = f(\Psi_{p,k}; \mu) - \sum_i a(b_i^{f;h}(\mu_i), \Psi_{p,k}; \mu), \tag{62}$$

for $1 \le k \le n^{\Gamma}_{\mathrm{A},p}$, $1 \le k' \le n^{\Gamma}_{\mathrm{A},p'}$, $1 \le p, p' \le n^{\Gamma}$. We then symmetrize as

$$\tilde{\mathbb{A}}^{\mathrm{PR}} \equiv \frac{1}{2}\tilde{\mathbb{A}}^{\mathrm{PR},*}(\mu) + \frac{1}{2}\tilde{\mathbb{A}}^{\mathrm{PR},*}(\mu)^{\mathrm{T}}, \tag{63}$$

we define the port-reduced SCRBE system as

$$\tilde{\mathbb{A}}^{\mathrm{PR}}(\mu)\tilde{\mathbb{U}}^{\mathrm{PR}}(\mu) = \tilde{\mathbb{F}}^{\mathrm{PR}}(\mu), \tag{64}$$

and we define the port-reduced SCRBE field approximation $\tilde{u}^{\mathrm{PR}}(\mu) \approx u^h(\mu)$ as

$$\tilde{u}^{\mathrm{PR}}(\mu) = \sum_{i=1}^{I} \tilde{b}_i^f(\mu_i) + \sum_{p=1}^{n^{\Gamma}} \sum_{k=1}^{n^{\Gamma}_{\mathrm{A},p}} \tilde{\mathbb{U}}^{\mathrm{PR}}_{p,k}(\mu)\tilde{\Phi}_{p,k}(\mu). \tag{65}$$

The associated port-reduced SCRBE compliance output approximation is $\tilde{s}^{\mathrm{PR}}(\mu) = f(\tilde{u}^{\mathrm{PR}}(\mu); \mu)$.

The purpose of port reduction is of course to reduce the size of the Schur complement system — and thus computational cost — while maintaining accuracy of the approximation. The size of the system (64), n_{A}, is equal to the total number of active port modes in the system. In practice, we shall typically invoke only a few port degrees of freedom on each port such that $n_{\mathrm{A}} \ll n_{\mathrm{SC}}$. A good choice for the port modes $\chi_{i,j,k}$ is key to the accuracy of the port-reduced SCRBE approximation, and is the focus of the next subsection.

Empirical port mode training

To ensure port compatibility we must for each port type develop our port basis on the associated reference port domain β as discussed under "Port compatibility" above. To this end we pursue a pairwise training algorithm that provides a port space tailored to the family of solutions associated with this port type. We shall develop bases for the full port spaces (6) and not merely the space spanned by "Active" modes; the remaining "Inactive" modes shall play a role in certification (for residual calculation), which we discuss further in the "Certification framework" section.

Our port spaces shall consist of three sets of modes. The first set of modes is explicitly specified and consists of the six modes associated with rigid-body motion.[f] We include these six modes for two reasons: first, it simplifies the procedure for specification of typical Dirichlet boundary conditions, and second, it ensures invertibility of the Schur complement operator associated with "Inactive" modes, which is a property we require for our non-conforming error estimation framework.

The second set of modes consists of the $n_{\text{pod}}^{\beta} \leq \mathcal{N}^{\beta} - 6$ modes which shall be the outcome of our pairwise training algorithm. The third set of modes consists of $\mathcal{N}^{\beta} - n_{\text{pod}}^{\beta} - 6$ singular Sturm-Liouville eigenmodes restricted to the orthogonal complement of the first $n_{\text{pod}}^{\beta} + 6$ empirical modes [12]. These modes serve to complete the discrete port space in a numerically stable fashion.

Recall that the total number of modes associated with the reference port β is $\mathcal{N}^{\beta}$. We consider here the case $d = 3$ and thus $\beta \subset \mathbb{R}^2$; each port mode $\hat{\chi}_i^{\beta}$, $1 \leq i \leq \mathcal{N}^{\beta}$, has the form $\hat{\chi}_i^{\beta} = (\hat{\chi}_i^{\beta,1}, \hat{\chi}_i^{\beta,2}, \hat{\chi}_i^{\beta,3})$, where the number of degrees of freedom associated with each field component $\hat{\chi}_i^{\beta,j}$ is $\mathcal{N}^{\beta}/3$.

In the case that β is the square $\beta = [-0.5, 0.5]^2$, the first six reference port modes are explicitly defined as

$$\hat{\chi}_1^{\beta} = (1,0,0), \quad \hat{\chi}_2^{\beta} = (0,1,0), \quad \hat{\chi}_3^{\beta} = (0,0,1), \tag{66}$$

for the three ports associated with translation; as

$$\hat{\chi}_4^{\beta}(\xi,\eta) = (-\eta,\xi,0) \tag{67}$$

for the mode associated with pure rotation; and as

$$\hat{\chi}_5^{\beta}(\xi,\eta) = (0,0,\xi), \qquad \hat{\chi}_6^{\beta} = (0,0,\eta), \tag{68}$$

for the two modes associated with flipping. Note these six modes are mutually $(L^2(\beta))^d$-orthonormal. (If β is not the square $\beta = [-0.5, 0.5]^2$ we apply Gram-Schmidt orthonormalization to these first six modes to recover $(L^2(\beta))^d$-orthonormality.)

The next n_{pod}^{β} port modes are the outcome of our pairwise empirical training algorithm. In this algorithm we exploit the fact that within any system, the solution on any global (shared, say) port is determined completely by the parameter values assigned to the pair of components sharing the port *and* the (typically relatively smooth) solution on all other ports associated with these two components. The purpose of our pairwise training algorithm is to explore the associated "solution manifold" induced by local parameter dependence and neighboring ports in a systematic fashion such that the empirical modes associated with each port type are tailored to all possible component connectivity and all admissible component parameter values.

For our empirical training algorithm we shall require discrete "Legendre polynomials" $L_i^{\beta}, 1 \leq i \leq \mathcal{N}^{\beta}/3$, such that the L_i^{β} are the eigenvectors of a scalar singular Sturm-Liouville eigenproblem [16] over β ordered according to increasing eigenvalue; we shall also require a univariate random variable r with uniform density; and we introduce an algorithm tuning parameter $\gamma > 1$ related to anticipated regularity. We then identify one or several pairs of components in the component library that may connect through a global port of the relevant port type β.

The empirical training procedure for each such pair is now given by Algorithm 1: we sample (solve) each pair N_{sample} times for different (random) parameters and different (random but smooth thanks to the parameter $\gamma > 1$) boundary conditions on all non-connected ports (note that we assign random boundary conditions independently to each vector component); for each such sample we extract the solution on the shared port Γ_{p^*} of the relevant type, map it to the reference port β, subtract from this mapped solution its orthogonal $(L^2(\beta))^d$-projection onto each of the six rigid body modes $\hat{\chi}_i^{\beta}$, $1 \leq i \leq 6$, and then finally include the result ζ in a snapshot set S_{pair} associated with the current pair. Note that in Algorithm 1 $(\cdot,\cdot)_{L^2(\beta)}$ refers to the vector $(L^2(\beta))^d$ inner product.

Algorithm 1 Pairwise training (two components connected at global port Γ_{p^*})

$S_{\text{pair}} = \emptyset$.
for $n = 1,\dots,N_{\text{samples}}$ **do**
  Assign random parameters $\mu_i \in \mathcal{D}_i$ to component $i = 1, 2$.
  On all non-shared ports Γ_p, assign random boundary conditions:

$$u^i|_{\Gamma_p} = \sum_{k=1}^{\mathcal{N}^\beta/3} r\frac{1}{k^\gamma}L_k^\beta, \quad i = 1,2,3.$$

  Solve the two-component system; extract solution on shared port Γ_{p^*} (mapped to β):

$$\zeta \leftarrow u|_{\Gamma_{p^*}}$$

  Subtract the orthogonal projection onto rigid-body modes:
  for $i = 1,\dots,6$ **do**

$$\zeta \leftarrow \zeta - \frac{(\zeta,\hat{\chi}_i^\beta)_{L^2(\beta)}}{\|\hat{\chi}_i^\beta\|_{L^2(\beta)}}\hat{\chi}_i^\beta,$$

  end for
  Include the result in the snapshot set:

$$S_{\text{pair}} \leftarrow S_{\text{pair}} \cup \zeta$$

end for

After pairwise training of all pairs relevant for one port type, we form the bigger snapshot set

$$S_{\text{type}} = \bigcup_{\text{pair}} S_{\text{pair}}. \tag{69}$$

We then perform a data compression step: we invoke the proper orthogonal decomposition (POD) [17] (with respect to the vector $(L^2(\beta))^d$ inner product). The output from the POD procedure is a set of n_{pod}^β mutually $(L^2(\beta))^d$-orthonormal POD modes which are also orthonormal to the six first modes $\hat{\chi}_i^\beta$, $1 \le i \le 6$, related to rigid-body motion. We choose these n_{pod}^β POD modes as our next reference port basis functions $\hat{\chi}_{6+i}^\beta$, $1 \le i \le n_{\text{pod}}^\beta$; we typically observe rapid (often exponential) convergence [12] of these POD modes with respect to the input snapshot set S_{type}.

We refer to all first $n_{\text{pod}}^\beta + 6$ port modes as our *empirical port modes*. If n_{pod}^β is chosen such that $n_{\text{pod}}^\beta + 6 < \mathcal{N}^\beta$, we now complete the discrete space with Sturm-Liouville singular eigenmodes restricted to the orthogonal complement space $(\text{span}\{\hat{\chi}_i^\beta\}_{i=1}^{n_{\text{pod}}^\beta+6})^\perp$ (of dimension $\mathcal{N}^\beta - n_{\text{pod}}^\beta - 6$) as discussed in detail in [12].

We finally note that for our pairwise training approach we may employ the (non-port-reduced) SCRBE framework or we may use standard FE approximations. The computational cost associated with empirical training is not critical as the procedure is performed offline. For our numerical results in this paper we have used the non-port-reduced SCRBE framework to calculate empirical modes.

Computational procedures

The computational procedures associated with our port-reduced SCRBE approximation framework naturally decouple into an offline preprosessing stage and an online evaluation stage, and we now discuss each in more detail. Note we provide here only descriptions of each of the offline and online steps involved; for detailed online operation counts we refer to [12].

Offline

The offline stage is the preprosessing stage — performed only once — in which we construct and prepare the archetype component library. This stage consists of the following steps.

Off1. Empirical pairwise training by Algorithm 1. For each port type we sample pairs of components to obtain efficient port space basis functions $\hat{\chi}_k^{\beta}$, $1 \leq k \leq \mathcal{N}^{\beta}$, associated with each reference port domain β. In the current implementation, we employ the non-port-reduced SCRBE [5] (rather than standard global FE) for the pairwise training.

Off2. RB space construction. For each archetype component m, $1 \leq m \leq M$, we must train $\sum_{j=1}^{n_m^{\gamma}} \mathcal{N}_{m,j}^{\gamma} + 1$ *different* RB spaces to accommodate the RB approximations (44) and (45). Each construction of an RB space requires a number of component-local FE solves (each associated with an RB space basis function), and thus this step is potentially rather expensive, depending on the component spatial discretization and parametric complexity $\hat{Q}_m^a$ and $\hat{Q}_m^f$ in the bilinear and linear form expansions (1).
Note, however, that the construction of the RB approximation spaces (subsequent to port space construction) is embarrassingly parallel. Also note that we do not consider parameters for spatial orientation (because of the mapping Jacobian cancellations in the archetype domain formulation), and furthermore recall that components often have "free" parameters such as component-wide thermal conductivity or Young's modulus, with which the solutions to (18) and (19) simply scale linearly. As a result, RB space dimensions are typically rather small (around ten basis functions often suffice for each RB space), and thus although this step typically dominates offline cost the computational effort is not onerous: typically a couple of CPU hours is required for each archetype component.

Off3. Online dataset preparation. For each archetype component we construct data to enable efficient assembly of the component-local Schur complement matrix and vector blocks in the subsequent online stage. The computation time depends stongly on component spatial discretization and parametric complexity, but is typically between minutes and hours (on a single CPU) for each component. The online dataset also contains all RB basis functions, which are required for online global field visualization, if desired.

Off4. Data loading. We finally read the online datasets (typically a few hundred Mb) for all library components into computer memory to prepare for the online stage.

Note that with our current implementation of the framework, since we employ the non-port-reduced SCRBE in step Off1 above, we must first perform a step Off0a (similar to Off2) and then a step Off0b (similar to Off3) in order to enable the necessary "online" pair evaluation in Off1.

Online

The online stage is the stage in which we instantiate archetype components, and assemble and solve our system. This stage consists of the following steps, which in the current implementation is performed on a single CPU.

On1. Component instantiation. Instantiate I components from the library, assign the relevant parameter values to each component, and connect components to other components through ports of the same type to form a system; this step is most easily effected through a graphical user interface [Additional file 1].

On2. Schur complement system formation. Perform component-local RB solves (of small RB dimension) associated with all "Active" degrees of freedom to obtain (RB coefficients for) the RB approximations $\tilde{\phi}_{i,j',k'}(\mu_i)$ and $\tilde{b}^f(\mu_i)$, assemble the associated matrix and vector blocks for each component, and assemble the Schur complement system (64) through a direct-stiffness procedure [5,12].
The entries in the component-local matrix blocks are of the form

$$\tilde{\mathbb{A}}^i_{\mathrm{A,A};(j,k),(j',k')}(\mu_i) = \frac{1}{2}a_i(\tilde{\phi}_{i,j',k'}(\mu_i), \psi_{i,j,k}; \mu_i) + \frac{1}{2}a_i(\tilde{\phi}_{i,j,k}(\mu_i), \psi_{i,j',k'}; \mu_i) \tag{70}$$

(the symmetrization is performed on the component level) and the entries in the component-local vector blocks are of the form $\tilde{\mathbb{F}}^i_{\mathrm{A};(j,k)}(\mu_i) = f(\psi_{i,j,k}; \mu) - a_i(\tilde{b}^f(\mu_i), \psi_{i,j,k}; \mu)$; the subscripts $_\mathrm{A}$ refer to assembly of "Active" component matrices and vectors. However, thanks to an efficient construction-evaluation procedure [6], which relies on the affine operator expansions (1), only the RB *coefficients* associated with $\tilde{\phi}_{i,j',k'}(\mu_i)$ and $\tilde{b}^f(\mu_i)$ are required for this assembly step. We emphasize in particular that the underlying component FE discretization is never invoked.
We recall that parameters related to spatial orientation (component "docking") do not appear in the (archetype) bilinear forms due to cancellation of the associated Jacobians (we demonstrate this for isotropic linear elasticity in the "Microtruss beam application" section); and moreover, certain parametric variations such as component-wide conductivity or Young's modulus are "free" in the sense that they enter as scalars outside the bilinear forms in (18) and (19). As a consequence, matrix and vector blocks associated with different component instantiations are in practice often identical (in the context of "free" parameters up to a multiplicative constant). We may thus in typical systems often consider only $I_{\mathrm{eff}} \ll I$ effectively different (or unique) component instantiations, for which we perform RB solves and assemble component-local matrices and vectors. The component-local matrices and vectors for the remaining $I - I_{\mathrm{eff}}$ component instantiations are then simply copies of the respective data from effectively identical components. This consideration of component "clones" together with

the realization of "docking" parameter cancellation and "free" parameters contribute significantly to the modest computational cost associated with On2.[g] The typical computation time is a few seconds.

On3. Evaluate. Solve the "Active" Schur complement system, and evaluate any relevant derived quantities from the solution vector (for example a compliance output). The typical computation time is a few seconds.

The computational cost associated with this online stage is dominated by On2 (when I_{eff} is close to I) or On3 (when $I_{\text{eff}} \ll I$). However, the offline and online stages above are only concerned with the port-reduced SCRBE *approximation*. We consider the computational procedures associated with *a posteriori* error estimation in the next section.

Certification framework

Our port-reduced SCRBE approximation is equipped with efficiently computable *a posteriori* error bounds and estimators that provide certificates for the error in the approximation with respect to the underlying global FE discretization. We employ in this paper the energy-norm and compliance output bound developed in [12], and we present the main ingredients and certain extensions below. We furthermore sharpen the bounds by consideration of a multi-reference parameter bound conditioner.

The error in our approximation derives from two sources: port reduction and RB approximation. Below we first address the error due to port reduction, that is to say, the case in which the error due to RB approximation is zero. In this case the error bound presentation simplifies significantly and in particular admits a pure functional interpretation. We then subsequently perturb the equivalent algebraic interpretation to provide a bound for the general case in which the error due to RB approximation is non-zero.

Port reduction error contribution

We assume in this subsection only that the only source of error is port reduction and hence that there is no RB-induced error. We introduce the function

$$u^{\text{PR}}(\mu) = \sum_{i=1}^{I} b^{f;h}(\mu_i) + \sum_{p=1}^{n^\Gamma} \sum_{k=1}^{n^\Gamma_{\text{A},p}} \mathbb{U}^{\text{PR}}_{p,k}(\mu)\Phi_{p,k}(\mu) \in X^h(\Omega), \tag{71}$$

which satisfies

$$a(u^{\text{PR}}(\mu), v; \mu) = f(v; \mu), \quad \forall v \in \mathcal{S}^{\text{PR}}(\Omega); \tag{72}$$

hence $u^{\text{PR}}(\mu)$ is the port-reduced approximation to $u^h(\mu)$ obtained in the absence of RB errors. We note that we may (as in (25)) replace the skeleton space $\mathcal{S}^{\text{PR}}(\Omega)$ in (72) by the skeleton space

$$\mathcal{S}^{\text{PR}}_{\text{symm}}(\Omega) = \text{span}\{\Phi_{p,k}(\mu), \quad 1 \le k \le n^\Gamma_{\text{A},p}, 1 \le p \le n^\Gamma\} \subset \mathcal{S}_{\text{symm}}(\Omega), \tag{73}$$

and thus $u^{\text{PR}}(\mu) \in X^h(\Omega)$ also satisfies

$$a(u^{\text{PR}}(\mu), v; \mu) = f(v; \mu), \quad \forall v \in \mathcal{S}^{\text{PR}}_{\text{symm}}(\Omega); \tag{74}$$

note that $u^{\text{PR}}(\mu) \notin \mathcal{S}^{\text{PR}}_{\text{symm}}(\Omega)$ because of the source bubble terms $b^{f;h}(\mu_i)$ in (71).

We define the associated (RB-error-free) error field as

$$e_0^h(\mu) \equiv u^h(\mu) - u^{\mathrm{PR}}(\mu)$$
$$= \sum_{p=1}^{n^\Gamma}\left(\sum_{k=1}^{n^\Gamma_{\mathrm{A},p}}(\mathbb{U}_{p,k}(\mu) - \mathbb{U}^{\mathrm{PR}}_{p,k}(\mu))\Phi_{p,k}(\mu) + \sum_{k=n^\Gamma_{\mathrm{A},p}+1}^{\mathcal{N}^\Gamma_p} \mathbb{U}_{p,k}(\mu)\Phi_{p,k}(\mu)\right), \tag{75}$$

(in which the subscript $_0$ refers to the case of zero RB error contribution) and we note that $e_0^h(\mu) \in \mathcal{S}_{\mathrm{symm}}(\Omega)$ because the source bubble contributions from $u^h(\mu)$ and $u^{\mathrm{PR}}(\mu)$ cancel. Our goal is to develop a bound for the energy $\|e_0^h(\mu)\|_\mu$, where

$$\|\cdot\|_\mu \equiv \sqrt{a(\cdot,\cdot;\mu)} \tag{76}$$

is the usual energy norm. From (25) and (74) we see that

$$a(e_0^h(\mu), v; \mu) = f(v) - a(u^{\mathrm{PR}}(\mu), v; \mu), \quad \forall v \in \mathcal{S}_{\mathrm{symm}}(\Omega); \tag{77}$$

this error-residual relationship is the point of departure for our error bound development.

Thanks to coercivity and symmetry of $a(\cdot,\cdot;\mu)$, the error field $e_0^h(\mu)$ admits the equivalent definition

$$e_0^h(\mu) = \arg \min_{v\in\mathcal{S}_{\mathrm{symm}}(\Omega)} \mathcal{J}(v;\mu), \tag{78}$$

where

$$\mathcal{J}(v;\mu) \equiv \frac{1}{2}a(v,v) - \big(f(v) - a(u^{\mathrm{PR}}(\mu), v; \mu)\big), \tag{79}$$

and furthermore $\|e_0^h(\mu)\|_\mu^2 = a(e_0^h(\mu), e_0^h(\mu);\mu) = -2\mathcal{J}(e_0^h(\mu);\mu)$. We now relax the minimization (78) by consideration of a discontinuous (non-conforming) skeleton space

$$\mathcal{S}^{\mathrm{NC}}_{\mathrm{symm}}(\Omega) \equiv \mathcal{S}^{\mathrm{PR}}_{\mathrm{symm}}(\Omega)$$
$$\oplus\ \mathrm{span}\{\phi_{i,j,k}(\mu), \quad (n^\gamma_{\mathrm{A},i,j}+1) \le k \le \mathcal{N}^\gamma_{i,j}, 1 \le j \le n^\gamma_i, 1 \le i \le I\}$$
$$\equiv \mathrm{span}\{\Phi'_i(\mu), \quad 1 \le i \le n_{\mathrm{NC}}\}, \tag{80}$$

in which the basis functions $\Phi'_i(\mu)$, $1 \le i \le n_{\mathrm{NC}}$, merely represent a re-indexing of the basis functions $\Phi_{p,k}(\mu)$, $1 \le k \le n^\Gamma_p$, $1 \le p \le n^\Gamma$, and $\phi_{i,j,k}(\mu)$, $(n^\gamma_{\mathrm{A},i,j}+1) \le k \le \mathcal{N}^\gamma_{i,j}$, $1 \le j \le n^\gamma_i$, $1 \le i \le I$. Note that the $\phi_{i,j,k}(\mu)$ represent independent (non-conforming) degrees of freedom local to component i. The dimension of $\mathcal{S}^{\mathrm{NC}}_{\mathrm{symm}}(\Omega)$ is

$$n_{\mathrm{NC}} = n_{\mathrm{A}} + \sum_{i=1}^{I}\sum_{j=1}^{n^\gamma_i} \mathcal{N}^\gamma_{i,j} - n^\gamma_{\mathrm{A},i,j} \ge n_{\mathrm{SC}}; \tag{81}$$

note that $\mathcal{S}^{\mathrm{NC}}_{\mathrm{symm}}(\Omega) \supseteq \mathcal{S}_{\mathrm{symm}}(\Omega)$. We also define a non-conforming skeleton space $\mathcal{S}^{\mathrm{NC}} \supseteq \mathcal{S}(\Omega)$ as

$$\mathcal{S}^{\mathrm{NC}}(\Omega) \equiv \mathcal{S}^{\mathrm{PR}}(\Omega)$$
$$\oplus\ \mathrm{span}\{\psi_{i,j,k}, \quad (n^\gamma_{\mathrm{A},i,j}+1) \le k \le \mathcal{N}^\gamma_{i,j}, 1 \le j \le n^\gamma_i, 1 \le i \le I\}$$
$$\equiv \mathrm{span}\{\Psi'_i, \quad 1 \le i \le n_{\mathrm{NC}}\}. \tag{82}$$

Hence for

$$e_0^{\mathrm{NC}}(\mu) \equiv \arg \min_{v\in\mathcal{S}^{\mathrm{NC}}_{\mathrm{symm}}(\Omega)} \mathcal{J}(v;\mu) \tag{83}$$

(recall the "broken" definition of $a(\cdot,\cdot;\mu)$ in (8)) we must have

$$\mathcal{J}(e_0^{\mathrm{NC}}(\mu);\mu) \leq \mathcal{J}(e_0^h(\mu);\mu) \tag{84}$$

and thus $a(e_0^{\mathrm{NC}}(\mu), e_0^{\mathrm{NC}}(\mu);\mu) \geq a(e_0^h(\mu), e_0^h(\mu);\mu)$. This first relaxation of (78) not only provides a bound on the energy of the error field, but also accommodates efficient bound calculation thanks to the non-conforming space $\mathcal{S}_{\mathrm{symm}}^{\mathrm{NC}}(\Omega)$.

A second relaxation step is required to obtain a computationally tractable error bound. To this end we introduce a bound conditioner, the bilinear form $b_\mu : \mathcal{S}_{\mathrm{symm}}^{\mathrm{NC}}(\Omega) \times \mathcal{S}_{\mathrm{symm}}^{\mathrm{NC}} \to \mathbb{R}$, defined as

$$b_\mu(\cdot,\cdot) \equiv a(\cdot,\cdot;\mu_{\mathrm{ref}}^\mu) \tag{85}$$

for a reference parameter value $\mu_{\mathrm{ref}}^\mu \in \mathcal{D}$. Note that here, $b_\mu(\cdot,\cdot)$ depends implicitly on μ through the parameter-dependent reference parameter μ_{ref}^μ. In fact, an important innovation of this paper is this multi-reference parameter bound conditioner: in the online stage, we optimally select μ_{ref}^μ from a database of a few candidate reference parameters (through a discrete enumeration procedure); we discuss the selection of μ_{ref}^μ further in the "Computational procedures" subsection below. We also define

$$\lambda_{\min}(\mu) \equiv \min_{v \in \mathcal{S}_{\mathrm{symm}}^{\mathrm{NC}}(\Omega)} \frac{a(v,v;\mu)}{b_\mu(v,v)}. \tag{86}$$

We then introduce a modified functional

$$\mathcal{J}_b(v;\mu) \equiv \frac{\lambda_{\min}(\mu)}{2} b_\mu(v,v) - \big(f(v) - a(u^{\mathrm{PR}}(\mu), v;\mu)\big), \quad \forall v \in \mathcal{S}_{\mathrm{symm}}^{\mathrm{NC}}(\Omega), \tag{87}$$

and we consider the minimization

$$\bar{e}_0^{\mathrm{NC}}(\mu) \equiv \arg \min_{v \in \mathcal{S}_{\mathrm{symm}}^{\mathrm{NC}}(\Omega)} \mathcal{J}_b(v;\mu). \tag{88}$$

By the definition of $\lambda_{\min}(\mu)$ in (86) it is clear that $\mathcal{J}_b(v;\mu) \leq \mathcal{J}(v;\mu)$ for all $v \in \mathcal{S}_{\mathrm{symm}}^{\mathrm{NC}}(\Omega)$. Thus in particular, since $\bar{e}_0^{\mathrm{NC}}(\mu)$ is the minimizer,

$$\mathcal{J}_b(\bar{e}_0^{\mathrm{NC}}(\mu);\mu) \leq \mathcal{J}_b(e_0^{\mathrm{NC}}(\mu);\mu) \leq \mathcal{J}(e_0^{\mathrm{NC}}(\mu);\mu) \leq \mathcal{J}(e_0(\mu);\mu), \tag{89}$$

where the last inequality follows from (84). Consequently, we obtain the energy-norm error bound

$$\lambda_{\min}(\mu) b_\mu(\bar{e}_0^{\mathrm{NC}}(\mu), \bar{e}_0^{\mathrm{NC}}(\mu)) \geq a(e_0^h(\mu), e_0^h(\mu);\mu) \tag{90}$$

where the field variable $\bar{e}_0^{\mathrm{NC}}(\mu) \in \mathcal{S}_{\mathrm{symm}}^{\mathrm{NC}}(\Omega)$ — a presumably rather good approximation to the original error field $e_0^h(\mu)$ [12] — satisfies the elliptic problem $b_\mu(\bar{e}_0^{\mathrm{NC}}(\mu), v) = \lambda_{\min}(\mu)^{-1}(f(v;\mu) - a(u^{\mathrm{PR}}(\mu), v;\mu))$ for all $v \in \mathcal{S}_{\mathrm{symm}}^{\mathrm{NC}}(\Omega)$. Equivalently, because of the Galerkin orthogonality in (19),

$$b_\mu(\bar{e}_0^{\mathrm{NC}}(\mu), v) = \frac{1}{\lambda_{\min}(\mu)} \big(f(v;\mu) - a(u^{\mathrm{PR}}(\mu), v;\mu)\big), \quad \forall v \in \mathcal{S}^{\mathrm{NC}}(\Omega). \tag{91}$$

Thanks to incorporation of the modes related to rigid-body motion in our port space bases (presuming $n_{\mathrm{A},p}^\Gamma \geq 6$ on all global ports Γ_p, $1 \leq p \leq n^\Gamma$) we expect in general (and for a particular system, we computationally verify) that (91) is well-posed; for the simpler class of problems with scalar-valued fields we demonstrate this well-posedness in [12]. The RB-error-free bound given in (90) (together with (91)) is the basis on which we in the next subsection extend our error estimation framework to the general case of non-zero RB errors and furthermore to certain outputs of interest.

In order to implement this error bound, and to facilitate incorporation of RB-induced error contributions, we now interpret the error bound (90) in terms of algebraic quantities. To this end, we first note that, for any $v(\mu) = \sum_{p=1}^{n^\Gamma} \sum_{k=1}^{\mathcal{N}_p^\Gamma} \mathbb{V}_{p,k}(\mu)\Phi_{p,k}(\mu)$ — that is, for any $v(\mu) \in \mathcal{S}_{\text{symm}}(\Omega)$ with coefficients $\mathbb{V}(\mu)$ — we have

$$a(v(\mu), v(\mu); \mu) = \mathbb{V}(\mu)^{\mathrm{T}}\mathbb{A}(\mu)\mathbb{V}(\mu); \tag{92}$$

we refer to the right-hand side of (92) as the "Schur energy" of $\mathbb{V}(\mu)$. It shall prove convenient to introduce the zero-extended solution vectors

$$\hat{\mathbb{U}}_0^{\text{PR}}(\mu) \equiv \begin{bmatrix} \mathbb{U}^{\text{PR}}(\mu) \\ \mathbf{0} \end{bmatrix} \in \mathbb{R}^{n_{\text{SC}}}, \qquad \text{and} \qquad \hat{\mathbb{U}}_0^{\text{PR,NC}}(\mu) \equiv \begin{bmatrix} \mathbb{U}^{\text{PR}}(\mu) \\ \mathbf{0} \end{bmatrix} \in \mathbb{R}^{n_{\text{NC}}}, \tag{93}$$

in which all but the first n_{A} entries are explicitly set to zero. We also define the error coefficient vector

$$\mathbb{E}_0(\mu) \equiv \mathbb{U}(\mu) - \hat{\mathbb{U}}_0^{\text{PR}}(\mu) \in \mathbb{R}^{n_{\text{SC}}} \tag{94}$$

such that the error (75) can be written $e_0^h(\mu) = \sum_{p=1}^{n^\Gamma} \sum_{k=1}^{\mathcal{N}_p^\Gamma} \mathbb{E}_{0;p,k}(\mu)\Phi_{p,k}(\mu)$. Note here, we tacitly interpret (without loss of generality) $\mathbb{U}(\mu)$ such that the first n_{A} entries correspond to the n_{A} active degrees of freedom. The algebraic version of the error residual equation (77) is

$$\mathbb{A}(\mu)\mathbb{E}_0(\mu) = \mathbb{R}_0(\mu), \tag{95}$$

where the residual vector is given as

$$\mathbb{R}_0(\mu) = \mathbb{F}(\mu) - \mathbb{A}(\mu)\hat{\mathbb{U}}_0^{\text{PR}}(\mu); \tag{96}$$

note that, thanks to (92) and the fact that $e_0^h(\mu) \in \mathcal{S}_{\text{symm}}(\Omega)$, (95) is equivalent to (77).

We now introduce a non-conforming matrix $\mathbb{A}^{\text{NC}}(\mu) \in \mathbb{R}^{n_{\text{NC}} \times n_{\text{NC}}}$ and vector $\mathbb{F}^{\text{NC}}(\mu) \in \mathbb{R}^{n_{\text{NC}}}$ as

$$\mathbb{A}_{i,j}^{\text{NC}}(\mu) = a(\Phi_j'(\mu), \Phi_i'(\mu); \mu), \tag{97}$$

$$\mathbb{F}_i^{\text{NC}}(\mu) = f(\Phi_i'(\mu); \mu) - \sum_{l=1}^{I} a(b_l^{f;h}(\mu_l), \Phi_i'(\mu); \mu), \tag{98}$$

for $1 \leq i,j \leq n_{\text{NC}}$. Note that $a(\Phi_j'(\mu), \Phi_i'(\mu); \mu) = a(\Phi_j'(\mu), \Psi_i'; \mu)$ because of the Galerkin orthogonality in (19), and thus $\mathbb{A}^{\text{NC}}(\mu)$ is indeed the non-conforming version of the Schur complement matrix $\mathbb{A}(\mu)$ in (26); similarly, note that $f(\Phi_i'(\mu); \mu) - \sum_{l=1}^{I} a(b_l^{f;h}(\mu_l), \Phi_i'(\mu); \mu) = f(\Psi_i'; \mu) - \sum_{l=1}^{I} a(b_l^{f;h}(\mu_l), \Psi_i'(\mu); \mu)$ because of (18) and the fact that $\Phi_i'(\mu) - \Psi_i'$ vanish on ports, and thus $\mathbb{F}^{\text{NC}}(\mu)$ is the non-conforming version of the vector $\mathbb{F}(\mu)$ in (26).

We further define a non-conforming reference matrix

$$\mathbb{B}_\mu^{\text{NC}} \equiv \mathbb{A}^{\text{NC}}(\mu_{\text{ref}}^\mu), \tag{99}$$

which corresponds to the bilinear form $b_\mu(\cdot,\cdot)$. We also introduce a non-conforming residual vector $\mathbb{R}_0^{\text{NC}}(\mu) \in \mathbb{R}^{n_{\text{NC}}}$ as

$$\mathbb{R}_{0;i}^{\text{NC}}(\mu) = f(\Phi_i'(\mu)) - a(u^{\text{PR}}(\mu), \Phi_i'(\mu); \mu), \quad 1 \leq i \leq n_{\text{NC}}; \tag{100}$$

note that $\mathbb{R}_0^{\text{NC}}(\mu) = \mathbb{F}^{\text{NC}}(\mu) - \mathbb{A}^{\text{NC}}(\mu)\hat{\mathbb{U}}_0^{\text{PR}}(\mu)$.

Next, we introduce a (unknown) coefficient vector $\bar{\mathbb{E}}_0^{\rm NC}(\mu) \in \mathbb{R}^{n_{\rm NC}}$ such that

$$\bar{e}_0^{\rm NC}(\mu) = \sum_{i=1}^{n_{\rm NC}} \bar{\mathbb{E}}_{0;i}(\mu)\Phi_i'(\mu). \tag{101}$$

Thus from (91), (99), and (100) we obtain

$$\bar{\mathbb{E}}_0^{\rm NC}(\mu) = \frac{1}{\lambda_{\rm min}(\mu)}(\mathbb{B}_\mu^{\rm NC})^{-1}\mathbb{R}_0^{\rm NC}(\mu). \tag{102}$$

Similarly to (92), we note that for any $v(\mu) = \sum_{i=1}^{n_{\rm NC}} \mathbb{V}_i(\mu)\Phi_i'(\mu)$ — that is, for any $v(\mu) \in \mathcal{S}_{\rm symm}^{\rm NC}(\Omega)$ — we have

$$a(v(\mu), v(\mu); \mu) = \mathbb{V}(\mu)^{\rm T}\mathbb{A}^{\rm NC}(\mu)\mathbb{V}(\mu). \tag{103}$$

Hence in particular, since $\bar{e}_0^{\rm NC}(\mu) \in \mathcal{S}_{\rm symm}^{\rm NC}(\Omega)$, we obtain

$$\begin{aligned}\lambda_{\rm min}(\mu)b_\mu(\bar{e}_0^{\rm NC}(\mu), \bar{e}_0^{\rm NC}(\mu)) &= \lambda_{\rm min}(\mu)\bar{\mathbb{E}}_0^{\rm NC}(\mu)^{\rm T}\mathbb{B}_\mu^{\rm NC}\bar{\mathbb{E}}_0^{\rm NC}(\mu)\\ &= \frac{1}{\lambda_{\rm min}(\mu)}\mathbb{R}_0^{\rm NC}(\mu)^{\rm T}(\mathbb{B}_\mu^{\rm NC})^{-1}\mathbb{R}_0^{\rm NC}(\mu).\end{aligned} \tag{104}$$

Further, since $e_0^h(\mu) \in \mathcal{S}_{\rm symm}$, we may invoke (92) and write

$$a(e_0(\mu), e_0(\mu); \mu) = \mathbb{E}_0(\mu)^{\rm T}\mathbb{A}(\mu)\mathbb{E}_0(\mu). \tag{105}$$

Finally, we note that $\lambda_{\rm min}(\mu)$ of (86) is the smallest eigenvalue associated with the generalized eigenproblem

$$\mathbb{A}^{\rm NC}(\mu)\mathbb{V}(\mu) = \lambda(\mu)\mathbb{B}_\mu^{\rm NC}. \tag{106}$$

The algebraic interpretation of the port reduction error bound (90) is thus

$$\frac{1}{\lambda_{\rm min}(\mu)}\mathbb{R}_0^{\rm NC}(\mu)^{\rm T}(\mathbb{B}_\mu^{\rm NC})^{-1}\mathbb{R}_0^{\rm NC}(\mu) \geq \mathbb{E}_0(\mu)^{\rm T}\mathbb{A}(\mu)\mathbb{E}_0(\mu). \tag{107}$$

We note that the bound (107) necessitates a solve $(\mathbb{B}_\mu^{\rm NC})^{-1}\mathbb{R}_0^{\rm NC}(\mu)$ of dimension $n_{\rm NC} \geq n_{\rm SC}$. However, this solve may be performed efficiently thanks to *i)* the non-conforming skeleton space $\mathcal{S}^{\rm NC}(\Omega)$ which in a natural way allows component-local elimination of all degrees of freedom that do not couple at shared global ports; and *ii)* the quasi parameter-independent bound conditioner matrix $\mathbb{B}_\mu^{\rm NC}$ associated with the bilinear form b_μ, which allows offline pre-factorization for all these component-local solves. And furthermore, in actual practice we invoke not $\lambda_{\rm min}(\mu)$ but rather a computationally tractable eigenvalue lower bound $\tilde{\lambda}_{\rm min,LB}(\mu) \leq \lambda_{\rm min}(\mu)$. We consider computational aspects of our error estimation framework in more detail in the "Computational procedures" subsection below.

RB error contribution — *A Posteriori* error estimators

We now modify (107) in order to obtain an efficiently computable *a posteriori* error bound which is also valid in the presence of RB error contributions. First, as we in the SCRBE context only have access to an approximation of the FE Schur complement system, the residual can not be computed exactly and we thus instead compute a residual approximation together with bounds on associated RB-error-induced residual perturbation terms. Second, we introduce a lower bound (valid under an eigenvalue proximity assumption) for the eigenvalue $\lambda_{\rm min}(\mu)$ which is based on the solution to a port-reduced eigenproblem, an approximate eigenproblem residual, and bounds on associated RB-error-induced eigenproblem residual perturbation terms.

Moreover, in the presence of RB error contributions the error in the Schur energy is not equal to the energy of the error in the field, and thus in addition to a bound on the former we require a bound on *additional* RB perturbation terms to obtain a bound for the latter. Further, we develop in this section, from our Schur energy error bound, a new bound on port-restricted compliance outputs. For this output bound we must take into account that we in this paper (in contrast to in [12]) employ $\mathcal{S}^{\mathrm{PR}}(\Omega)$ rather than $\tilde{\mathcal{S}}^{\mathrm{PR}}_{\mathrm{symm}}(\Omega) \subset \tilde{\mathcal{S}}_{\mathrm{symm}}(\Omega)$ (the former being a port-reduced version of the latter, which is defined in (51)) as our skeleton space. Finally, we introduce *asymptotically rigorous* error estimators, by which we reduce computational cost by neglecting typically very small quadratic RB error bound contributions.

To begin, we define the error field as

$$e^h(\mu) \equiv u^h(\mu) - \tilde{u}^{\mathrm{PR}}(\mu). \tag{108}$$

It is again convenient to introduce the zero-extended solution vectors,

$$\hat{\mathbb{U}}^{\mathrm{PR}}(\mu) \equiv \begin{bmatrix} \tilde{\mathbb{U}}^{\mathrm{PR}}(\mu) \\ \mathbf{0} \end{bmatrix} \in \mathbb{R}^{n_{\mathrm{SC}}}, \qquad \text{and} \qquad \hat{\mathbb{U}}^{\mathrm{PR,NC}}(\mu) \equiv \begin{bmatrix} \tilde{\mathbb{U}}^{\mathrm{PR}}(\mu) \\ \mathbf{0} \end{bmatrix} \in \mathbb{R}^{n_{\mathrm{NC}}}, \tag{109}$$

in which the solution $\tilde{\mathbb{U}}^{\mathrm{PR}}(\mu)$ of (64) is extended by $n_{\mathrm{SC}} - n_{\mathrm{A}}$ and $n_{\mathrm{NC}} - n_{\mathrm{A}}$ zeros, respectively. We may then write

$$e^h(\mu) = \sum_{i=1}^{I} (b_i^{h;f}(\mu_i) - \tilde{b}_i^{h;f}(\mu_i)) + \sum_{p=1}^{n^\Gamma} \sum_{k=1}^{\mathcal{N}_p^\Gamma} \left(\mathbb{U}_{p,k}(\mu)\Phi_{p,k}(\mu) - \hat{\mathbb{U}}^{\mathrm{PR}}_{p,k}(\mu)\tilde{\Phi}_{p,k}(\mu) \right), \tag{110}$$

and we note that $e^h(\mu)$ is *not* a member of $\mathcal{S}_{\mathrm{symm}}(\Omega)$ because of the errors in the RB bubble approximations. We also define a vector of error coefficients as

$$\mathbb{E}(\mu) \equiv \mathbb{U}(\mu) - \hat{\mathbb{U}}^{\mathrm{PR}}(\mu). \tag{111}$$

We first develop a bound for the error in the Schur energy norm, $\sqrt{\mathbb{E}(\mu)^{\mathrm{T}}\mathbb{A}(\mu)\mathbb{E}(\mu)}$, through perturbations of the left-hand side of (107). We subsequently modify this bound to obtain a bound on $\|e^h(\mu)\|_\mu$; note the former is not equivalent to the latter because $e^h(\mu)$ is not a member of $\mathcal{S}_{\mathrm{symm}}(\Omega)$.

The usual error-residual relationship still holds in the presence of RB error contributions. In this case the relevant error-residual equation is

$$\mathbb{A}(\mu)\mathbb{E}(\mu) = \mathbb{R}(\mu), \tag{112}$$

where the residual vector is given as

$$\mathbb{R}(\mu) = \mathbb{F}(\mu) - \mathbb{A}(\mu)\hat{\mathbb{U}}^{\mathrm{PR}}(\mu). \tag{113}$$

The difference between (95) and (112) is rather subtle: the former features the residual associated with the RB-error-free solution vector $\hat{\mathbb{U}}_0^{\mathrm{PR}}(\mu)$ (never computationally realized), while the latter features the residual associated with the RB-error-affected SCRBE solution vector $\hat{\mathbb{U}}^{\mathrm{PR}}(\mu)$ (computed in practice). The non-conforming version of the residual is

$$\mathbb{R}^{\mathrm{NC}}(\mu) \equiv \mathbb{F}^{\mathrm{NC}}(\mu) - \mathbb{A}^{\mathrm{NC}}(\mu)\hat{\mathbb{U}}^{\mathrm{PR}}(\mu). \tag{114}$$

Next, we *redefine* our quasi parameter-independent (due to online reference parameter selection) bound conditioner matrix $\mathbb{B}_{\bar{\mu}}^{\mathrm{NC}}$ from the previous subsection as $\mathbb{B}_{\bar{\mu}}^{\mathrm{NC}} =$

$\tilde{\mathbb{A}}^{\mathrm{NC}}(\mu^{\mu}_{\mathrm{ref}})$; note that any SPD matrix may serve as our bound conditioner, and thus the RB approximations now present in $\mathbb{B}^{\mathrm{NC}}_{\mu}$ do not necessitate modifications to the error bound expression (and therefore the $\mathbb{B}^{\mathrm{NC}}_{\mu}$ of the previous subsection did not bear a subscript $_0$). Henceforth, the eigenproblem (106) is interpreted with this redefined $\mathbb{B}^{\mathrm{NC}}_{\mu}$ as the right-hand side matrix, and $\lambda_{\min}(\mu)$ is interpreted as the associated smallest eigenvalue. In the presence of RB error contributions, (107) now becomes

$$\frac{\mathbb{R}^{\mathrm{NC}}(\mu)^{\mathrm{T}}(\mathbb{B}^{\mathrm{NC}}_{\mu})^{-1}\mathbb{R}^{\mathrm{NC}}(\mu)}{\lambda_{\min}(\mu)} \geq \mathbb{E}(\mu)^{\mathrm{T}}\mathbb{A}(\mu)\mathbb{E}(\mu). \tag{115}$$

To bound the error in the Schur energy, we must thus, based on residual and eigenvalue approximations, develop upper and lower bounds for the numerator and denominator, respectively, of the left-hand side of (115).

We first consider the approximation to the non-conforming residual $\mathbb{R}^{\mathrm{NC}}(\mu)$. As we do not have access to $\mathbb{F}^{\mathrm{NC}}(\mu)$ and $\mathbb{A}^{\mathrm{NC}}(\mu)$ as defined in (97) and (98), but rather to RB-approximated versions $\tilde{\mathbb{F}}^{\mathrm{NC}}(\mu) \approx \mathbb{F}^{\mathrm{NC}}(\mu)$ and $\tilde{\mathbb{A}}^{\mathrm{NC}}(\mu) \approx \mathbb{A}^{\mathrm{NC}}(\mu)$, we introduce our approximation based on $\tilde{\mathbb{F}}^{\mathrm{NC}}(\mu) \approx \mathbb{F}^{\mathrm{NC}}(\mu)$ and $\tilde{\mathbb{A}}^{\mathrm{NC}}(\mu) \approx \mathbb{A}^{\mathrm{NC}}(\mu)$ as

$$\tilde{\mathbb{R}}^{\mathrm{NC}}(\mu) = \tilde{\mathbb{F}}^{\mathrm{NC}}(\mu) - \tilde{\mathbb{A}}^{\mathrm{NC}}(\mu)\hat{\mathbb{U}}^{\mathrm{PR,NC}}(\mu) \tag{116}$$

such that $\tilde{\mathbb{R}}^{\mathrm{NC}}(\mu) = \mathbb{R}^{\mathrm{NC}}(\mu) + \delta\mathbb{R}^{\mathrm{NC}}(\mu)$. Here,

$$\delta\mathbb{R}^{\mathrm{NC}}(\mu) = \tilde{\mathbb{F}}^{\mathrm{NC}}(\mu) - \mathbb{F}^{\mathrm{NC}}(\mu) + (\mathbb{A}^{\mathrm{NC}}(\mu) - \tilde{\mathbb{A}}^{\mathrm{NC}}(\mu))\hat{\mathbb{U}}^{\mathrm{PR,NC}}(\mu) \tag{117}$$

is an RB-error-induced perturbation term. We may readily from standard RB error bounds [5,6] develop bounds on these perturbation quantities; we introduce a vector $\boldsymbol{\sigma}(\mu)$ such that, for any $\mu \in \mathcal{D}$,

$$\boldsymbol{\sigma}_i(\mu) \geq |\delta\mathbb{R}^{\mathrm{NC}}_i(\mu)|, \quad 1 \leq i \leq n_{\mathrm{NC}}. \tag{118}$$

We next consider the approximation to the eigenvalue $\lambda_{\min}(\mu)$. Again, as we do not in practice have access to $\mathbb{A}^{\mathrm{NC}}(\mu)$, and furthermore as we wish to avoid solution of a full eigenproblem of dimension n_{NC}, we consider an approximation $\tilde{\lambda}^{\mathrm{PR}}_{\min}(\mu)$ to $\lambda_{\min}(\mu)$ given as the smallest eigenvalue associated with the port-reduced SCRBE eigenproblem

$$\tilde{\mathbb{A}}^{\mathrm{PR}}(\mu)\mathbb{V}(\mu) = \tilde{\lambda}^{\mathrm{PR}}(\mu)\mathbb{B}^{\mathrm{PR}}_{\mu}\mathbb{V}(\mu); \tag{119}$$

here, $\mathbb{B}^{\mathrm{PR}}_{\mu}$ denotes the block of $\mathbb{B}^{\mathrm{NC}}_{\mu}$ associated with "Active" degrees of freedom. We denote by $\mathbb{V}^{\mathrm{PR}}_{\min}(\mu)$ the eigenvector associated with $\tilde{\lambda}^{\mathrm{PR}}_{\min}(\mu)$, and we assume the normalization $\mathbb{V}^{\mathrm{PR}}_{\min}(\mu)^{\mathrm{T}}\mathbb{B}^{\mathrm{PR}}_{\mu}\mathbb{V}^{\mathrm{PR}}_{\min}(\mu) = 1$. We also introduce an approximate eigenproblem residual

$$\tilde{\mathbb{R}}^{\mathrm{NC}}_{\mathrm{eig}}(\mu) = \tilde{\mathbb{A}}^{\mathrm{NC}}(\mu)\hat{\mathbb{V}}^{\mathrm{PR}}_{\min}(\mu) - \tilde{\lambda}^{\mathrm{PR}}_{\min}(\mu)\mathbb{B}^{\mathrm{NC}}_{\mu}\hat{\mathbb{V}}^{\mathrm{PR}}_{\min}(\mu), \tag{120}$$

in which $\hat{\mathbb{V}}^{\mathrm{PR}}_{\min}(\mu) \in \mathbb{R}^{n_{\mathrm{NC}}}$ is a zero-expanded version of $\mathbb{V}^{\mathrm{PR}}_{\min}(\mu) \in \mathbb{R}^{n_{\mathrm{A}}}$. Note that the *exact* eigenproblem residual is given as $\mathbb{R}^{\mathrm{NC}}_{\mathrm{eig}}(\mu) = \mathbb{A}^{\mathrm{NC}}\hat{\mathbb{V}}^{\mathrm{PR}}_{\min}(\mu) - \tilde{\lambda}^{\mathrm{PR}}_{\min}(\mu)\mathbb{B}^{\mathrm{NC}}_{\mu}\hat{\mathbb{V}}^{\mathrm{PR}}_{\min}(\mu)$, and we may thus define a vector of RB perturbation terms $\delta\mathbb{R}^{\mathrm{NC}}_{\mathrm{eig}}(\mu)$ such that $\mathbb{R}^{\mathrm{NC}}_{\mathrm{eig}}(\mu) = \tilde{\mathbb{R}}^{\mathrm{NC}}_{\mathrm{eig}}(\mu) + \delta\mathbb{R}^{\mathrm{NC}}_{\mathrm{eig}}(\mu)$. We may then develop bounds on these RB-error-induced perturbation quantities — we introduce a vector $\boldsymbol{\sigma}_{\mathrm{eig}}(\mu)$ such that, for any $\mu \in \mathcal{D}$,

$$\boldsymbol{\sigma}_{\mathrm{eig},i}(\mu) \geq |\delta\mathbb{R}^{\mathrm{NC}}_{\mathrm{eig},i}(\mu)|, \quad 1 \leq i \leq n_{\mathrm{NC}}. \tag{121}$$

We now obtain a *computable* eigenvalue lower bound in

Lemma 1. *Let $C > 0$ be such that*

$$\delta\mathbb{R}_{\text{eig}}^{\text{NC}}(\mu)^{\text{T}}(\mathbb{B}_{\mu}^{\text{NC}})^{-1}\delta\mathbb{R}_{\text{eig}}^{\text{NC}}(\mu) \leq C\|\delta\mathbb{R}_{\text{eig}}^{\text{NC}}(\mu)\|_2^2, \tag{122}$$

assume that

$$|\lambda_{\min}^{\text{PR}}(\mu) - \lambda_{\min}(\mu)| \leq |\lambda_{\min}^{\text{PR}}(\mu) - \lambda(\mu)|, \tag{123}$$

for all $\lambda(\mu)$ which satisfy (106) (with the redefined $\mathbb{B}_{\mu}^{\text{NC}}$), and let

$$\lambda_{\min,\text{LB}}(\mu; C) \equiv \tilde{\lambda}_{\min}^{\text{PR}}(\mu) - \sqrt{\tilde{\mathbb{R}}_{\text{eig}}^{\text{NC}}(\mu)^{\text{T}}(\mathbb{B}_{\mu}^{\text{NC}})^{-1}\tilde{\mathbb{R}}_{\text{eig}}^{\text{NC}}(\mu) + 2\boldsymbol{\sigma}_{\text{eig}}(\mu)^{\text{T}}|(\mathbb{B}_{\mu}^{\text{NC}})^{-1}\tilde{\mathbb{R}}_{\text{eig}}^{\text{NC}}(\mu)| + C\|\boldsymbol{\sigma}_{\text{eig}}(\mu)\|_2^2}. \tag{124}$$

Then

$$\lambda_{\min,\text{LB}}(\mu; C) \leq \lambda_{\min}(\mu). \tag{125}$$

Proof. We refer to ([12], Proposition 1) for the proof, and we note that a similar residual-based eigenvalue bound has been developed in [18] for the standard eigenproblem. □

With the residual approximation $\tilde{\mathbb{R}}(\mu)$, associated RB error bounds $\boldsymbol{\sigma}(\mu)$, and the eigenvalue lower bound $\lambda_{\min,\text{LB}}(\mu; C)$ above, we may now obtain a *computable* bound for the left-hand side of (115) and thus the error in the Schur energy norm in

Proposition 1. *Let $C > 0$ be a computable constant such that*

$$\delta\mathbb{R}^{\text{NC}}(\mu)^{\text{T}}(\mathbb{B}_{\mu}^{\text{NC}})^{-1}\delta\mathbb{R}^{\text{NC}}(\mu) \leq C\|\delta\mathbb{R}^{\text{NC}}(\mu)\|_2^2, \tag{126}$$

$$\delta\mathbb{R}_{\text{eig}}^{\text{NC}}(\mu)^{\text{T}}(\mathbb{B}_{\mu}^{\text{NC}})^{-1}\delta\mathbb{R}_{\text{eig}}^{\text{NC}}(\mu) \leq C\|\delta\mathbb{R}_{\text{eig}}^{\text{NC}}(\mu)\|_2^2. \tag{127}$$

Then define

$$\Delta^{\mathbb{U}}(\mu; C) \equiv \sqrt{\frac{\tilde{\mathbb{R}}^{\text{NC}}(\mu)^{\text{T}}(\mathbb{B}_{\mu}^{\text{NC}})^{-1}\tilde{\mathbb{R}}^{\text{NC}}(\mu) + 2\boldsymbol{\sigma}(\mu)^{\text{T}}|(\mathbb{B}_{\mu}^{\text{NC}})^{-1}\tilde{\mathbb{R}}^{\text{NC}}(\mu)| + C\|\boldsymbol{\sigma}(\mu)\|_2^2}{\lambda_{\min,\text{LB}}(\mu; C)}}. \tag{128}$$

Then if the assumption (123) holds, we have

$$\sqrt{\mathbb{E}(\mu)^{\text{T}}\mathbb{A}(\mu)\mathbb{E}(\mu)} \leq \Delta^{\mathbb{U}}(\mu; C). \tag{129}$$

Proof. We merely note here that the numerator in (128) is an upper bound for the numerator in (115), and that $\lambda_{\min,\text{LB}}(\mu; C) \leq \lambda_{\min}(\mu)$ is a lower bound for the denominator in (115). We refer to ([12], Appendix A) for the detailed proof. □

We proceed to bound the energy of the error in the field. Since $e^h(\mu)$ is not a member of $\mathcal{S}_{\text{symm}}(\Omega)$, a small modification to (128) is necessary to obtain a bound for $\|e^h(\mu)\|_{\mu}$. To this end, we introduce additional RB perturbation terms

$$\Delta b^f(\mu) \equiv \sum_{i=1}^{I}(b_i^{f;h}(\mu) - \tilde{b}_i^f(\mu)) \tag{130}$$

$$\Delta\Phi_{\text{A}}(\mu) \equiv \sum_{p=1}^{n^\Gamma}\sum_{k=1}^{n_{\text{A},p}^\Gamma}\tilde{\mathbb{U}}_{\text{A},p,k}(\mu)(\Phi_{p,k}(\mu) - \tilde{\Phi}_{p,k}(\mu)); \tag{131}$$

we also introduce an RB error bound [6] $\kappa(\mu)$ such that, for any $\mu \in \mathcal{D}$,

$$\kappa(\mu) \geq \|\Delta b^f(\mu) + \Delta\Phi_{\mathrm{A}}(\mu)\|_\mu. \tag{132}$$

We then introduce our bound for the energy of the error field in

Proposition 2. *Define* $\Delta^u(\mu; C)$ *as*

$$\Delta^u(\mu; C) \equiv \sqrt{\left(\Delta^{\mathbb{U}}(\mu; C)\right)^2 + \kappa(\mu)^2}. \tag{133}$$

where $\kappa(\mu)$ *is given in* (132). *Then if the assumption* (123) *holds, we have*

$$\|e^h(\mu)\|_\mu \leq \Delta^u(\mu; C). \tag{134}$$

Proof. We refer to ([12], Appendix A) for the proof. □

Next, we develop a bound for the error in *port-restricted* compliance outputs. To this end we introduce a matrix $\boldsymbol{\sigma}_{\mathbb{A}}(\mu) \in \mathbb{R}^{n_{\mathrm{A}} \times n_{\mathrm{A}}}$ such that

$$\boldsymbol{\sigma}_{\mathbb{A},i,j}(\mu) \geq |\mathbb{A}_{i,j}(\mu) - \tilde{\mathbb{A}}_{i,j}(\mu)|, \qquad 1 \leq i, j \leq n_{\mathrm{A}}. \tag{135}$$

We then state

Proposition 3. *Let*

$$\Delta^s(\mu; C) \equiv \left(\Delta^{\mathbb{U}}(\mu; C)\right)^2 + |\tilde{\mathbb{U}}^{\mathrm{PR}}(\mu)|^{\mathrm{T}} \boldsymbol{\sigma}_{\mathbb{A}}(\mu) |\tilde{\mathbb{U}}^{\mathrm{PR}}(\mu)| \tag{136}$$

(in which $|\cdot|$ *denotes entry-wise absolute value and not vector modulus). Assume that the source* $f(\cdot; \mu)$ *is restricted to ports such that* $b_i^{f;h}(\mu_i) = 0,\ 1 \leq i \leq I$. *The error in a port-restricted compliance output* $\tilde{s}^{\mathrm{PR}}(\mu) = f(\tilde{u}^{\mathrm{PR}}(\mu); \mu)$ *can then be bounded as*

$$|s^h(\mu) - \tilde{s}^{\mathrm{PR}}(\mu)| \leq \Delta^s(\mu; C) \tag{137}$$

Proof. We provide here a full proof as in the present paper (skeleton space $\mathcal{S}^{\mathrm{PR}}(\Omega)$) the proof is different from a related proof in [12] (skeleton space $\tilde{\mathcal{S}}_{\mathrm{symm}}(\Omega)$).

We first note that

$$e^h(\mu) = \Delta b^f(\mu) + \Delta\Phi_{\mathrm{A}}(\mu) + \sum_{p=1}^{n^\Gamma} \sum_{k=1}^{\mathcal{N}_p^\Gamma} \mathbb{E}_{p,k}(\mu)\Phi_{p,k}(\mu); \tag{138}$$

note in the port-restricted output case considered here, $\Delta b^f(\mu) = 0$. For the compliance output error, we may then write (using symmetry of $a(\cdot, \cdot; \mu)$)

$$\begin{aligned} s^h(\mu) - \tilde{s}^{\mathrm{PR}}(\mu) &= a(u^h(\mu), e^h(\mu); \mu) \\ &= a(e^h(\mu), u^h(\mu); \mu) = a(e^h(\mu), e^h(\mu); \mu) + a(e^h(\mu), \tilde{u}^{\mathrm{PR}}(\mu); \mu), \end{aligned} \tag{139}$$

and thus by (138) (and again symmetry of $a(\cdot, \cdot; \mu)$)

$$\begin{aligned} s^h(\mu) - \tilde{s}^{\mathrm{PR}}(\mu) &= \mathbb{E}(\mu)^{\mathrm{T}} \mathbb{A}(\mu) \mathbb{E}(\mu) + a(\Delta\Phi_{\mathrm{A}}(\mu), \Delta\Phi_{\mathrm{A}}(\mu); \mu) \\ &\quad + 2 \sum_{p=1}^{n^\Gamma} \sum_{k=1}^{\mathcal{N}_p^\Gamma} a(\Phi_{p,k}(\mu), \Delta\Phi_{\mathrm{A}}(\mu); \mu) + a(e^h(\mu), \tilde{u}^{\mathrm{PR}}(\mu); \mu). \end{aligned} \tag{140}$$

We note that $e^h(\mu)$ is *not* Galerkin-orthogonal to $\tilde{u}^{\mathrm{PR}}(\mu)$ because $\tilde{u}^{\mathrm{PR}}(\mu)$ (even in the case $b_i^{f;h}(\mu_i) = 0$) is not a member of the skeleton test space $\mathcal{S}^{\mathrm{PR}}(\Omega)$. We thus do not obtain equality between the compliance output error and the squared energy of the error field in (139). This is the key difference between the compliance output error bound result here and in [12]; in [12], we invoke the skeleton space $\tilde{\mathcal{S}}^{\mathrm{PR}}_{\mathrm{symm}}(\Omega) \subset \tilde{\mathcal{S}}^{\mathrm{PR}}(\Omega)$ (the latter is defined in (51)) of which $\tilde{u}^{\mathrm{PR}}(\mu)$ *is* a member (for port-restricted compliance such that $\tilde{b}^f_i(\mu) = 0$), and thus we directly obtain this equality.

We next note that $\Delta\Phi_{\mathrm{A}}(\mu)|_{\Omega_i}$ vanish on all ports and thus is a member of the FE bubble space associated with instantiated component i. From the Galerkin orthogonality (19) we then conclude that the third term on the right-hand side of (140) is equal to zero, and we obtain

$$s^h(\mu) - \tilde{s}^{\mathrm{PR}}(\mu) = \mathbb{E}(\mu)^{\mathrm{T}}\mathbb{A}(\mu)\mathbb{E}(\mu) + a(\Delta\Phi_{\mathrm{A}}(\mu), \Delta\Phi_{\mathrm{A}}(\mu); \mu) + a(e^h(\mu), \tilde{u}^{\mathrm{PR}}(\mu); \mu) \quad (141)$$

We now consider the two right-most terms on the right-hand side of (141) (we omit the μ-dependence for simplicity of exposition). We first obtain

$$\begin{aligned}
&a(\Delta\Phi_{\mathrm{A}}, \Delta\Phi_{\mathrm{A}}) + a(e^h, \tilde{u}^{\mathrm{PR}}) \\
&\quad = \sum_{p=1}^{n^\Gamma}\sum_{k=1}^{n^\Gamma_{\mathrm{A},p}}\sum_{p'=1}^{n^\Gamma}\sum_{k'=1}^{n^\Gamma_{\mathrm{A},p}} \tilde{\mathbb{U}}^{\mathrm{PR}}_{p,k}\tilde{\mathbb{U}}^{\mathrm{PR}}_{p',k'} a(\Phi_{p,k} - \tilde{\Phi}_{p,k}, \Phi_{p',k'} - \tilde{\Phi}_{p',k'}) \\
&\qquad + \sum_{p=1}^{n^\Gamma}\sum_{k=1}^{n^\Gamma_{\mathrm{A},p}}\sum_{p'=1}^{n^\Gamma}\sum_{k'=1}^{n^\Gamma_{\mathrm{A},p}} \tilde{\mathbb{U}}^{\mathrm{PR}}_{p,k}\tilde{\mathbb{U}}^{\mathrm{PR}}_{p',k'} a(\Phi_{p,k} - \tilde{\Phi}_{p,k}, \tilde{\Phi}_{p',k'}) \\
&\qquad + \sum_{p=1}^{n^\Gamma}\sum_{k=1}^{\mathcal{N}^\Gamma_p}\sum_{p'=1}^{n^\Gamma}\sum_{k'=1}^{n^\Gamma_{\mathrm{A},p}} \mathbb{E}_{p,k}\tilde{\mathbb{U}}^{\mathrm{PR}}_{p',k'} a(\Phi_{p,k}, \tilde{\Phi}_{p',k'}),
\end{aligned} \quad (142)$$

by the expression for $e^h(\mu)$ in (138) (for $\Delta b^f(\mu) = 0$) and the definition of $\Delta\Phi_{\mathrm{A}}(\mu)$ in (131). For the first two terms on the right-hand side of (142) we obtain

$$\begin{aligned}
&\sum_{p=1}^{n^\Gamma}\sum_{k=1}^{n^\Gamma_{\mathrm{A},p}}\sum_{p'=1}^{n^\Gamma}\sum_{k'=1}^{n^\Gamma_{\mathrm{A},p}} \tilde{\mathbb{U}}^{\mathrm{PR}}_{p,k}\tilde{\mathbb{U}}^{\mathrm{PR}}_{p',k'} a(\Phi_{p,k} - \tilde{\Phi}_{p,k}, \Phi_{p',k'} - \tilde{\Phi}_{p',k'}) \\
&\qquad + \sum_{p=1}^{n^\Gamma}\sum_{k=1}^{n^\Gamma_{\mathrm{A},p}}\sum_{p'=1}^{n^\Gamma}\sum_{k'=1}^{n^\Gamma_{\mathrm{A},p}} \tilde{\mathbb{U}}^{\mathrm{PR}}_{p,k}\tilde{\mathbb{U}}^{\mathrm{PR}}_{p',k'} a(\Phi_{p,k} - \tilde{\Phi}_{p,k}, \tilde{\Phi}_{p',k'}) \\
&\quad = \sum_{p=1}^{n^\Gamma}\sum_{k=1}^{n^\Gamma_{\mathrm{A},p}}\sum_{p'=1}^{n^\Gamma}\sum_{k'=1}^{n^\Gamma_{\mathrm{A},p}} \tilde{\mathbb{U}}^{\mathrm{PR}}_{p,k}\tilde{\mathbb{U}}^{\mathrm{PR}}_{p',k'} a(\Phi_{p,k} - \tilde{\Phi}_{p,k}, \Phi_{p',k'}) \\
&\quad = \sum_{p=1}^{n^\Gamma}\sum_{k=1}^{n^\Gamma_{\mathrm{A},p}}\sum_{p'=1}^{n^\Gamma}\sum_{k'=1}^{n^\Gamma_{\mathrm{A},p}} \tilde{\mathbb{U}}^{\mathrm{PR}}_{p,k}\tilde{\mathbb{U}}^{\mathrm{PR}}_{p',k'} a(\Phi_{p',k'}, \Phi_{p,k} - \tilde{\Phi}_{p,k}) = 0,
\end{aligned} \quad (143)$$

where in the second step we invoke symmetry of $a(\cdot, \cdot; \mu)$ and in the final step the Galerkin orthogonality (19). For the last term on the right-hand side of (142) we obain

$$\begin{aligned}
&\sum_{p=1}^{n^\Gamma}\sum_{k=1}^{\mathcal{N}^\Gamma_p}\sum_{p'=1}^{n^\Gamma}\sum_{k'=1}^{n^\Gamma_{\mathrm{A},p}} \mathbb{E}_{p,k}\tilde{\mathbb{U}}^{\mathrm{PR}}_{p',k'} a(\Phi_{p,k}, \tilde{\Phi}_{p',k'}) \\
&\qquad = \sum_{p=1}^{n^\Gamma}\sum_{k=1}^{\mathcal{N}^\Gamma_p}\sum_{p'=1}^{n^\Gamma}\sum_{k'=1}^{n^\Gamma_{\mathrm{A},p}} \mathbb{E}_{p,k}\tilde{\mathbb{U}}^{\mathrm{PR}}_{p',k'} a(\Phi_{p,k}, \Phi_{p',k'}) = \mathbb{E}^{\mathrm{T}}\mathbb{A}\hat{\mathbb{U}}^{\mathrm{PR}},
\end{aligned} \quad (144)$$

where we again exploit Galerkin orthogonality with respect to $\Phi_{p',k'}(\mu) - \tilde{\Phi}_{p',k'}(\mu)$.

We note that $\mathbb{A}(\mu)\mathbb{E}(\mu) = \mathbb{A}(\mu)(\mathbb{U}(\mu) - \hat{\mathbb{U}}^{\mathrm{PR}}(\mu)) = \mathbb{F}(\mu) - \mathbb{A}(\mu)\hat{\mathbb{U}}^{\mathrm{PR}}(\mu)$. With (141), (142), (143), and (144) (and symmetry of $\mathbb{A}(\mu)$) we then obtain

$$\begin{aligned} s^h(\mu) - \tilde{s}^{\mathrm{PR}}(\mu) &= \mathbb{E}(\mu)^{\mathrm{T}}\mathbb{A}(\mu)\mathbb{E}(\mu) + \mathbb{E}(\mu)^{\mathrm{T}}\mathbb{A}(\mu)\hat{\mathbb{U}}^{\mathrm{PR}}(\mu) \\ &= \mathbb{E}(\mu)^{\mathrm{T}}\mathbb{A}(\mu)\mathbb{E}(\mu) + \left(\mathbb{F}(\mu) - \mathbb{A}(\mu)\hat{\mathbb{U}}^{\mathrm{PR}}(\mu)\right)^{\mathrm{T}}\hat{\mathbb{U}}^{\mathrm{PR}}(\mu). \end{aligned} \tag{145}$$

In the case of a port-restricted compliance, we have $b_i^{f;h}(\mu_i) = \tilde{b}_i^f(\mu_i) = 0$ and thus also $\mathbb{F}(\mu) = \tilde{\mathbb{F}}(\mu)$. It is furthermore straightforward to show that $(\tilde{\mathbb{F}}(\mu) - \tilde{\mathbb{A}}(\mu)\hat{\mathbb{U}}^{\mathrm{PR}}(\mu))^{\mathrm{T}}\hat{\mathbb{U}}^{\mathrm{PR}}(\mu) = 0$ because the port-reduced SCRBE solution vector $\tilde{\mathbb{U}}^{\mathrm{PR}}(\mu)$ (that is, the non-zero coefficients of $\hat{\mathbb{U}}^{\mathrm{PR}}(\mu)$) satisfies (64) exactly. We thus obtain in this case

$$\begin{aligned} \left(\mathbb{F}(\mu) - \mathbb{A}(\mu)\hat{\mathbb{U}}^{\mathrm{PR}}(\mu)\right)^{\mathrm{T}}\hat{\mathbb{U}}^{\mathrm{PR}}(\mu) &= \left(\tilde{\mathbb{F}}(\mu) - \mathbb{A}(\mu)\hat{\mathbb{U}}^{\mathrm{PR}}(\mu)\right)^{\mathrm{T}}\hat{\mathbb{U}}^{\mathrm{PR}}(\mu) \\ &= \left(\tilde{\mathbb{F}}(\mu) - \tilde{\mathbb{A}}(\mu)\hat{\mathbb{U}}^{\mathrm{PR}}(\mu) + \tilde{\mathbb{A}}(\mu)\hat{\mathbb{U}}^{\mathrm{PR}}(\mu) \right. \\ &\quad \left. - \mathbb{A}(\mu)\hat{\mathbb{U}}^{\mathrm{PR}}(\mu)\right)^{\mathrm{T}}\hat{\mathbb{U}}^{\mathrm{PR}}(\mu) \\ &= \left((\tilde{\mathbb{A}}(\mu) - \mathbb{A}(\mu))\hat{\mathbb{U}}^{\mathrm{PR}}(\mu)\right)^{\mathrm{T}}\hat{\mathbb{U}}^{\mathrm{PR}}(\mu). \end{aligned} \tag{146}$$

From (145) and (146) (and symmetry of $\tilde{\mathbb{A}}(\mu)$ and $\mathbb{A}(\mu)$) we then conclude that

$$s^h(\mu) - \tilde{s}^{\mathrm{PR}}(\mu) = \mathbb{E}(\mu)^{\mathrm{T}}\mathbb{A}(\mu)\mathbb{E}(\mu) + \hat{\mathbb{U}}^{\mathrm{PR}}(\mu)^{\mathrm{T}}\left(\tilde{\mathbb{A}}(\mu) - \mathbb{A}(\mu)\right)\hat{\mathbb{U}}^{\mathrm{PR}}(\mu), \tag{147}$$

which, with the triangle inequality and (135), yields the desired result. □

We do not in the present paper consider bounds on more general outputs.

We reiterate that Lemma 1, Proposition 1, Proposition 2, and Proposition 3 all provide rigorous bounds under the eigenvalue proximity assumption given in (123). These bounds necessitate computation of a bound C for Rayleigh quotients associated with $(\mathbb{B}_\mu^{\mathrm{NC}})^{-1}$, for which we may choose $C = 1/\lambda_{\min}(\mathbb{B}_\mu^{\mathrm{NC}})$, where $\lambda_{\min}(\mathbb{B}_\mu^{\mathrm{NC}})$ is the smallest eigenvalue associated with $\mathbb{B}_\mu^{\mathrm{NC}}$. Unfortunately, this choice for C is typically a rather pessimistic Rayleigh quotient bound,[h] and furthermore calculation of $\lambda_{\min}(\mathbb{B}_\mu^{\mathrm{NC}})$ requires considerable (albeit, as discussed in the next section, not onerous) computational cost. However, we note that the terms which multiply C in (124) and (128), as well as the term $\kappa(\mu)^2$ in (133), are quadratic in RB error bound contributions and thus presumably small compared to the terms that are linear in RB error bound contributions.

We thus introduce *asymptotically rigorous* error *estimators*, in which we choose to neglect these terms: we set $C = 0$ in (124) and (128) to obtain a Schur energy error estimator $\Delta^{\mathrm{U}}(\mu; 0)$; we then obtain an estimator for the energy of the error field as

$$\Delta^{u;0}(\mu) \equiv \Delta^{\mathrm{U}}(\mu; 0), \tag{148}$$

in which we also neglect the term $\kappa(\mu)^2$ in (133); and finally we obtain an estimator for the port-restricted compliance output error as

$$\Delta^{s;0}(\mu) \equiv \Delta^{s}(\mu; 0). \tag{149}$$

In actual practice, RB errors are typically rather small, and we shall thus for our large-scale numerical results in this paper employ the error estimators (148) and (149).

Computational procedures

The main computational costs associated with our *a posteriori* error estimation framework derive from the two non-conforming solves $(\mathbb{B}_\mu^{\mathrm{NC}})^{-1}\tilde{\mathbb{R}}_{\mathrm{eig}}^{\mathrm{NC}}(\mu)$ and $(\mathbb{B}_\mu^{\mathrm{NC}})^{-1}\tilde{\mathbb{R}}^{\mathrm{NC}}(\mu)$ required in (124) and in the numerator of (128), respectively, and from the calculation of the smallest eigenvalue $\tilde{\lambda}_{\min}^{\mathrm{PR}}(\mu)$ of (119). We now discuss the former in more detail; for the latter we employ an implementation of a Krylov-Schur (inverted spectrum) iterative solver from the SLEPc library [19].

For our discussion here it is convenient to first introduce a particular interpretation of the non-port-reduced SCRBE system matrix and right-hand side as

$$\tilde{\mathbb{A}}(\mu) = \begin{bmatrix} \tilde{\mathbb{A}}_{\mathrm{A,A}}(\mu) & \tilde{\mathbb{A}}_{\mathrm{A,I}}(\mu) \\ \tilde{\mathbb{A}}_{\mathrm{I,A}}(\mu) & \tilde{\mathbb{A}}_{\mathrm{I,I}}(\mu) \end{bmatrix}, \qquad \tilde{\mathbb{F}}(\mu) = \begin{bmatrix} \tilde{\mathbb{F}}_{\mathrm{A}}\ (\mu) \\ \tilde{\mathbb{F}}_{\mathrm{I}}\ (\mu) \end{bmatrix}. \tag{150}$$

Here, the matrix block $\tilde{\mathbb{A}}_{\mathrm{A,A}}(\mu) = \tilde{\mathbb{A}}^{\mathrm{PR}}(\mu)$ is the "Active" matrix block which we invoke for our port-reduced SCRBE approximation, the blocks $\tilde{\mathbb{A}}_{\mathrm{I,A}}(\mu)$ and $\tilde{\mathbb{A}}_{\mathrm{I,A}}(\mu)$ correspond to couplings between the "Active" and "Inactive" degrees of freedom, and the block $\tilde{\mathbb{A}}_{\mathrm{I,I}}(\mu)$ is associated only with "Inactive" degrees of freedom. Note that the interpretation (150) simply corresponds to a particular ordering of (54).

In the particular case of only two instantiated components, the system matrix $\tilde{\mathbb{A}}(\mu)$ may be written as

$$\tilde{\mathbb{A}}(\mu) = \begin{bmatrix} \tilde{\mathbb{A}}^1_{\mathrm{A,A}}(\mu_1) + \mathbb{A}^2_{\mathrm{A,A}}(\mu_2) & \tilde{\mathbb{A}}^1_{\mathrm{A,I}}(\mu_1) + \tilde{\mathbb{A}}^2_{\mathrm{A,I}}(\mu_2) \\ \tilde{\mathbb{A}}^1_{\mathrm{I,A}}(\mu_1) + \mathbb{A}^2_{\mathrm{I,A}}(\mu_2) & \tilde{\mathbb{A}}^1_{\mathrm{I,I}}(\mu_1) + \tilde{\mathbb{A}}^2_{\mathrm{I,I}}(\mu_2) \end{bmatrix}, \tag{151}$$

where each submatrix $\tilde{\mathbb{A}}^i_{*,*'}(\mu)$ is a matrix block associated with instantiated component i, $i = 1, 2$. The non-conforming matrix $\tilde{\mathbb{A}}^{\mathrm{NC}}(\mu)$ for this two-component system is then

$$\tilde{\mathbb{A}}^{\mathrm{NC}}(\mu) = \begin{bmatrix} \tilde{\mathbb{A}}^1_{\mathrm{A,A}}(\mu_1) + \tilde{\mathbb{A}}^2_{\mathrm{A,A}}(\mu_2) & \tilde{\mathbb{A}}^1_{\mathrm{A,I}}(\mu_1) & \tilde{\mathbb{A}}^2_{\mathrm{A,I}}(\mu_2) \\ \tilde{\mathbb{A}}^1_{\mathrm{I,A}}(\mu_1) & \tilde{\mathbb{A}}^1_{\mathrm{I,I}}(\mu_1) & \mathbf{0} \\ \tilde{\mathbb{A}}^2_{\mathrm{I,A}}(\mu_2) & \mathbf{0} & \tilde{\mathbb{A}}^2_{\mathrm{I,I}}(\mu_2) \end{bmatrix}. \tag{152}$$

Note that the difference between (151) and (152) is that the latter does not couple "Inactive" port degrees of freedom.

For the computation of the residual approximation $\tilde{\mathbb{R}}^{\mathrm{NC}}(\mu)$ in (116) we note that

$$\begin{aligned} \tilde{\mathbb{R}}^{\mathrm{NC}}(\mu) &= \begin{bmatrix} \tilde{\mathbb{R}}_{\mathrm{A}}(\mu) \\ \tilde{\mathbb{R}}^1_{\mathrm{I}}(\mu) \\ \tilde{\mathbb{R}}^2_{\mathrm{I}}(\mu) \end{bmatrix} \\ &= \begin{bmatrix} \tilde{\mathbb{F}}_{\mathrm{A}}(\mu) \\ \tilde{\mathbb{F}}^1_{\mathrm{I}}(\mu_1) \\ \tilde{\mathbb{F}}^2_{\mathrm{I}}(\mu_2) \end{bmatrix} - \begin{bmatrix} \tilde{\mathbb{A}}_{\mathrm{A,A}}(\mu) & \tilde{\mathbb{A}}^1_{\mathrm{A,I}}(\mu_1) & \tilde{\mathbb{A}}^2_{\mathrm{A,I}}(\mu_2) \\ \tilde{\mathbb{A}}^1_{\mathrm{I,A}}(\mu_1) & \tilde{\mathbb{A}}^1_{\mathrm{I,I}}(\mu_1) & \mathbf{0} \\ \tilde{\mathbb{A}}^2_{\mathrm{I,A}}(\mu_2) & \mathbf{0} & \tilde{\mathbb{A}}^2_{\mathrm{I,I}}(\mu_2) \end{bmatrix} \begin{bmatrix} \tilde{\mathbb{U}}^{\mathrm{PR}}(\mu) \\ \mathbf{0} \\ \mathbf{0} \end{bmatrix} \\ &= \begin{bmatrix} \tilde{\mathbb{F}}_{\mathrm{A}}(\mu) - \tilde{\mathbb{A}}_{\mathrm{A,A}}(\mu)\tilde{\mathbb{U}}^{\mathrm{PR}}(\mu) \\ \tilde{\mathbb{F}}^1_{\mathrm{I}}(\mu_1) - \tilde{\mathbb{A}}^1_{\mathrm{I,A}}(\mu_1)\tilde{\mathbb{U}}^{\mathrm{PR}}(\mu) \\ \tilde{\mathbb{F}}^2_{\mathrm{I}}(\mu_2) - \tilde{\mathbb{A}}^2_{\mathrm{I,A}}(\mu_2)\tilde{\mathbb{U}}^{\mathrm{PR}}(\mu) \end{bmatrix} = \begin{bmatrix} \mathbf{0} \\ \tilde{\mathbb{F}}^1_{\mathrm{I}}(\mu_1) - \tilde{\mathbb{A}}^1_{\mathrm{I,A}}(\mu_1)\tilde{\mathbb{U}}^{\mathrm{PR},1}(\mu) \\ \tilde{\mathbb{F}}^2_{\mathrm{I}}(\mu_2) - \tilde{\mathbb{A}}^2_{\mathrm{I,A}}(\mu_2)\tilde{\mathbb{U}}^{\mathrm{PR},2}(\mu) \end{bmatrix}, \end{aligned} \tag{153}$$

where $\tilde{\mathbb{U}}^{\mathrm{PR},i}(\mu)$ is extracted from $\tilde{\mathbb{U}}^{\mathrm{PR}}(\mu)$ for the degrees of freedom associated with component i. Note that the first n_{A} entries in the residual vector are zero, and that we may obtain the local residuals $\tilde{\mathbb{R}}^i_{\mathrm{I}}(\mu)$ by component-local evaluation. The eigenproblem residual approximation $\tilde{\mathbb{R}}^{\mathrm{NC}}_{\mathrm{eig}}(\mu)$ admits a similar procedure.

We now consider the system $\mathbb{B}_{\mu}^{\mathrm{NC}} z(\mu) = \tilde{\mathbb{R}}^{\mathrm{NC}}(\mu)$, which we may write as

$$\begin{bmatrix} \mathbb{B}^1_{\mathrm{A,A}} + \mathbb{B}^2_{\mathrm{A,A}} & \mathbb{B}^1_{\mathrm{A,I}} & \mathbb{B}^2_{\mathrm{A,I}} \\ \mathbb{B}^1_{\mathrm{I,A}} & \mathbb{B}^1_{\mathrm{I,I}} & \mathbf{0} \\ \mathbb{B}^2_{\mathrm{I,A}} & \mathbf{0} & \mathbb{B}^2_{\mathrm{I,I}} \end{bmatrix} \begin{bmatrix} z_{\mathrm{A}}(\mu) \\ z^1_{\mathrm{I}}(\mu) \\ z^2_{\mathrm{I}}(\mu) \end{bmatrix} = \begin{bmatrix} \tilde{\mathbb{R}}_{\mathrm{A}}(\mu) \\ \tilde{\mathbb{R}}^1_{\mathrm{I}}(\mu) \\ \tilde{\mathbb{R}}^2_{\mathrm{I}}(\mu) \end{bmatrix}, \tag{154}$$

and we note that

$$\begin{aligned} &[\mathbb{B}^1_{\mathrm{A,A}} + \mathbb{B}^2_{\mathrm{A,A}} - \mathbb{B}^1_{\mathrm{A,I}}(\mathbb{B}^1_{\mathrm{I,I}})^{-1}\mathbb{B}^1_{\mathrm{I,A}} - \mathbb{B}^2_{\mathrm{A,I}}(\mathbb{B}^2_{\mathrm{I,I}})^{-1}\mathbb{B}^2_{\mathrm{I,A}}] z_{\mathrm{A}}(\mu) \\ &\qquad = \tilde{\mathbb{R}}_{\mathrm{A}}(\mu) - \mathbb{B}^1_{\mathrm{A,I}}(\mathbb{B}^1_{\mathrm{I,I}})^{-1}\tilde{\mathbb{R}}^1_{\mathrm{I}}(\mu) - \mathbb{B}^2_{\mathrm{A,I}}(\mathbb{B}^2_{\mathrm{I,I}})^{-1}\tilde{\mathbb{R}}^2_{\mathrm{I}}(\mu). \end{aligned} \tag{155}$$

We may thus obtain $z(\mu)$ by consideration of a second Schur complement: we first solve smaller local problems associated with each of the two components, and then a global problem of size n_{A} for $z_{\mathrm{A}}(\mu)$; we finally recover $z(\mu)$ by standard back-substitution as $z^i_{\mathrm{I}}(\mu) = (\mathbb{B}^i_{\mathrm{I,I}})^{-1}(\tilde{\mathbb{R}}^i_{\mathrm{I}}(\mu) - \mathbb{B}^i_{\mathrm{I,A}} z_{\mathrm{A}}(\mu))$. The extension of this procedure to a system with an arbitrary number of components and ports is straightforward.

An important innovation of this paper for our error bound framework is a multi-reference parameter bound conditioner. In fact, the system reference parameter value μ^{μ}_{ref} shall be chosen online, based on a database of component-local reference parameter values $\hat{\mu}^t_{m,\mathrm{ref}}$, $1 \le t \le n_{m,\mathrm{ref}}$, $1 \le m \le M$. The component-local reference matrices $\mathbb{B}^i_{*,*}$ in (154) and (155) are thus chosen online from a database of $n_{m,\mathrm{ref}}$ precomputed component-local matrices $\hat{\mathbb{B}}^{m,t}_{*,*}$ associated with the parameter values $\hat{\mu}^t_{m,\mathrm{ref}} \in \hat{\mathcal{D}}_m$. For our numerical results of this paper, we choose the component reference parameters to minimize the Euclidean distance between μ^{μ}_{ref} and μ. This multi-reference parameter bound conditioner procedure significantly sharpens our error bound through a closer-to-unity smallest eigenvalue $\tilde{\lambda}^{\mathrm{PR}}_{\min}(\mu)$ (and associated eigenvalue bound) at only minor additional computational cost (note a related approach is considered in [14] in a different context).

The computational efficacy of our error bound framework is thus realized largely through the quasi parameter-independent and non-conforming operator $\mathbb{B}^{\mathrm{NC}}_{\mu}$. As for the SCRBE approximation framework, the computational procedures associated with the error bound framework naturally decouple into offline and online stages. We consider these stages as extensions of the offline and online *approximation* computational stages discussed earlier, and we now discuss each in more detail (we again refer to [12] for detailed online operation counts).

Offline

Off5. Online dataset preparation. For each archetype component we construct data to enable efficient assembly of the matrix blocks $\tilde{\mathbb{A}}^i_{\mathrm{I,A}}(\mu)$ required for residual calculation in (153) (the matrix blocks $\tilde{\mathbb{A}}^i_{\mathrm{A,A}}(\mu)$ are also required for residual calculation; however the associated data is already constructed in Off3). Note that the blocks $\tilde{\mathbb{A}}^i_{\mathrm{A,I}}(\mu)$ and, more importantly, the blocks $\tilde{\mathbb{A}}^i_{\mathrm{I,I}}(\mu)$, are *not* required for residual calculation. Hence the cost of this stage scales quadratically in n_{A} but only linearly in n_{I}.

Off6. Bound conditioner preparation. For each archetype component m, $1 \le m \le M$, we choose (manually) $n_{m,\mathrm{ref}}$ (typically only a few) reference parameter values $\hat{\mu}^t_{m,\mathrm{ref}}$ and compute associated bound conditioner reference matrices $\hat{\mathbb{B}}^{m,t}_{\mathrm{I,I}}$, $\hat{\mathbb{B}}^{m,t}_{\mathrm{AA}}$, and $\hat{\mathbb{B}}^{m,t}_{\mathrm{A,I}}$, $1 \le t \le n_{m,\mathrm{ref}}$.

We also perform and store the Cholesky factorization of each of the $\hat{\mathbb{B}}^{m,t}_{\mathrm{I,I}}$, and we precompute the terms $\hat{\mathbb{B}}^{m,t}_{\mathrm{A,I}}(\hat{\mathbb{B}}^{m,t}_{\mathrm{I,I}})^{-1}\hat{\mathbb{B}}^{m,t}_{\mathrm{I,A}}$ required for assembly of the left-hand side of (155).

Off7. Data loading. We finally read the online datasets and error bound conditioner data (typically a few Gb combined) for all library components into computer memory to prepare for the online stage.

Online

On4. Port-reduced eigenproblem. We compute the smallest eigenvalue and associated eigenvector associated with (119) using a Krylov-Schur algorithm [19].

On5. Matrix and vector block assembly. Assemble component matrix and vector blocks $\tilde{\mathbb{A}}^i_{\mathrm{I,A}}(\mu)$ and vectors $\tilde{\mathbb{F}}^i_{\mathrm{I}}(\mu)$ for each unique component instantiation; note that the "Active" component matrix and vector blocks $\tilde{\mathbb{A}}^i_{\mathrm{A,A}}(\mu)$ are already assembled in On2. As in On2, we exploit "cloned" component instantiations to effectively reduce the number of component instantiations to $I_{\mathrm{eff}} \ll I$.

On6. Residual calculation. Given the solution vector $\hat{\mathbb{U}}(\mu)$, the eigenvalue $\tilde{\lambda}^{\mathrm{PR}}_{\min}(\mu)$, and the associated (normalized and zero-expanded) eigenvector $\hat{\mathbb{V}}^{\mathrm{PR}}_{\min}(\mu)$, we calculate $\tilde{\mathbb{R}}^{\mathrm{NC}}(\mu)$ and $\tilde{\mathbb{R}}^{\mathrm{NC}}_{\mathrm{eig}}(\mu)$ locally on each component.

On7. Non-conforming solves. We first choose the reference parameter value

$$\mu^{\mu}_{\mathrm{ref}} = (\hat{\mu}^{t_1}_{\mathcal{M}(1),\mathrm{ref}}, \ldots, \hat{\mu}^{t_I}_{\mathcal{M}(I),\mathrm{ref}}) \tag{156}$$

(where $1 \leq t_i \leq n_{\mathcal{M}(i),\mathrm{ref}}$) from the database of candidate component reference parameter values such that the Euclidean distance between each μ_i and $\hat{\mu}^{t_i}_{\mathcal{M}(i),\mathrm{ref}}$ is minimized. We then compute $(\mathbb{B}^{\mathrm{NC}}_{\mu})^{-1}\mathbb{R}^{\mathrm{NC}}(\mu)$ and $(\mathbb{B}^{\mathrm{NC}}_{\mu})^{-1}\mathbb{R}^{\mathrm{NC}}_{\mathrm{eig}}(\mu)$ through component-local elimination of "Inactive" degrees of freedom as indicated in (155). Note that this step is particularly efficient thanks to the preparation in Off6.

On8. Calculation of $\lambda_{\min}(\mathbb{B}^{\mathrm{NC}}_{\mu})$. In the case that we wish to employ a rigorous error bound (we choose $C = 1/\lambda_{\min}(\mathbb{B}^{\mathrm{NC}}_{\mu})$ rather than $C = 0$), we must also compute $\lambda_{\min}(\mathbb{B}^{\mathrm{NC}}_{\mu})$. Note that we may compute $\lambda_{\min}(\mathbb{B}^{\mathrm{NC}}_{\mu})$ rather efficiently through (typically) a few inverse power iterations, and hence only a few additional non-conforming solves. This procedure is applicable for $\lambda_{\min}(\mathbb{B}^{\mathrm{NC}}_{\mu})$ but not for $\lambda_{\min}(\tilde{\mathbb{A}}^{\mathrm{NC}}(\mu))$ because the latter would have required expensive online formation of the $\tilde{\mathbb{A}}^i_{\mathrm{I,I}}(\mu)$ component-local matrix blocks.

The computational cost associated with this online stage is typically dominated by On4 and On7. However for systems in which almost all components are unique — that is, I_{eff} close to I — the cost of matrix assembly in On5 is considerable. In any event, the *error estimation* online computational cost discussed here is typically larger than the *approximation* online computational cost discussed earlier (we report actual timings in the next section).

Microtruss beam application

We consider here application of our port-reduced SCRBE framework to structural analysis of a microtruss beam. The particular beam we consider is in practice manufactured from microcylinders that are welded together in a three-dimensional square array configuration to form a larger but light-weight truss structure; see Figure 3. Many examples

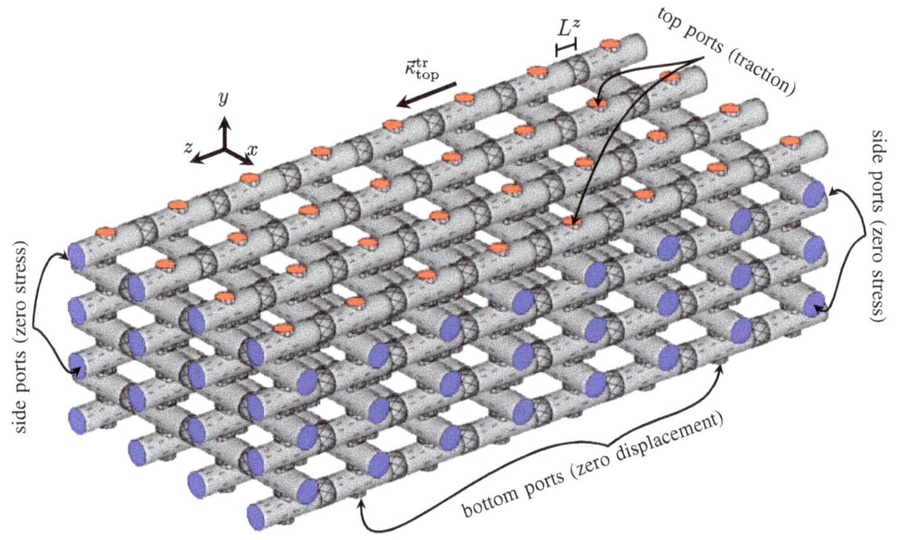

Figure 3 Microtruss structure. The system has $I = 408$ instantiated components, 224 of which are of type `component 1` and 184 of which are of type `component 2`.

of microtruss structures exist in literature and in engineering, and our choice here is only one of numerous possibilities. We refer to [20-22] for analyses and manufacturing considerations for such structures, including the particular type we consider here.

The microtruss structure is a good fit for our methodology. First, the structure admits a very natural decomposition into components, and the macroscale beam is comprised of many identical or similar instantiations of the same component archetypes; thus typically we may obtain $I_{\mathrm{eff}} \ll I$ which implies particularly effective treatment by the port-reduced SCRBE. Second, the behavior of the macroscale beam as a function of component parameters and system topology is non-trivial, and furthermore the solution may exhibit large localized stresses within the components; hence the fidelity of a full FE discretization — provided by the port-reduced SCRBE framework at a fraction of the cost — is desired. Third, it is often of interest to assess performance in off-design conditions in particular in the presence of inevitable flaws, in which not just natural periodicity but departures from periodicity — well within the capabilities of the SCRBE — are important.

Archetype component library

Before we introduce our components, we consider the non-dimensionalization of the equations of isotropic linear elasticity for a "generic" archetype (and thus entities below bear ˆs). To this end we first define the non-dimensional tensor $\hat{C}$ as

$$\hat{C}_{ijkl} \equiv \frac{\nu}{(1+\nu)(1-2\nu)}\delta_{ij}\delta_{kl} + \frac{1}{2(1+\nu)}(\delta_{ik}\delta_{jl} + \delta_{il}\delta_{jk}), \quad 1 \le i,j,k,l \le 3, \tag{157}$$

in which ν is the Poisson ratio (we choose $\nu = 0.3$ for steel); the dimensional elasticity tensor is then given as the product $\hat{E}^{\mathrm{dim}}\hat{C}_{ijkl}$, where $\hat{E}^{\mathrm{dim}}$ is the Young's modulus. The associated stress tensor $\hat{\sigma}^{\mathrm{dim}}(\hat{u}^{\mathrm{dim}})$, given the dimensional displacement $\hat{u}^{\mathrm{dim}}$, is defined as $\hat{\sigma}^{\mathrm{dim}}_{ij}(\hat{u}^{\mathrm{dim}}) = \hat{E}^{\mathrm{dim}}\hat{C}_{ijkl}\partial\hat{u}^{\mathrm{dim}}_k/\partial\hat{x}^{\mathrm{dim}}_l$.

We shall consider either homogeneous Dirichlet boundary conditions, or (port) tractions. In the latter case the boundary conditions are enforced through the stress tensor as $\hat{\sigma}_{ij}^{\text{dim}}\hat{\mathbf{e}}_j^{\text{dim}} = \hat{\kappa}_i^{\text{tr;dim}}$ ($\hat{\mathbf{e}}_j^{\text{dim}}$ denotes the canonical vectors) for a specified traction vector $\hat{\kappa}^{\text{tr;dim}}$.

To derive non-dimensional equations we introduce the dimensionless variables $\hat{x} = \hat{x}^{\text{dim}}/\hat{L}^{\text{dim},0}$, $\hat{u} = \hat{u}^{\text{dim}}/\hat{L}^{\text{dim},0}$, $\hat{E} = \hat{E}^{\text{dim}}/\hat{E}^{\text{dim},0}$, $\hat{\sigma} = \hat{\sigma}^{\text{dim}}/\hat{E}^{\text{dim},0}$, and $\hat{\kappa}^{\text{tr}} = \hat{\kappa}^{\text{tr;dim}}/\hat{E}^{\text{dim},0}$, where $\hat{L}^{\text{dim},0}$ is a characteristic length, and $\hat{E}^{\text{dim},0}$ is a characteristic Young's modulus. The non-dimensional traction boundary conditions then become $\sigma_{ij}\mathbf{n}_j = \kappa_i^{\text{tr}}$. Below, all our equations take a non-dimensional form.

We now introduce our (non-dimensional) archetype component library, which consists of the two three-dimensional isotropic linear-elastic components illustrated in Figure 4; note $L^{\text{dim},0}$ in Figure 4 is the characteristic length used in our non-dimensionalization.

The first archetype, `component 1`, is a "Steinmetz cylinder," and has four circular ports: the left and right ports are of type `port 1` and the top and bottom ports are of type `port 2`. For both port types, the reference port space dimension is $\mathcal{N}^{\beta} = 219$ (73 mesh nodes). The FE discretization for `component 1` has $\mathcal{N}_1 = 115{,}443$ degrees of freedom in linear hexahedral elements. Note in Figure 4 that the mesh is significantly refined where the weld stub meets the cylinder base in order to resolve potentially high stress concentrations in this area.

The archetype parameter vector for this component is

$$\hat{\mu}_1 = (\hat{E}_1, \hat{\kappa}^{\text{tr}}_{1,\text{top}}), \tag{158}$$

where $\hat{E}_1 = \hat{E}_1^{\text{dim}}/\hat{E}_1^{\text{dim},0}$ is a Young's modulus scaling parameter and $\hat{\kappa}^{\text{tr}}_{1,\text{top}}$ is a directional traction applied on the top port. The archetype bilinear and linear forms associated with `component 1` are, for all $\hat{w}, \hat{v} \in \hat{X}_1^h$, given as

$$\hat{a}_1(\hat{w}, \hat{v}; \hat{\mu}_1) = \hat{E}_1 \int_{\hat{\Omega}_1} \frac{\partial \hat{w}^i}{\partial \hat{x}_j} \hat{C}_{ijkl} \frac{\partial \hat{v}^k}{\partial \hat{x}_l}, \tag{159}$$

$$\hat{f}_1(\hat{v}; \hat{\mu}_1) = \hat{\kappa}^{\text{tr}}_{1,\text{top},i} \int_{\hat{\gamma}_{1,\text{top}}} \hat{v}^i, \tag{160}$$

where $\hat{\gamma}_{1,\text{top}}$ denotes the boundary associated with the top port. For the bound conditioner reference matrix blocks we consider a single reference parameter value $\hat{\mu}^1_{1,\text{ref}} = \hat{E}^1_{1,\text{ref}} = 1$ (thus $n_{1,\text{ref}} = 1$).

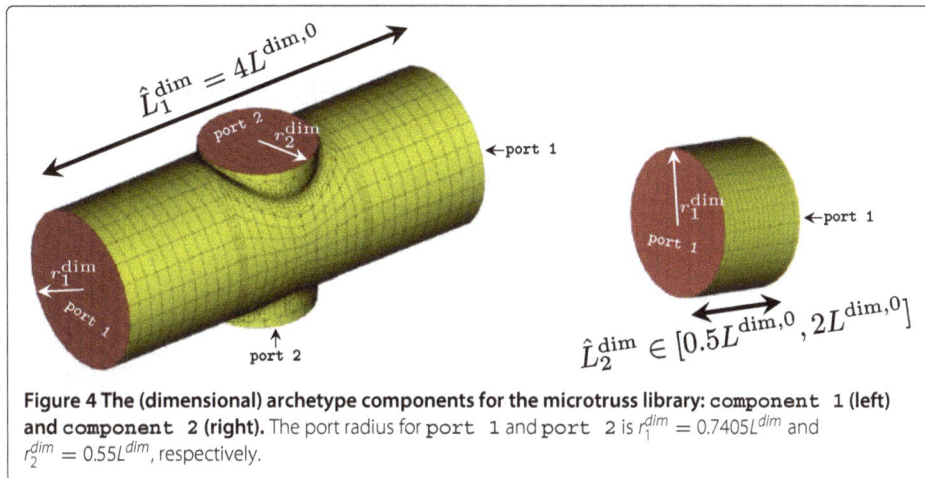

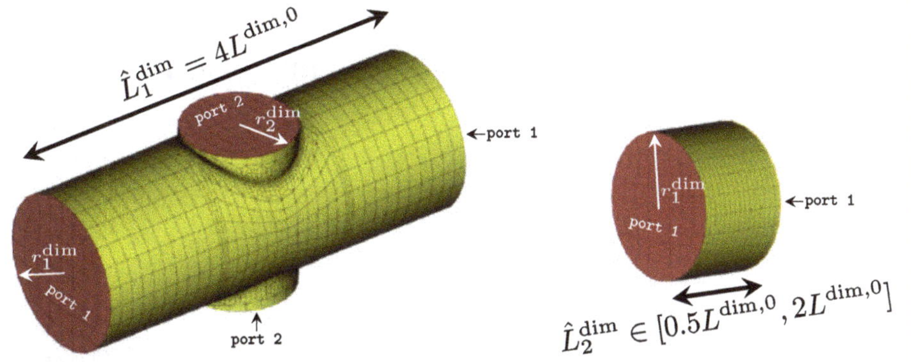

Figure 4 The (dimensional) archetype components for the microtruss library: `component 1` (left) and `component 2` (right). The port radius for `port 1` and `port 2` is $r_1^{dim} = 0.7405L^{dim}$ and $r_2^{dim} = 0.55L^{dim}$, respectively.

We note that $\hat{Q}_1^a = 1$; the Young's modulus parameter $\hat{E}_1$ is "free" in the sense that it enters outside the integral of (19). As a result, we may consider any value of $\hat{E}_1$ with only a single RB basis function in each of the RB approximation spaces. For $\hat{f}_1$ we note that $\hat{Q}_1^f = 3$; however, as $\hat{f}_1$ is port-restricted, we obtain $\hat{b}_1^{f;h} = 0$.

We also note that $\hat{a}_1$ in (159) does not reflect the rigid-body mapping parameters required to "dock" instantiations of `component 1` to the correct position in the system frame. To demonstrate this property more explicitly we consider the mapping $\mathcal{T}_1 = \mathcal{T}_1^{\mathrm{rot}}\mathcal{T}_1^{\mathrm{def}}$, in which, for `component 1`, $\mathcal{T}_1^{\mathrm{def}}$ is pure translation; we introduce an associated rotation matrix $\mathbb{Q} \in \mathbb{R}^{3\times 3}$ and a translation vector $\mathbb{T} \in \mathbb{R}^3$. For any coordinate $\hat{x} \in \hat{\Omega}_1$, we thus have $x_i = \mathbb{Q}_{ij}(\hat{x} + \mathbb{T})_j$, $1 \le i \le 3$, where $x = (x_1, x_3, x_3) \in \Omega_1$ and Ω_1 is the instantiated component domain; note that the Jacobian of the mapping, $\mathbb{Q}$, is unitary and thus $\det\mathbb{Q} = 1$.

On Ω_1, the instantiated component bilinear form reads, for any $w, v \in X_1$,

$$a_1(w, v; \mu) = E_1 \int_{\Omega_1} \frac{\partial w^i}{\partial x_j} C_{ijkl} \frac{\partial v^k}{\partial x_l}\, \mathrm{d}\Omega_1, \quad 1 \le i, j, k, l \le 3. \tag{161}$$

Here, C is the elasticity tensor on the instantiated domain (i.e., in the system coordinates) such that [23]

$$C_{ijkl} = \mathbb{Q}_{ii'}\mathbb{Q}_{jj'}\mathbb{Q}_{kk'}\mathbb{Q}_{ll'}\hat{C}_{i'j'k'l'} \tag{162}$$

for $\hat{C}$ defined in (157). We recall from the Section "Component-based static condensation" that we apply $\mathcal{T}^{\mathrm{rot}}$ to the dependent variables, and thus for any function $v \in X_1$ we write $v^i = \mathbb{Q}_{ij}\hat{v}^j$, where $\hat{v} = (\hat{v}^1, \hat{v}^2, \hat{v}^3) \in \hat{X}_1$. Starting from (161), we obtain in this case, for $w, v \in X_1$,

$$a_1(w, v; \mu_1) = E_1 \int_{\hat{\Omega}} \left(\mathbb{Q}_{jj'}^{-1}\frac{\partial w^i}{\partial \hat{x}_j}\right) C_{ij'kl'} \left(\mathbb{Q}_{ll'}^{-1}\frac{\partial v^k}{\partial \hat{x}_l}\right)(\det\mathbb{Q})\, \mathrm{d}\hat{\Omega} \tag{163}$$

$$= E_1 \int_{\hat{\Omega}} \left(\mathbb{Q}_{jj'}^{-1}\mathbb{Q}_{i'i}\frac{\partial \hat{w}^i}{\partial \hat{x}_j}\right) C_{i'j'k'l'} \left(\mathbb{Q}_{ll'}^{-1}\mathbb{Q}_{k'k}\frac{\partial \hat{v}^k}{\partial \hat{x}_l}\right) \mathrm{d}\hat{\Omega} \tag{164}$$

$$= E_1 \int_{\hat{\Omega}} \left(\mathbb{Q}_{jj'}^{-1}\mathbb{Q}_{i'i}\frac{\partial \hat{w}^i}{\partial \hat{x}_j}\right) \mathbb{Q}_{i'i''}\mathbb{Q}_{j'j''}\mathbb{Q}_{k'k''}\mathbb{Q}_{l'l''}\hat{C}_{i''j''k''l''} \left(\mathbb{Q}_{ll'}^{-1}\mathbb{Q}_{k'k}\frac{\partial \hat{v}^k}{\partial \hat{x}_l}\right) \mathrm{d}\hat{\Omega} \tag{165}$$

$$= E_1 \int_{\hat{\Omega}} \frac{\partial \hat{w}^i}{\partial \hat{x}_j} (\mathbb{Q}_{i'i}\mathbb{Q}_{i'i''})(\mathbb{Q}_{jj'}^{-1}\mathbb{Q}_{j'j''})(\mathbb{Q}_{k'k}Q_{k'k''})(\mathbb{Q}_{ll'}^{-1}\mathbb{Q}_{l'l''})\hat{C}_{i''j''k''l''}\frac{\partial \hat{v}^k}{\partial \hat{x}_l}\, \mathrm{d}\hat{\Omega} \tag{166}$$

$$= E_1 \int_{\hat{\Omega}} \frac{\partial \hat{w}^i}{\partial \hat{x}_j} (\mathbb{Q}^{\mathrm{T}}\mathbb{Q})_{ii''}(\mathbb{Q}^{-1}\mathbb{Q})_{jj''}(\mathbb{Q}^{\mathrm{T}}\mathbb{Q})_{kk''}(\mathbb{Q}^{-1}\mathbb{Q})_{ll''}\hat{C}_{i''j''k''l''}\frac{\partial \hat{v}^k}{\partial \hat{x}_l}\, \mathrm{d}\hat{\Omega} \tag{167}$$

$$= E_1 \int_{\hat{\Omega}} \frac{\partial \hat{w}^i}{\partial \hat{x}_j}\hat{C}_{ijkl}\frac{\partial \hat{v}^k}{\partial \hat{x}_l}\, \mathrm{d}\hat{\Omega} \tag{168}$$

$$= \hat{a}_1(\hat{w}, \hat{v}; \mu_1). \tag{169}$$

The key point in (163)–(169) is that the representation of the instantiated bilinear form in archetype coordinates does not require parameters related to the rotation Jacobian $\mathbb{Q}$ due to the cancellations in (167) and the fact that $\det\mathbb{Q} = 1$. For our RB approximations we employ the archetype domain for all computations and thus the RB spaces do not need to take these "docking" parameters into account. Furthermore the matrix and vector blocks for component instantiations that differ only in spatial orientation are identical, which thus contributes to the realization of $I_{\text{eff}} \ll I$ in an instantiated system.

We next consider our second archetype, `component 2`, which is a short cylinder stub. It has two ports of type `port 1` and may thus connect to the left and right ports of `component 1`. The reference port space dimension is again $\mathcal{N}^{\beta} = 219$. The FE discretization for `component 2` has $\mathcal{N}_2 = 3{,}504$ degrees of freedom in linear hexahedral elements. The parameter vector for this component is

$$\hat{\mu}_2 = (\hat{E}_2, \hat{L}_2, \hat{\boldsymbol{\kappa}}^{\text{tr}}_{2,\text{left}}, \hat{\boldsymbol{\kappa}}^{\text{tr}}_{2,\text{right}}), \tag{170}$$

where $\hat{E}_2 = \hat{E}_2^{\text{dim}}/\hat{E}_2^{\text{dim},0}$ is a Young's modulus scaling parameter, $\hat{L}_2 = \hat{L}_2^{\text{dim}}/\hat{L}^{\text{dim},0} \in [0.5, 2]$ is a length scaling parameter, and $\hat{\boldsymbol{\kappa}}^{\text{tr}}_{2,\text{left}}$ and $\hat{\boldsymbol{\kappa}}^{\text{tr}}_{2,\text{right}}$ are directional traction applied on the left and right ports, respectively. The archetype bilinear and linear forms associated with `component 2` are, for all $\hat{w}, \hat{v} \in \hat{X}_2(\hat{\Omega}_2)$, given as

$$\begin{aligned} \hat{a}_2(\hat{w}, \hat{v}; \hat{\mu}_2) &= \hat{E}_2\left(\int_{\hat{\Omega}_2} \frac{\partial \hat{w}^i}{\partial \hat{x}_j}\hat{C}_{ijk3}\frac{\partial \hat{v}^k}{\partial \hat{x}_3} + \int_{\hat{\Omega}_2} \frac{\partial \hat{w}^i}{\partial \hat{x}_3}\hat{C}_{i3kl}\frac{\partial \hat{v}^k}{\partial \hat{x}_l}\right) \\ &\quad + \frac{\hat{E}_2}{\hat{L}_2}\int_{\hat{\Omega}_2} \frac{\partial \hat{w}^i}{\partial \hat{x}_3}\hat{C}_{i3k3}\frac{\partial \hat{v}^k}{\partial \hat{x}_3} + \hat{E}_2\hat{L}_2\int_{\hat{\Omega}_2} \frac{\partial \hat{w}^i}{\partial \hat{x}_j}\hat{C}_{ijkl}\frac{\partial \hat{v}^k}{\partial \hat{x}_l}, \end{aligned} \tag{171}$$

$$\hat{f}_2(\hat{v}; \hat{\mu}_2) = \hat{\kappa}^{\text{tr}}_{2,\text{left},i}\int_{\hat{\gamma}_{2,\text{left}}} \hat{v}^i + \hat{\kappa}^{\text{tr}}_{2,\text{right},i}\int_{\hat{\gamma}_{2,\text{right}}} \hat{v}^i, \tag{172}$$

where, in (171), j and l take only the values 1, 2, and where, in (172) $\hat{\gamma}_{2,\text{left}}$ and $\hat{\gamma}_{2,\text{right}}$ are the boundaries associated with the left and right port, respectively. We note that $\hat{Q}_2^a = 3$ and that $\hat{Q}_2^f = 6$. Note that the bilinear form depends on the dilation parameter $\hat{L}_2$, but not on spatial orientation of the component; we may show this by reverse application of the arguments in (163)–(169) to each of three terms in (171). For the bound conditioner reference matrix blocks we consider three parameter values $\hat{\mu}^1_{2,\text{ref}} = (\hat{E}^1_{2,\text{ref}}, \hat{L}^1_{2,\text{ref}}) = (1, 0.75)$, $\hat{\mu}^2_{2,\text{ref}} = (\hat{E}^2_{2,\text{ref}}, \hat{L}^2_{2,\text{ref}}) = (1, 1)$, and $\hat{\mu}^3_{2,\text{ref}} = (\hat{E}^3_{2,\text{ref}}, \hat{L}^3_{2,\text{ref}}) = (1, 1.5)$ (thus $n_{2,\text{ref}} = 3$).

Pairwise empirical port mode training

We now discuss the pairwise empirical port mode training for our library components. For the `port 1` type we consider the three component pairs shown in

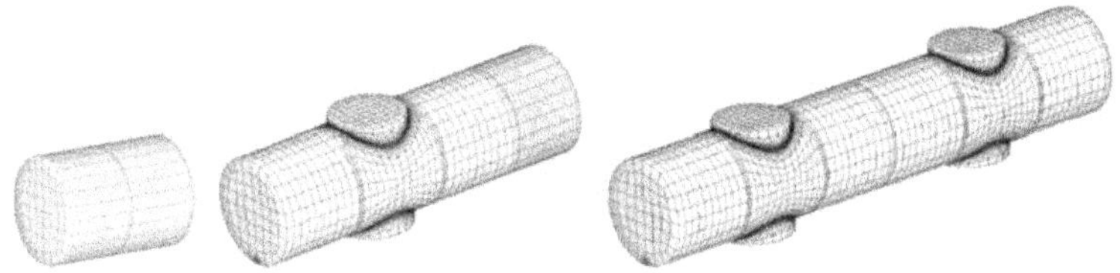

Figure 5 Component pairs used for empirical training of `port 1`.

Figure 5: a pair with two instantiations of `component 2`, a pair with one instantiation of `component 1` and one instantiation of `component 2`, and a pair with two instantiations of `component 1` connected via side ports (`port 1`). For the `port 2` type we consider the single pair of instantiations of `component 1` connected via a `port 2` type port as shown in Figure 6; note that when we consider a large microtruss structure we shall always "weld" our cylinders in this particular cross configuration.

We then execute Algorithm 1 for each pair; for the boundary condition regularity parameter in Algorithm 1 we choose $\gamma = 3$. For the training of `port 1` we extract $N_{\text{samples}} = 150$ different port samples in S_{pair} from each of the three pairs; recall that we subtract the projection onto the six modes $\hat{\chi}_i^{\beta}$, $1 \leq i \leq 6$, related to rigid-body motion from all snapshots. We then combine all 450 modes in S_{type}, and perform a POD over these 450 modes to compress the data to $n_{\text{pod}}^{\beta} = 44$ POD modes. We then obtain $n_{\text{pod}}^{\beta} + 6 = 50$ empirical modes, which we complement by $\mathcal{N}^{\beta} - 50 = 169$ eigenmodes (restricted to the orthogonal complement space) to complete the discrete space (note in practice we shall always use less than 50 modes for the port-reduced SCRBE approximation).

The approach for the training of `port 2` is identical except we perform POD over $N_{\text{samples}} = 300$ different port samples (with the projections onto the rigid body modes subtracted) extracted from the single component pair. We choose the same number of POD modes ($n_{\text{POD}}^{\beta} = 44$) and thus empirical modes for this port type.

We shall use these empirical port modes for most of our numerical results below. However we shall also compare these results to results obtained using more standard (and in

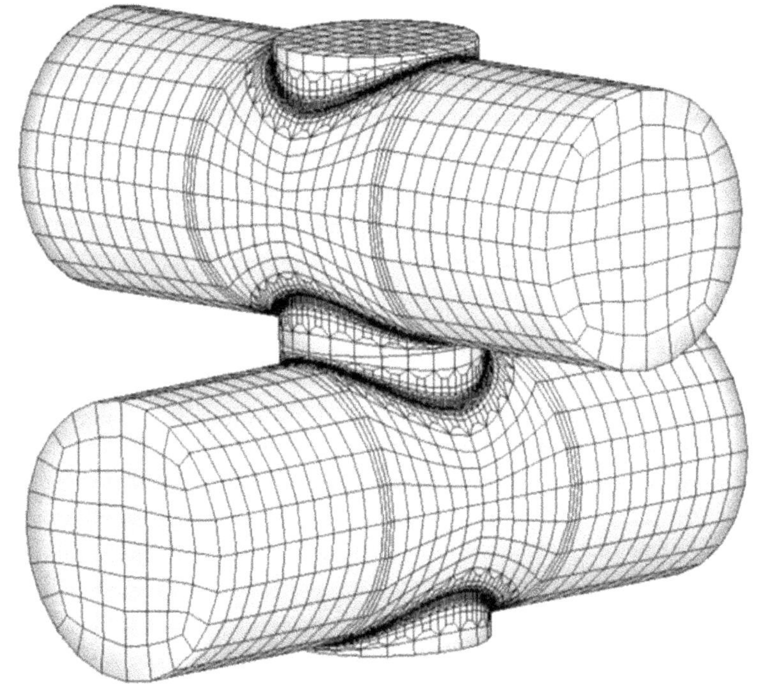

Figure 6 Component pairs used for empirical training of `port 2`.

particular non-empirical) "Legendre" port eigenmodes. In this latter case, the reference port modes $\hat{\chi}_k^\beta$ are given as

$$\hat{\chi}_1^\beta = (L_1^\beta, 0, 0),\ \hat{\chi}_2^\beta = (0, L_1^\beta, 0),\ \hat{\chi}_3^\beta = (0, 0, L_1^\beta),$$
$$\hat{\chi}_4^\beta = (L_2^\beta, 0, 0),\ \hat{\chi}_5^\beta = (0, L_2^\beta, 0),\ \hat{\chi}_6^\beta = (0, 0, L_2^\beta), \ldots \quad (173)$$

where the L_i^β, $1 \le i \le \mathcal{N}^\beta/3$, are the eigenvectors of a scalar singular Sturm-Liouville eigenproblem over β ordered according to increasing eigenvalue.

Numerical results

We now present numerical results for our three-dimensional linear-elastic microtruss library to demonstrate our port-reduced SCRBE approximation and error estimation framework. Our implementation is in C++ and is based on the library libMesh [24,25]. In our current implementation offline calculations are performed in parallel, while online calculations are limited to a single core.

The offline computation time for our microtruss library is about five hours using up to 24-cores on an AMD Opteron 6238 workstation computer. In offline stages Off4 and Off7 we load all required data into memory to prepare for the online stage. An (upper bound for) the online memory footprint for this library is 1.5Gb.

Cylindrical cantilever beam

We shall first consider a cylindrical cantilever beam system, for which we may compare our compliance output results to standard (Euler-Bernoulli) beam theory [26]. Hence this system provides an opportunity to confirm both the validity of the SCRBE framework — in terms both of approximation and certification — as well as the fidelity of the underlying FE "truth" component discretization.

Our cantilever system is of total length $l = 8L$ and consists of $I = 8$ instantiations of `component 2` of individual length $L_i = L$, $1 \le i \le I$; we consider $E_i = 1$, $1 \le i \le I$, and thus here $I_{\text{eff}} = 1$. We prescribe zero Dirichlet conditions on the left-most port of the system and we apply a unity-magnitude tangential traction on the right-most port as shown in Figure 7; the deformations in Figure 7 show the displacement field, and the colors indicate the Von Mises stresses[i] with higher stresses in red. The output for this system is the average displacement over the right-most port in the direction of the

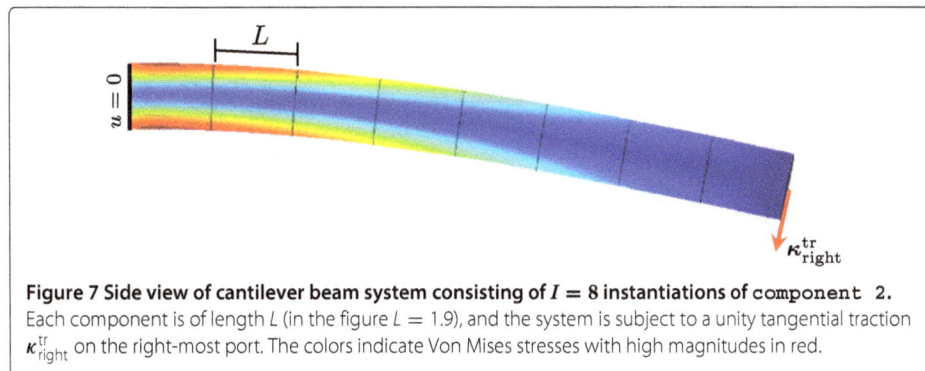

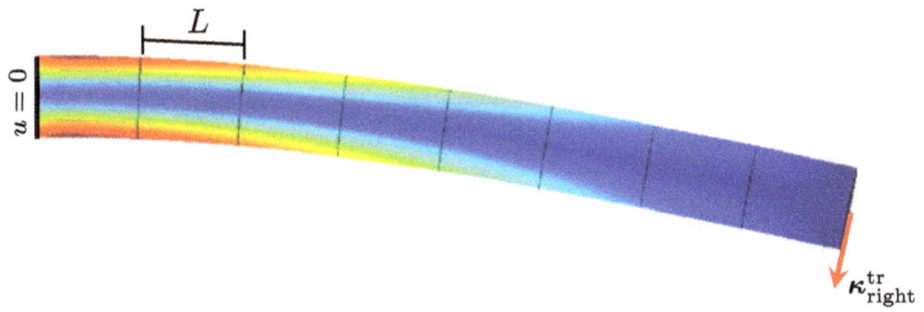

Figure 7 Side view of cantilever beam system consisting of $I = 8$ instantiations of `component 2`. Each component is of length L (in the figure $L = 1.9$), and the system is subject to a unity tangential traction $\boldsymbol{\kappa}_{\text{right}}^{\text{tr}}$ on the right-most port. The colors indicate Von Mises stresses with high magnitudes in red.

specified traction and is thus equal to compliance normalized by port area. We use $n^{\Gamma}_{\mathrm{A},p} = 20$ empirical port modes on all global ports Γ_p, $1 \leq p \leq n^{\Gamma}$.

We report in the second, third, and fourth columns of Table 1 the port-reduced SCRBE compliance output approximation $\tilde{s}^{\mathrm{PR}}(\mu)$, the relative error in $\tilde{s}^{\mathrm{PR}}(\mu)$ with respect to the FE "truth" compliance output $s^h(\mu)$, and the effectivity of the compliance output estimator, $\Delta^{s;0}(\mu)$, respectively. We note that the error in the output approximation is very small, and that the error estimator is relatively sharp; note that for large values of L the term $|\tilde{\mathbb{U}}^{\mathrm{PR}}(\mu)|^{\mathrm{T}}\boldsymbol{\sigma}_{\mathbb{A}}(\mu)|\tilde{\mathbb{U}}^{\mathrm{PR}}(\mu)|$ dominates in (136) and causes the effectivity to increase somewhat. We emphasize that our error estimator is for all these cases indeed an error upper bound: the effectivities are greater than unity.

We report in the fourth and fifth columns of Table 1 the theoretical maximum deflection $s^{\mathrm{EB}}(\mu)$ as predicted by classical Euler-Bernoulli beam theory,[j] and the relative difference between $\tilde{s}^{\mathrm{PR}}(\mu)$ and $s^{\mathrm{EB}}(\mu)$. The theoretical predictions match the computational results reasonably well, and in particular become increasingly accurate for larger L (the analytical results are valid in the limit of a long cantilever). Furthermore the discrepancy is for larger L sufficiently small that we deem our component FE discretization sufficiently rich.

Next, we consider the behavior of our port-reduced SCRBE compliance output approximation and associated error estimator as functions of $n^{\Gamma}_{\mathrm{A},p}$ *empirical* port modes for a fixed length parameter $L = 1.3$. In Figure 8 we report the relative compliance error $|s^h(\mu) - \tilde{s}^{\mathrm{PR}}(\mu)|/\tilde{s}^{\mathrm{PR}}(\mu)$, the relative error estimator $\Delta^{s;0}(\mu)/\tilde{s}^{\mathrm{PR}}(\mu)$ given in (149), and the relative error bound $\Delta^{s}(\mu; C)/\tilde{s}^{\mathrm{PR}}(\mu)$ given in Proposition 3 realized for $C = 1/\lambda_{\min}(\mathbb{B}^{\mathrm{NC}}_{\mu})$. We make several observations: first, the relative error decreases very fast and is of order 10^{-4} already for $n^{\Gamma}_{\mathrm{A},p} = 10$. Second, the error *estimator* is always greater than the error and is furthermore reasonably sharp — the effectivity is $\mathcal{O}(10)$ — for $n^{\Gamma}_{\mathrm{A},p} \leq 18$; at $n^{\Gamma}_{\mathrm{A},p} = 18$ the RB error bound contribution $|\tilde{\mathbb{U}}(\mu)|^{\mathrm{T}}\boldsymbol{\sigma}_{\mathbb{A}}(\mu)|\tilde{\mathbb{U}}(\mu)|$ becomes the dominating term in (136) and thus adding additional port modes will not reduce the error estimator.[k] Third, the rigorous error *bound* is reasonably sharp only for small $n^{\Gamma}_{\mathrm{A},p}$: the term $\|\boldsymbol{\sigma}(\mu)\|_2^2/\lambda_{\min}(\mathbb{B}^{\mathrm{NC}}_{\mu})$ in (128) dominates from an early point not because of large RB error bound contributions *per se* but because $C = 1/\lambda_{\min}(\mathbb{B}^{\mathrm{NC}}_{\mu})$ is a pessimistic estimate for the Rayleigh quotient associated with $\mathbb{B}^{\mathrm{NC}}_{\mu}$ and $\boldsymbol{\sigma}(\mu)$.

We also compare our empirical port approximation to the more standard eigenmode (Legendre) port approximation introduced in (173). In Figure 9 we report for the

Table 1 Results for variable L for the cylindrical cantilever beam system using $n^{\Gamma}_{\mathrm{A},p} = 20$ empirical port modes on each port

$l = 8L$	$\tilde{s}^{\mathrm{PR}}(\mu)$	$\frac{s^h(\mu)-\tilde{s}^{\mathrm{PR}}(\mu)}{\tilde{s}^{\mathrm{PR}}(\mu)}$	$\frac{\Delta^{s;0}(\mu)}{\lvert s^h(\mu)-\tilde{s}^{\mathrm{PR}}(\mu)\rvert}$	$s^{\mathrm{EB}}(\mu)$	$\frac{\tilde{s}^{PR}(\mu)-s^{EB}(\mu)}{\tilde{s}^{\mathrm{PR}}(\mu)}$
4.0	1.6504e+2	8.4e-5	1.7e+1	1.5562e+2	5.7e-2
5.6	4.3969e+2	3.8e-5	6.8e+0	4.2702e+2	2.8e-2
7.2	9.2362e+2	2.8e-5	2.3e+1	9.0758e+2	1.7e-3
8.8	1.6767e+3	2.4e-5	4.5e+1	1.6571e+3	1.2e-3
10.4	2.7584e+3	2.2e-5	3.3e+1	2.7352e+3	8.4e-3
12.0	4.2281e+3	1.8e-5	2.5e+1	4.2018e+3	6.2e-3
13.6	6.1450e+3	1.3e-5	1.7e+2	6.1165e+3	4.6e-3
15.2	8.5671e+3	2.0e-5	3.5e+2	8.5392e+3	3.3e-3

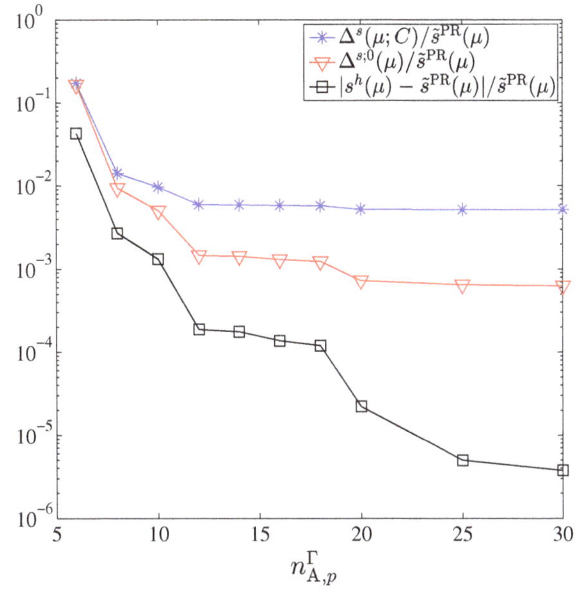

Figure 8 Relative compliance output error, relative output error estimator, and relative output error bound as functions of n^{Γ}_{A} for the cantilever beam ($L = 1.3$), using empirical port spaces.

Legendre case the relative compliance error, the relative error estimator, and the relative error bound superposed on the results for the empirical case (in gray). From the two error curves (squares) we note that the empirical port mode approximation is more than an order of magnitude better than the Legendre approximation for small $n^{\Gamma}_{A,p}$, and for larger $n^{\Gamma}_{A,p}$ the error in the Legendre approximation decreases significantly only for certain eigenmodes whereas the empirical approximation converges in a more regular

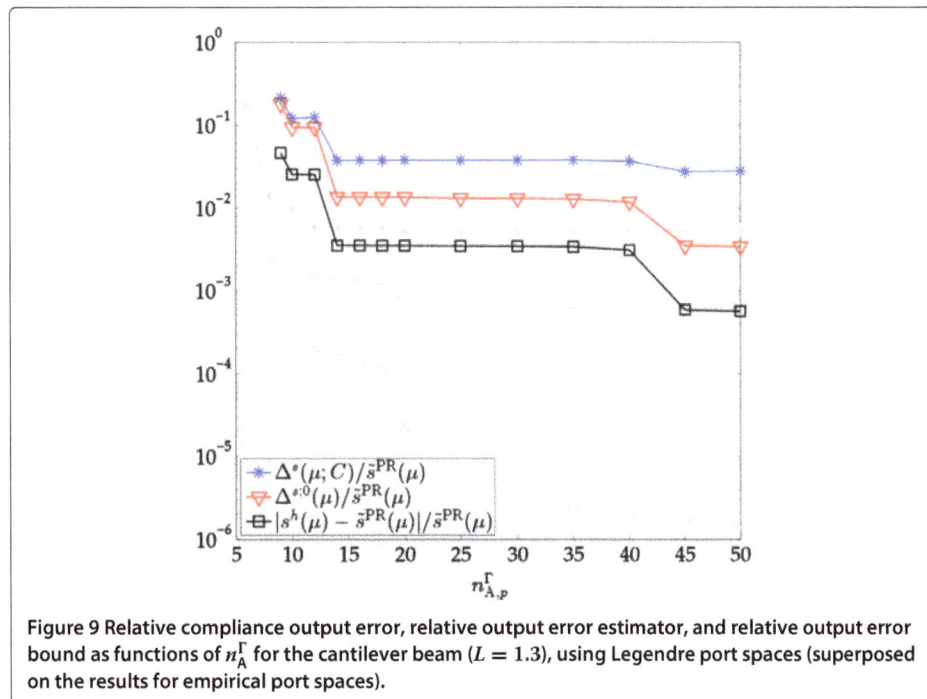

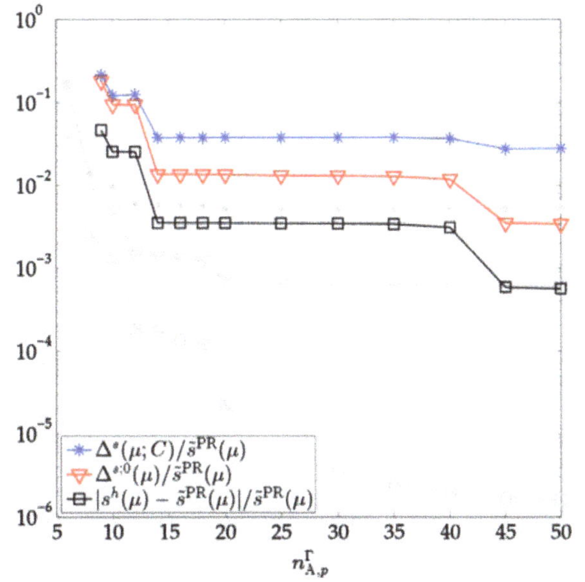

Figure 9 Relative compliance output error, relative output error estimator, and relative output error bound as functions of n^{Γ}_{A} for the cantilever beam ($L = 1.3$), using Legendre port spaces (superposed on the results for empirical port spaces).

fashion. We also note that the error estimator (triangles) and bound (asterisk) for the empirical approximation is about an order of magnitude smaller than the estimator and bound for the Legendre approximation, respectively.

Finally, we consider in Figure 10 and Figure 11 the relative compliance output error estimator $\Delta^{s;0}(\mu)/\tilde{s}^{\mathrm{PR}}(\mu)$ for empirical and Legendre port modes as functions of L for three different values of $n^{\Gamma}_{\mathrm{A},p}$; note the results for the latter case are superposed on the results for the former case in Figure 11. We note that for L far from the reference parameter values (recall $\hat{L}^1_{2,\mathrm{ref}} = 0.75$, $\hat{L}^2_{2,\mathrm{ref}} = 1.0$, $\hat{L}^3_{2,\mathrm{ref}} = 1.5$) only the empirical port modes provide a good approximation; in fact using empirical port modes we obtain even for $n^{\Gamma}_{\mathrm{A},p} = 10$ a relative error estimator smaller than 0.013 for all sampled values of L. Again, we emphasize that these error estimates indeed provide bounds on the error: for all cases the relative error with respect to the FE discretization is smaller than 10^{-4} as reported in the second column of Table 1.

For the remainder of our numerical results we exclusively employ the error estimator (136) or (148) rather than the respective rigorous bound.

Microtruss structure

We shall now consider a larger microtruss beam. Our first microtruss system, `system 1`, is an array of of $I = 408$ components (224 of which are of archetype `component 1` and 184 of which are of archetype `component 2`). We illustrate the system assembly process in Figure 12 and Figure 3; note that this procedure is efficient thanks to a graphical user interface that allows "cloning" of smaller subsystems which we may interconnect to form the final system [Additional file 1]. Note in actual (engineering) practice, this microtruss beam may be manufactured from $N_{\mathrm{rods}} = 40$ rods that are welded together.

To the final system shown in Figure 3 we apply zero Dirichlet boundary conditions on the 32 bottom ports; we apply homogeneous Neumann boundary conditions on the 80 side ports; we apply a unity-magnitude tangential traction (Neumann) $\kappa^{\mathrm{tr}}_{\mathrm{top}}$ on the 32 top (red) ports in the z-direction. The size of the non-port-reduced Schur complement system

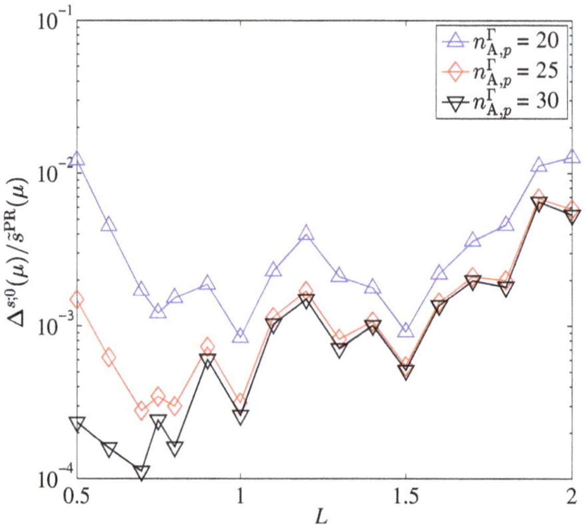

Figure 10 Cantilever beam relative compliance error estimator as a function of L for different $n^{\Gamma}_{\mathrm{A},p}$ using empirical port spaces.

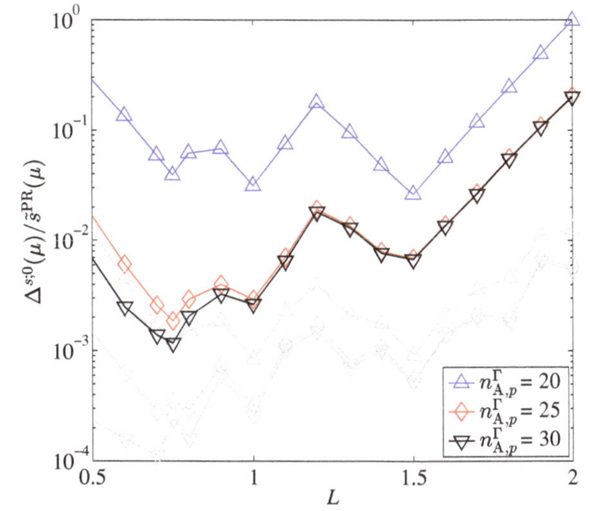

Figure 11 Cantilever beam relative compliance error estimator as a function of L for different $n^{\Gamma}_{A,p}$ using Legendre port spaces (superposed on the results for empirical port spaces).

is in this case $n_{SC} = 147,168$. Note that `system 1` refers to a particular topology configured with particular Dirichlet boundary conditions; we shall thus consider `system 1` for many different system parameter values. In particular, we denote by L^z and L^x the length of all component instantiations of `component 2` which are oriented in the z-direction and x-direction, respectively, as indicated for L^z in Figure 3.

We first demonstrate the ability of the port-reduced SCRBE framework to provide FE-fidelity field approximations at low computational cost. For our first calculation we consider the parameter values $E_i = 1$, $1 \le i \le I$, $L^z = 1.1$, and $L^x = 1$. We show (qualitatively) the solution fields in Figure 13 and Figure 14: the displacement field is shown in Figure 13 as a deformation of the original geometry (compare to the original geometry in Figure 3); a closeup of the Von Mises stress field near a "weld" is shown in Figure 14. The high-stress concentrations (red) are typically isolated to areas where a weld meets the cylinder base. Note that this high-stress, near-singular, area of the field is located somewhat close to the ports but nevertheless well within the interior of the

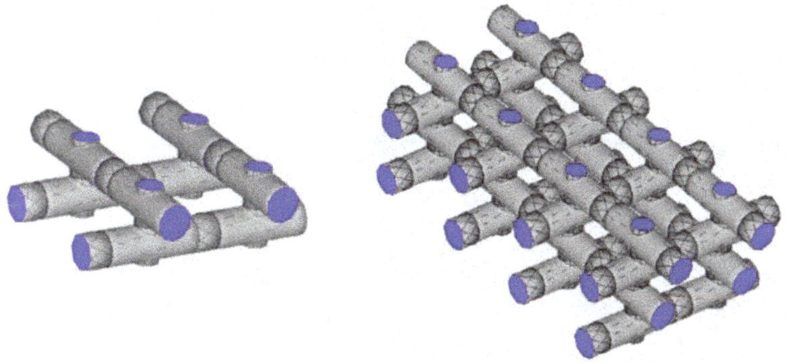

Figure 12 Assembly of the microtruss structure by component and subsystem "cloning" using a graphical user interface.

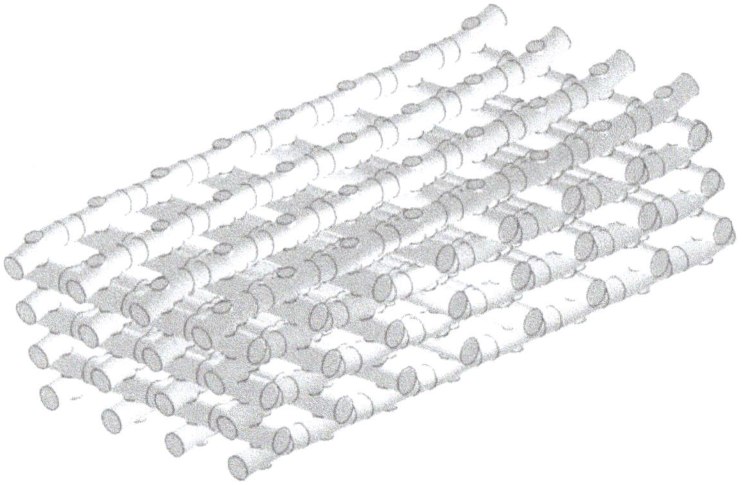

Figure 13 Displacement field shown as deformation for parameter values $L^x = 1, L^z = 1.1, E_i = 1, 1 \leq i \leq I$.

components. Placement of singular or more rapid behavior within the interior of a component, when possible, can reduce the number of port degrees of freedom required as provided by the pairwise training algorithm.

For $n^{\Gamma}_{A,p} = 20$ and $n^{\Gamma}_{A,p} = 25$ "Active" port modes we obtain the relative energy-norm error estimators

$$\frac{\|e^h(\mu)\|_\mu}{\|\tilde{u}^{\mathrm{PR}}(\mu)\|_\mu} \leq \frac{\Delta^{u;0}(\mu)}{\|\tilde{u}^{\mathrm{PR}}(\mu)\|_\mu} = 0.1139, \quad \frac{\|e^h(\mu)\|_\mu}{\|\tilde{u}^{\mathrm{PR}}(\mu)\|_\mu} \leq \frac{\Delta^{u;0}(\mu)}{\|\tilde{u}^{\mathrm{PR}}(\mu)\|_\mu} = 0.05641, \quad (174)$$

Figure 14 Von Mises stress field for parameter values $L^x = 1, L^z = 1.1, E_i = 1, 1 \leq i \leq I$.

respectively (note the inequalities are not confirmed but valid under the assumption that the error estimators indeed provide error bounds). The port-reduced SCRBE system sizes are $n_{\mathrm{A}} = 13,440$ and $n_{\mathrm{A}} = 16,800$, respectively, and we thus realize in both cases $n_{\mathrm{A}} \ll n_{\mathrm{SC}}$. For these calculations $I_{\mathrm{eff}} = 4$ — there are only two unique instantiations of `component 1`[1] and only two unique instantiations of `component 2` — and we thus realize very efficient online computations. The total (for solution and error estimate) online CPU time is approximately 12.9 seconds for the $n^{\Gamma}_{\mathrm{A},p} = 20$ calculation and approximately 18 seconds for the $n^{\Gamma}_{\mathrm{A},p} = 25$ calculation. We report detailed online timing results in the left and middle columns of Table 2, and we note that for both computations the certification dominates online cost. In particular, the calculation of the minimum eigenvalue (On4) together with the non-conforming solves (On7) contribute roughly 8/10 of total cost. Note that as we consider the error estimator rather than the error bound, we do not execute On8.

We next demonstrate the ability of the SCRBE framework to handle different topological configurations, here in the form of a simulated material flaw. To this end, we introduce a second microtruss system, `system 2`, which is identical to `system 1` except we assume that three "random" welds are broken such that now we do not couple the corresponding `port 2` ports (top or bottom local ports of `component 1`). These three shared global ports are thus split into six non-shared global ports, on which we impose homogeneous Neumann (zero-stress) boundary conditions. On all other ports the boundary conditions are the same as for `system 1`. The size of the non-port-reduced Schur complement system is in this case slightly larger: the non-port-reduced system is of size $n_{\mathrm{SC}} = 147{,}825$, and the port-reduced system for $n^{\Gamma}_{\mathrm{A},p} = 20$ active port modes is of size $n_{\mathrm{A}} = 13{,}500$. A closeup of the solution field near a broken weld is shown in Figure 15; note the low stress concentration at the failed weld compared to neighboring intact welds.

We now consider the compliance output and associated error estimators for `system 1` and `system 2`. The compliance is for these systems the integrated displacement in the z-direction over all top (red in Figure 3) ports, and is thus effectively a measure of the microtruss beam directional stiffness. For `system 1` and `system 2` we then compute solutions and corresponding compliance outputs for different values of $L^z \in [0.5, 2]$; we consider $L_x = 1$ and $E_i = 1$, $1 \le i \le I$. We thus effectively consider the directional stiffness of the microtruss beams as a function of the spacing between rods oriented in the x-direction.

The results for $n^{\Gamma}_{\mathrm{A},p} = 20$ are shown in Figure 16. The solid blue and solid red lines indicate the port-reduced SCRBE `system 1` and `system 2` output approximation, respectively; the dashed lines indicate the estimated bounds on the output as provided by

Table 2 Breakdown of majority of online computational cost in seconds for `system 1` for indicated $n^{\Gamma}_{\mathrm{A},p}$ active port modes and I_{eff} unique component instantiations

	$I_{eff} = 4, n^{\Gamma}_{\mathrm{A},p} = 20$	$I_{eff} = 4, n^{\Gamma}_{\mathrm{A},p} = 25$	$I_{\mathrm{eff}} = 80, n^{\Gamma}_{\mathrm{A},p} = 25$
On2	0.5	0.6	8.3
On3	1.3	2.5	2.5
On4	6.5	7.8	9.2
On5	0.8	0.9	18.1
On7	3.8	6.2	6.2
Total	12.9	18	44.3

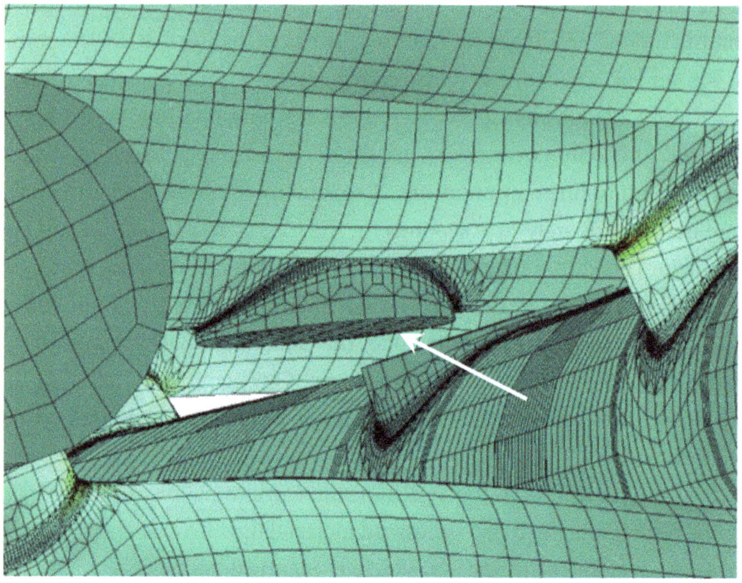

Figure 15 Solution (displacement and Von Mises stress (colors)) near a failed weld for `system 2`.

the error estimator (149). As expected, `system 2` (with the weld failure) is less stiff and thus exhibits larger top-port displacements than `system 1`. However, we can not in this case for larger L^z distinguish between `system 1` and `system 2` with any confidence because the (estimated) output bounds overlap.

We next consider the same "parameter sweep," but now using $n^{\Gamma}_{\mathrm{A},p} = 25$ empirical port modes. For `system 1` this corresponds to a port-reduced SCRBE system of size $n_{\mathrm{A}} = 16{,}800$ and for `system 2` a system of size $n_{\mathrm{A}} = 16{,}875$. In Figure 17, we show the outputs and output bounds for `system 1` and `system 2`, and we note that we are now able to easily distinguish the two systems.

So far we have for `system 1` and `system 2` considered only a single system parameter L_z and thus $I_{\mathrm{eff}} \ll I$. We now consider for `system 1` a somewhat more demanding case in which we also assign "random" Young's modulus $E^j_{\mathrm{rod}} \in [0.9, 1.1]$, $1 \leq j \leq N_{\mathrm{rods}}$, to

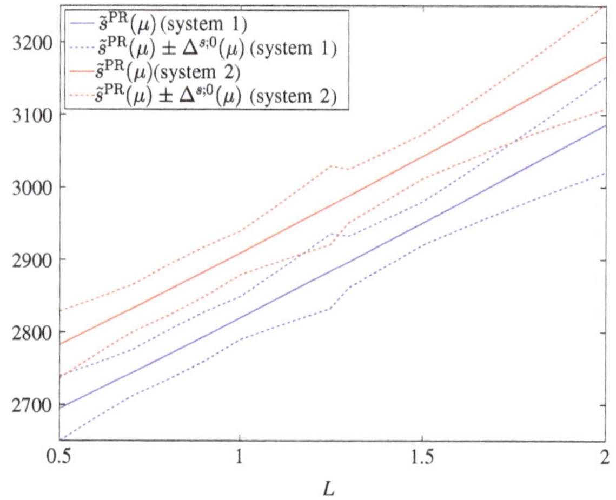

Figure 16 Parameter sweeps — compliance outputs and (estimated) compliance output bounds — over $L^z \in [0.5, 2]$ for `system 1` and `system 2` using $n^{\Gamma}_{\mathrm{A},p} = 20$ "Active" port modes.

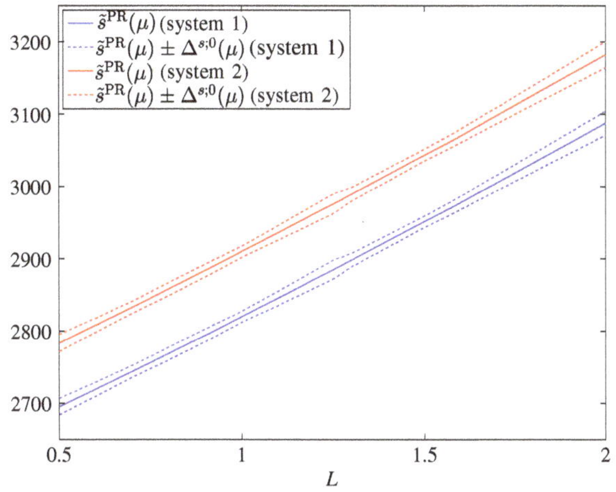

Figure 17 Parameter sweeps — compliance outputs and (estimated) compliance output bounds — over $L^z \in [0.5, 2]$ for `system 1` and `system 2` using $n^{\Gamma}_{A,p} = 25$ "Active" port modes.

each of the $N_{\rm rods} = 40$ rods of the system; we thus consider here $P = 40$ system parameters (one of the $E^j_{\rm rod}$ scales out) and we obtain in this case $I_{\rm eff} = 80$. Because of the larger $I_{\rm eff}$, computational cost increases somewhat as reported in the rightmost column of Table 2.[m] For this particular simulation we obtain a relative error field energy estimator $\Delta^{u;0}(\mu)/\|\tilde{u}^{\rm PR}(\mu)\|_{\mu} = 0.0712$ using $n^{\Gamma}_{A,p} = 25$ empirical port modes.

Finally, we close this section with comparison to standard global FE analysis of `system 1` for system parameters $L^x = 1$, $L^z = 1.1$ and $E_i = 1$, $1 \le i \le I$. We consider $n^{\Gamma}_{A,p} = 20$, $n^{\Gamma}_{A,p} = 25$, and $n^{\Gamma}_{A,p} = 30$, and we report in Table 3 for each case the relative output error, $(s^h(\mu) - \tilde{s}^{\rm PR}(\mu))/\tilde{s}^{\rm PR}(\mu)$, and the output error estimator effectivity, $\Delta^{s;0}(\mu)/(s^h(\mu) - \tilde{s}^{\rm PR}(\mu))$. For all calculations the relative output error is indeed small, and certainly within acceptable tolerances in an engineering context. The estimator effectivities are furthermore greater than unity — our error estimators are indeed error upper bounds — and moreover, the efficivities show that our estimators are relatively sharp.

The global FE space $X^h(\Omega)$ for `system 1` is of dimension $\mathcal{N}_{\rm FE} = 26{,}381{,}328$. The computation time for a single global FE simulation on a workstation with eight AMD Opteron 6238 cores is 59 minutes for 93 conjugate gradient iterations using an algebraic multigrid preconditioner [27] (we employ the BoomerAMG [28] parallel algebraic multigrid implementation provided by the hypre [29] linear solver library). For $n^{\Gamma}_{A,p} = 25$, the port-reduced SCRBE approximation and error estimation requires about 18 seconds on a single core and we thus obtain a speedup of almost 200 — note that significantly larger speedup is possible through parallelization of the SCRBE online stage. We finally note that to compute the results in Figure 17, we have for each parameter sweep performed seventeen evaluations of the port-reduced SCRBE compliance output and associated output bound

Table 3 Relative output error and output error estimator effectivity for `system 1` for parameter values $L^x = 1$, $L^z = 1.1$ and $E_i = 1$, $1 \le i \le I$

	$n^{\Gamma}_{A,p} = 20$	$n^{\Gamma}_{A,p} = 25$	$n^{\Gamma}_{A,p} = 35$
$\frac{s^h(\mu) - \tilde{s}^{\rm PR}(\mu)}{\tilde{s}^{\rm PR}(\mu)}$	4.67978e-4	2.31664e-4	2.5646e-5
$\frac{\Delta^{s;0}(\mu)}{s^h(\mu) - \tilde{s}^{\rm PR}(\mu)}$	28.3	14.8	67.6

estimators in only about five minutes total CPU time. A similar parametric analysis using a classical FE approach is clearly not equally tractable.

Conclusions

In this paper we have extended the port-reduced static condensation reduced basis element method to analysis of large-scale component-based structures. In particular we have demonstrated the applicability and efficacy of the procedure in three-dimensional linear elasticity analysis of a microtruss structure with hundreds of components.

Through a combination of *i)* component-interior reduced basis approximations and *ii)* port reduction using empirical modes tailored to the component library, we are able to obtain an accurate online approximation for any component parameter values and any system topology using very few global degrees of freedom. Moreover, we may estimate (and rigorously bound in the limit of small reduced basis error contributions) the error in this port-reduced SCRBE approximation with respect to the underlying global finite element discretization through efficiently computable *a posteriori* errorl estimators.

For the microtruss application we consider in this paper, more than twenty-six million degrees of freedom in the alternative global FE discretization is reduced to a few thousand degrees of freedom in the port-reduced SCRBE approximation. The online computation time is accordingly reduced from about an hour to only seconds, and thus the approach enables large-scale computation in many-query contexts such as interactive design or optimization. Further, our computational results for the microtruss structure indicate applications in stochastic homogenization and material failure identification, which may require many simulations for (say) random parameters and topology [30]. Another application is vibration analysis of structures as considered in [31].

The presented approach is an alternative to standard FE analysis of large component-based structures such as bridges, microtrusses, or vehicle or building frames. However, we may consider any linear elliptic or parabolic [13] parameter-dependent partial differential equation, and thus problems in (say) heat transfer [32], acoustics [33], and electromagnetics may be considered as well.

Endnotes

[a] For non-symmetric, non-coercive, complex-valued, or parabolic problems additional elements are required for our *a posteriori* error estimation framework.

[b] We first apply the inverse map to physical coordinates to obtain reference coordinates, and then evaluate the function on the reference domain.

[c] To illustrate this latter application of the mapping, consider for example a vector field $(0, 1)$ defined on $\hat{\Omega}_1$ in the frame $(\hat{x}_1, \hat{y}_1)$ in Figure 1. We then consider this same vector field over Ω_1 in Figure 2: by application of $\mathcal{T}_1^{\mathrm{rot}}$ to the field $(0, 1)$ we obtain an interpretation in the system frame (x, y) which is consistent with the interpretation on the archetype domain — the field is parallel to the original $\hat{x}_1$ axis (for Ω_1), and not parallel to the system x axis.

[d] We start with the strong formulation on each component; we multiply by a test function and integrate by parts; we then add the equations on adjacent components and invoke flux continuity to cancel the corresponding port integral terms. In practice this is automatically accommodated by the variational formulation (23).

[e] In this paper, we consider for $\mathcal{R}_{m,j}$ only rigid-body transformations; more general mappings and parametrized port deformations are also possible but is subject of future work.

[f] Note in the scalar-field case this simplifies to only the constant mode.
[g] Our current implementation does not recognize "free" parameters (Young's modulus, conductivity) and thus each set of component clones will contain components with different spatial orientation but identical ("non-docking") parameters.
[h] With the current $(L^2(\beta))^{d-1}$ orthogonalization of our port space bases, the ℓ_2 norm of the residual coefficients is rather strong. We conjecture that the constant C in (126) and (127) can be improved by consideration of an orthogonalization which provides a global Riesz basis (in the limit as the FE discretization parameter $h \to 0$) with respect to the $(H^{-1}(\Omega))^d$ semi-norm.
[i] The Von Mises stresses are calculated as $\sigma_{\mathrm{VM}} = \left(\frac{1}{2}\left((\sigma_{11}-\sigma_{22})^2 + (\sigma_{22}-\sigma_{33})^2 + (\sigma_{33}-\sigma_{11})^2 + 6(\sigma_{12}^2+\sigma_{23}^2+\sigma_{31}^2)\right)\right)^{1/2}$.
[j] The formula for the maximum deflection $d(l)$ of a cantilever beam of length l, Young's modulus E, second moment of inertia I, and subject to tangential force P at one end is $d(l) = Pl^3/(3EI)$.
[k] To reduce the error estimator in this case we would have to reduce the values in $\boldsymbol{\sigma}_{\mathbb{A}}(\mu)$ by adding additional RB snapshots to the RB bubble spaces in the offline stage.
[l] For `component 1` all instantiations have identical parameters, but there are two different component-local matrix blocks because we consider Dirichlet boundary conditions on all bottom ports of components located at the bottom of the microtruss structure.
[m] The E_i correspond to component-wide Young's modulus, and hence these parameters are "free" in the sense discussed in the "Model reduction" section. Thus with a more complete treatment of effectively identical components we would have recovered $I_{\mathrm{eff}} = 4$ for this case.

Additional file

Additional file 1: A short video which illustrates the methodology of this paper is published together with this paper as `prscrbe_movie.mp4`.

Competing interests
The authors declare that they have no competing interests

Authors' contributions
JLE developed computational procedures and the associated C++ implementation, contributed to the theoretical results, and drafted the manuscript. ATP developed computational procedures and theoretical results. All authors participated in the writing, review, and revision of the manuscript.

Acknowledgements
We are grateful to Dr. D. J. Knezevic for development of SCRBE library code, to Dr. D. B. P. Huynh for graphical system assembly and visualization software, and to Dr. S. Vallaghé for fruitful discussion. This work has been sponsored by the Research Council of Norway and ONR Grant N00014-11-0713.

References
1. Abdelal GF, Abuelfoutouh N, Gad AH (2013) Finite element analysis for satellite structures. Springer, London
2. Egeland O, Haraldsen PO (1974) SESAM-69 — a general purpose finite element method program. Comput Struct 4: 41–68
3. Craig R, Bampton M (1968) Coupling of substructures for dynamic analyses. AIAA J 6(7): 1313–1319
4. Hurty WC (1964) On the dynamic analysis of structural systems using component modes In: First AIAA Annual Meeting. AIAA (American Institute of Aeronautics and Astronautics), Washington. AIAA paper, no. 64-487
5. Huynh DBP, Knezevic DJ, Patera AT (2013) A static condensation reduced basis element method: approximation and a posteriori error estimation. ESAIM: Math Model Numerical Anal 47(1): 213–251
6. Rozza G, Huynh DBP, Patera AT (2008) Reduced basis approximation and a posteriori error estimation for affinely parametrized elliptic coercive partial differential equations: application to transport and continuum mechanics. Arch Comput Methods Eng 15(3): 229–275
7. Binev P, Cohen A, Dahmen W, DeVore R, Petrova G, Wojtaszczyk P (2011) Convergence rates for greedy algorithms in reduced basis methods. SIAM J Math Anal 43(3): 1457–1472

8. Haasdonk B (2013) Convergence rates of the pod–greedy method. ESAIM: Math Model Numerical Anal 47: 859–873
9. Bourquin F (1992) Component mode synthesis and eigenvalues of second order operators: discretization and algorithm. Math Model Numerical Anal 26(3): 385–423
10. Hetmaniuk UL, Lehoucq RB (2010) A special finite element method based on component mode synthesis. ESAIM: Math Model Numerical Anal 44(3): 401–420
11. Jakobsson H, Bengzon F, Larson MG (2011) Adaptive component mode synthesis in linear elasticity. Internat J Numer Methods Engrg 86(7): 829–844
12. Eftang JL, Patera AT (2013) Port reduction in parametrized component static condensation: approximation and a posteriori error estimation. Int J Numerical Methods Eng 96(5): 269–302
13. Vallaghé S (2013) The static condensation reduced basis element method for parabolic problems. M3AS: Math Models Methods Appl Sci. http://augustine.mit.edu/methodology/papers/SV_M3AS_2013.pdf
14. Veroy K, Rovas DV, Patera AT (2002) A posteriori error estimation for reduced-basis approximation of parametrized elliptic coercive partial differential equations: "convex inverse" bound conditioners. ESAIM: Control, Optimisation Calculus Variations 8: 1007–1028
15. Quarteroni A, Valli A (1994) Numerical approximation of partial differential equations Springer Series in Computational Mathematics, vol. 23. Springer, Berlin
16. Bernardi C, Maday Y (1997) Spectral methods In: Handbook of Numerical Analysis, North-Holland, Amsterdam, pp 209–485
17. Kunisch K, Volkwein S (2002) Galerkin proper orthogonal decomposition methods for a general equation in fluid dynamics. SIAM J Numer Anal 40(2): 492–515
18. Isaacson E, Keller HB (1994) Computation of eigenvalues and eigenvectors, analysis of numerical methods
19. Hernández V, Román JE, Tomás A, Vidal V (2007) Krylov-Schur Methods in SLEPc. Technical report, Universidad Politecnica De Valencia. http://www.grycap.upv.es/slepc
20. Queheillalt DT, Wadley HNG (2005) Cellular metal lattices with hollow trusses. Acta Materialia 53: 303–313
21. Wadley HNG (2006) Multifunctional periodic cellular metals. Philos Trans R Soc A 364: 31–68
22. Wadley HNG, Fleck NA, Evans AG (2003) Fabrication and structural performance of periodic cellular metal sandwich structures. Composites Sci Technol 63: 2331–2343
23. Flügge W (1972) Tensor analysis and continuum mechanics. Springer, Berlin
24. Kirk BS, Peterson JW, Stogner RH, Carey GF (2006) `libMesh`: A C++ library for parallel adaptive mesh refinement/coarsening simulations. Eng Comput 22(3–4): 237–254
25. Knezevic DJ, Peterson JW (2011) A high-performance parallel implementation of the certified reduced basis method. Comput Methods Appl Mech Eng 200(13–16): 1455–1466
26. Timoshenko SP (1953) History of strength of materials. McGraw-Hill, New York
27. Saad Y (2003) Iterative methods for sparse linear systems, 2nd edn. Society for Industrial and Applied Mathematics, Philadelphia
28. Henson VE, Yang UM (2002) BoomerAMG: a parallel algebraic multigrid solver and preconditioner. Appl Numer Math 41(1): 155–177. Developments and trends in iterative methods for large systems of equations—in memoriam Rüdiger Weiss (Lausanne, 2000)
29. hypre: Scalable linear solvers. http://computation.llnl.gov/casc/linear_solvers/sls_hypre.html
30. Anantharaman A, Le Bris C (2011) A numerical approach related to defect-type theories for some weakly random problems in homogenization. Multiscale Model Simul 9(2): 513–544
31. Vallaghé S, Huynh DBP, Knezevic DJ, Patera AT (2013) Component-based reduced basis for eigenproblems. Comput Struct. http://augustine.mit.edu/methodology/papers/VHKP_CS_July2013.pdf
32. Vallaghé S, Patera AT (2012) The static condensation reduced basis element method for a mixed-mean conjugate heat exchanger model. SIAM J Sci Comput. http://augustine.mit.edu/methodology/papers/VP_SISC_revised_May2013.pdf
33. Huynh DBP, Knezevic DJ, Patera AT (2013) A static condensation reduced basis element method: complex problems. Comput Methods Appl Mech Eng 259(0): 197–216

Permissions

All chapters in this book were first published in AMSES, by Springer; hereby published with permission under the Creative Commons Attribution License or equivalent. Every chapter published in this book has been scrutinized by our experts. Their significance has been extensively debated. The topics covered herein carry significant findings which will fuel the growth of the discipline. They may even be implemented as practical applications or may be referred to as a beginning point for another development.

The contributors of this book come from diverse backgrounds, making this book a truly international effort. This book will bring forth new frontiers with its revolutionizing research information and detailed analysis of the nascent developments around the world.

We would like to thank all the contributing authors for lending their expertise to make the book truly unique. They have played a crucial role in the development of this book. Without their invaluable contributions this book wouldn't have been possible. They have made vital efforts to compile up to date information on the varied aspects of this subject to make this book a valuable addition to the collection of many professionals and students.

This book was conceptualized with the vision of imparting up-to-date information and advanced data in this field. To ensure the same, a matchless editorial board was set up. Every individual on the board went through rigorous rounds of assessment to prove their worth. After which they invested a large part of their time researching and compiling the most relevant data for our readers.

The editorial board has been involved in producing this book since its inception. They have spent rigorous hours researching and exploring the diverse topics which have resulted in the successful publishing of this book. They have passed on their knowledge of decades through this book. To expedite this challenging task, the publisher supported the team at every step. A small team of assistant editors was also appointed to further simplify the editing procedure and attain best results for the readers.

Apart from the editorial board, the designing team has also invested a significant amount of their time in understanding the subject and creating the most relevant covers. They scrutinized every image to scout for the most suitable representation of the subject and create an appropriate cover for the book.

The publishing team has been an ardent support to the editorial, designing and production team. Their endless efforts to recruit the best for this project, has resulted in the accomplishment of this book. They are a veteran in the field of academics and their pool of knowledge is as vast as their experience in printing. Their expertise and guidance has proved useful at every step. Their uncompromising quality standards have made this book an exceptional effort. Their encouragement from time to time has been an inspiration for everyone.

The publisher and the editorial board hope that this book will prove to be a valuable piece of knowledge for researchers, students, practitioners and scholars across the globe.

List of Contributors

Alain G Combescure
LaMCoS, INSA-Lyon, Université de Lyon, CNRS UMR 5259, 18-20 rue des Sciences, F69621 Villeurbanne, France

Vikram V Garg
Massachusetts Institute of Technology, 77 Massachusetts Avenue, Cambridge, MA 02139, USA

Serge Prudhomme
Ecole Polytechnique de Montreal, C.P. 6079, succ. Centre-Ville, Montreal, Canada

Kris G van der Zee
The University of Nottingham, Nottingham, UK

Graham F Carey
The University of Texas at Austin, 1 University Station, C0200 Austin, USA

Brice Bognet
GeM UMR CNRS-Centrale Nantes, 1 rue de la Noe, F-44300 Nantes, France

Adrien Leygue
GeM UMR CNRS-Centrale Nantes, 1 rue de la Noe, F-44300 Nantes, France

Francisco Chinesta
GeM UMR CNRS-Centrale Nantes, 1 rue de la Noe, F-44300 Nantes, France

John Tinsley Oden
Institute for Computational Engineering and Sciences The University of Texas at Austin, 201 East 24th St, Stop C0200 POB 4.102, 78712 Austin, TX, USA

Kathryn Farrell
Institute for Computational Engineering and Sciences The University of Texas at Austin, 201 East 24th St, Stop C0200 POB 4.102, 78712 Austin, TX, USA

Danial Faghihi
Institute for Computational Engineering and Sciences The University of Texas at Austin, 201 East 24th St, Stop C0200 POB 4.102, 78712 Austin, TX, USA

Thomas Horger
Institute for Numerical Mathematics, Technische Universität München, Boltzmannstraße 3, 85748 Garching b, München, Germany

Stefan Kollmannsberger
Chair of Computation in Engineering, Technische Universität München, Arcisstraße 21, 80290 München, Germany

Felix Frischmann
Chair of Computation in Engineering, Technische Universität München, Arcisstraße 21, 80290 München, Germany

Ernst Rank
Chair of Computation in Engineering, Technische Universität München, Arcisstraße 21, 80290 München, Germany

Barbara Wohlmuth
Institute for Numerical Mathematics, Technische Universität München, Boltzmannstraße 3, 85748 Garching b, München, Germany

Elias Cueto
Aragon Institute of Engineering Research, Universidad de Zaragoza, Zaragoza, Spain

Francisco Chinesta
Ecole Centrale de Nantes, Nantes, France
Institut Universitaire de France, Paris, France

Jens L Eftang
Department of Mechanical Engineering, Massachusetts Institute of Technology, 77 Massachusetts Avenue, Cambridge
MA-02139, USA 2Current address: DNV GL - Software, P.O.Box 300, NO-1322, Høvik, Norway

Anthony T Patera
Department of Mechanical Engineering, Massachusetts Institute of Technology, 77 Massachusetts Avenue, Cambridge

www.ingramcontent.com/pod-product-compliance
Lightning Source LLC
Chambersburg PA
CBHW080258230326
41458CB00097B/5111
9781682850343